Advanced Many-body and Statistical Methods in Mesoscopic Systems 2011

Journal of Physics: Conference Series Volume 338

Constanta, Romania
27 June – 2 July 2011

ISBN: 978-1-62276-684-0
ISSN: 1742-6588

Printed from e-media with permission by:

Curran Associates, Inc.
57 Morehouse Lane
Red Hook, NY 12571

proceedings.com

Some format issues inherent in the e-media version may also appear in this print version.

Printed by Curran Associates, Inc. (2013)

Institute of Physics
Dirac House, Temple Back
Bristol BS1 6BE UK

Phone: 44 1 17 929 7481
Fax: 44 1 17 920 0979

techtracking@iop.org

Additional copies of this publication are available from:

Curran Associates, Inc.
57 Morehouse Lane
Red Hook, NY 12571 USA
Phone: 845-758-0400
Fax: 845-758-2634
Email: curran@proceedings.com
Web: www.proceedings.com

Preface

Mesoscopic nanoscale physics

Nuclear Physics

Zero bias anomalies in the Kondo regime of single and double quantum dots

A. Aldea, M. Ţolea, I.V. Dinu
National Institute of Materials Physics, POB MG-7, Bucharest-Magurele, Romania.

Abstract. The zero bias anomaly of the differential conductance of mesoscopic systems is a fingerprint of the electron-electron interaction. The common example is the peak of the differential conductance in Kondo mesoscopic systems.

We show that in complex mesoscopic systems, in particular in the side-coupled double dot, the dip of the differential conductance at zero bias is also possible due to simultaneous effects of interference and correlations. The external parameters that control this effect are the temperature and the gate potential on the lateral dot. We argue that even in single dots, in a multiple lead configuration, the shift from suppression to enhancement of dI/dV with increasing bias is allowed if the bias applied on the leads is not symmetric. The differential conductance exhibits a peak-dip crossover, the effect being controlled by the strength of the asymmetry and the ratio of the dot-lead couplings.

The scaling of the differential conductance for the double-dot system is also discussed. The results are obtained using an extended Anderson model, the Keldysh transport formalism and the equation of motion technique extended to the non-equilibrium.

1. Introduction

The Coulomb repulsion leads to important electronic correlations, resulting in strong modifications of the density of states, the (perhaps) most significant such case being illustrated by the mesoscopic Kondo effect [1]. In the classical single-dot case, the Kondo effect manifests in the differential conductance dI/dV as a pronounced peak at zero bias ($V = 0$), known as the zero bias anomaly (ZBA), meaning that the conductance is maximum close to equilibrium and is decays rapidly when the bias is increased. The reason for the depletion of the differential conductance is that the applied bias kills the Kondo correlations.

A more subtle problem regards the bias dependence of the differential conductance in complex systems. The three lead-Kondo problem -for instance- was theoretically considered by, e.g., [2, 3, 4, 5] or experimentally realized by Leturcq et. al [6, 7]. It can be shown that the presence of the third lead allows for a conductance suppression around zero bias, instead of the usual enhancement, in conditions of asymmetrically applied bias [5]. The result is interesting due to its counter-intuitive aspect, but, more importantly, one can extract the ratio of dot-leads coupling from the threshold values of the asymmetry parameter corresponding to the peak-dip crossover.

Another complex Kondo problem arises when considering a second side-coupled dot, a set-up intensively addressed in literature [8, 9, 10, 11, 12, 13, 14]. The system is typical for exhibiting Fano resonances when a gate potential V_g applied on the side dot is changed continuously. Our numerical results indicate that a dip-type ZBA can develop for values of V_g around the Fano zero.

Next, we address the scaling of the differential conductance $G = dI/dV$ for the double-dot system when the temperature and the bias are varied. The scaling properties resemble those for the single-dot case, provided that a suitable definition is used for the Kondo temperature of the double-dot system.

2. Three leads Kondo problem

A quantum dot connected to three leads is sketched in Fig.1. The electron current through one of the leads (say α) is given by [2]:

$$I_\alpha = -\frac{4e\Gamma_\alpha}{\pi\hbar\Gamma}\int_{-\infty}^{\infty} ImG_{dd}(\omega)\sum_{\beta\neq\alpha}\Gamma_\beta[f(\omega-\mu_\alpha)-f(\omega-\mu_\beta)]d\omega, \tag{1}$$

where Γ_α is the coupling of the lead α to the dot, $\Gamma = \sum\Gamma_\alpha$, and $f(\omega - \mu_\alpha)$ is the Fermi distribution in the α- lead. G_{dd} is the retarded Green function of the dot in the presence of the leads. Usually, one needs also the "lesser" Green function, however when the leads are all connected to the same site, the lesser function can be eliminated excepting for the case of magnetic leads [15]. The same simplification occurs for a geometrically symmetric many-site system [16].

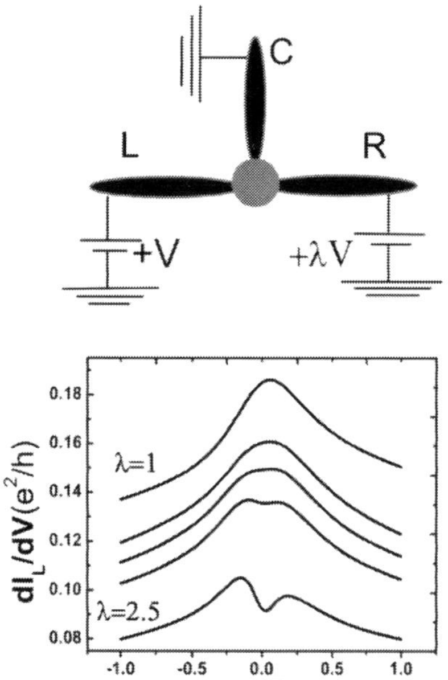

Figure 1. (above) Sketch of a quantum dot connected to three leads. The bias is applied asymmetrically on the left and right leads ($V_L = V$ and $V_R = \lambda V$; the usual symmetric case corresponds to $\lambda = -1$). (below) The differential conductance through the left lead for different values of the asymmetry parameter showing a change from peak to dip (around *zero* bias) ; the crossover occurs at $\lambda_0 = 1.58$; the other parameters: $\Gamma_R = \Gamma_L = 0.075, \Gamma_C/\Gamma_L = 3$, $T_K = 3.2 \cdot 10^{-4}$ (measured in the lead bandwidth), and $T = T_K/10$.[5]

If one applies the potential bias $V_L - V_R = V(1-\lambda)$ the differential conductance through the lead "L" can be calculated from eq.1:

$$\mathcal{G}_L(V) \;=\; \frac{dI_L}{dV} = \frac{4e^2\Gamma_L}{\pi\hbar\Gamma}\int_{-\infty}^{\infty}[(\Gamma_R+\Gamma_C)ImG_{dd}(\omega)\frac{df(\omega-eV)}{d\omega}$$

$$- \quad \lambda\Gamma_R ImG_{dd}(\omega)\frac{df(\omega-\lambda eV)}{d\omega}]d\omega. \tag{2}$$

At low temperatures, approximating the derivatives of the Fermi functions by delta-functions, and under the assumption that the spectral function depends weakly on the applied bias, the above formula becomes :

$$\mathcal{G}_L(V) = -\frac{4e^2\Gamma_L}{\pi\hbar\Gamma}[(\Gamma_R+\Gamma_C)ImG_{dd}(eV) - \lambda\Gamma_R ImG_{dd}(\lambda eV)]. \tag{3}$$

In order to prove the peak- or dip-like behaviour of the differential conductance, one has to compute the second derivative:

$$\mathcal{G}_L''(0) = -\frac{4e^4\Gamma_L\Gamma_R}{\pi\hbar\Gamma}ImG''_{dd}(0)[(1+\frac{\Gamma_C}{\Gamma_R}) - \lambda^3] \tag{4}$$

At this stage , one can make the following remarks:

i) in the above formulas $-ImG''_{dd}(0)$ is the second derivative of the spectral function at equilibrium (i.e. at $V=0$), and it is always negative for $T<T_K$ indicating the presence of the Kondo peak in the density of states (also, one has $ImG'_{dd}(0)=0$). This property will be used to determine the peak or dip behavior of the conductance.

ii) eq.(4) shows that the peak or dip aspect of the differential conductance is completely determined by the sign of the quantity $[1+\Gamma_C/\Gamma_R - \lambda^3]$.

iii) the symmetric bias corresponds to $\lambda=-1$, in which case the second derivative of the spectral function is positive. In other words, for the usual case of symmetric bias, the differential conductance is always enhanced at $V=0$ showing a peak-like behaviour.

iv) there is a crossover value of λ when the peak changes into dip; this value is $\lambda_0 = \sqrt[3]{1+\Gamma_C/\Gamma_R}$. This fact can be of interest for experimentalists: once the crossover value λ_0 is identified from experiment, the ratio Γ_C/Γ_R of the two dot-lead couplings can be immediately extracted. Otherwise, the ratio of the two couplings is inaccessible to direct measurement.

The above general argumentation will be illustrated by calculating the spectral function of a single quantum dot connected to three leads; the system is described by the Anderson Hamiltonian:

$$H_1 = \sum_{k,\sigma,\alpha}(\epsilon_k - \mu_\alpha)c^\dagger_{k\sigma,\alpha}c_{k\sigma,\alpha} + \sum_\sigma E_d d^\dagger_\sigma d_\sigma + U_H n_{d\uparrow}n_{d\downarrow} + \sum_{k,\sigma,\alpha} t_{kd}(c^\dagger_{k\sigma,\alpha}d_\sigma + h.c.) + H.c., \tag{5}$$

where the first term describes the three leads, E_d is the energy of the dot level, U_H is the Hubbard repulsion interaction inside the dot , and the last term describes the coupling to leads.

For the calculation of the spectral function we use the equation of motion method in the version of Ref.17 extended to the non-equilibrium situation. Details of this approach in the case of many-lead systems are given in Ref.5. By the use of eq.(3) the function $\mathcal{G}_L(V)$ can be obtained numerically. The differential conductance in Fig.1 is calculated for various λ and the crossover from the peak- to dip-like behaviour can be noticed with increasing asymmetry of the applied bias.

3. Non-equilibrium behaviour of the side-coupled double dot

The double-dot set-up can be realized by replacing the central lead in Fig.1 with a second dot. Only the the Kondo dot is directly coupled to leads, however the side dot influences the transport by allowing multiple electron paths resulting in non-trivial interference effects. We shall consider a multilevel non-interacting side-coupled dot, and focus on the Fano-Kondo regime of the system. In Kondo systems, due to the bias induced dephasing, the differential conductance is suppressed around the zero bias, giving rise to a peak-like *zero bias anomaly*. However, the side-coupled double dot is characterized also by Fano interference which compete with the Kondo correlations, so that the behaviour of the differential conductance is no more obvious. The process is controlled by the gate potential applied on the lateral dot and temperature.

Qualitatively, the physical process can be described as follows. The charge current through the device is the result of the superposition of two transport channels: the *resonant* channel through the side dot and the *continuous* channel passing through the Kondo dot. The first channel is controlled by the gate potential V_g applied on the lateral dot, while the Kondo channel is controlled mainly by temperature, but also by internal parameters describing the Kondo dot (e.g., Γ, E_d). The result of the interference depends on the transmittance amplitudes along the two channels, but also on the phase difference which can give rise to constructive or destructive interference depending on external parameters T and V_g. This can be the origin of peak or dip-like aspect of the differential conductance in different ranges of the parameters space $\{T, V_g\}$. In [14] we show using the equation of motion method for the Green function that indeed the dip-like behavior of the ZBA is also possible in the vicinity of the Fano dips, where the interference effects are most pronounced. The numerical results are shown in Fig.2 which exhibits the peak-dip crossover when V_g is varied at fixed temperature.

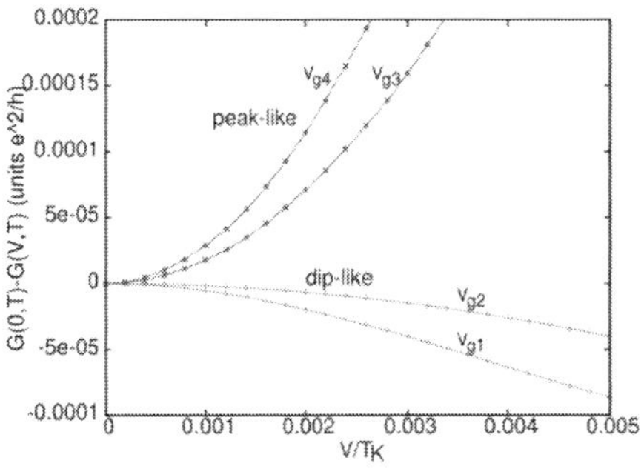

Figure 2. $\Delta G(V,T) = G(0,T) - G(V,T)$ as function of the bias (measured in units T_k) proves the peak-dip crossover when the gate potential V_g is changed at the fixed temperature $T = T_K/100$ (T_K= Kondo temperature).

The Hamiltonian of the double dot system can be written:

$$H = \sum_{k,\sigma,\alpha} (\epsilon_k - \mu_\alpha) c^\dagger_{k\sigma,\alpha} c_{k\sigma,\alpha} + \sum_\sigma E_d d^\dagger_\sigma d_\sigma + U_H n_{d\uparrow} n_{d\downarrow} + \sum_{i\sigma} E_i c^\dagger_{i\sigma} c_{i\sigma} + \frac{e^2}{2C} N^2$$
$$+ \sum_{k,\sigma,\alpha} t_{kd} (c^\dagger_{k\sigma,\alpha} d_\sigma + h.c.) + \sum_i \tau_{id} (c^\dagger_{i\sigma} d_\sigma + h.c.), \tag{6}$$

where one may identify the Hamiltonian of the leads ($\alpha = L(left), R(right)$), the Hamiltonian of the small dot including the Hubbard term, and that one of the multilevel lateral dot which contains also the Coulomb repulsion in the orthodox (capacitance) model; the operator of the total occupation number reads $N = \sum_{l\sigma} n_{l\sigma} = \sum_{l\sigma} c^\dagger_{l\sigma} c_{l\sigma}$. The last two terms stand for the coupling of the small dot to the leads and to the multilevel side dot.

Since only the Kondo dot is coupled to leads, the conductance takes the simple form (the symmetric bias will be considered in this case):

$$\mathcal{G}(V) = dI/dV = \frac{e^2}{2h} \sum_\sigma \Gamma_\sigma \; [Im G_{dd,\sigma}(-\frac{eV}{2}) + Im G_{dd,\sigma}(\frac{eV}{2})] \tag{7}$$

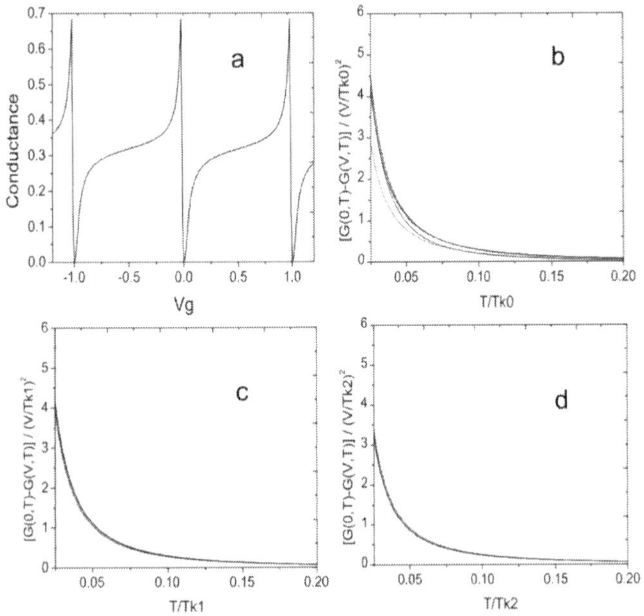

Figure 3. (a) Two Fano-Kondo resonances as function on V_g. The scaling function ΔG will be plotted for different values of Vg ($iV_g = -0.95, -0.7, -0.5, 0.1, 0.5, 0.95, 1.05, 1.1$) and considering three possible definitions for the Kondo temperature: (b) T_{K0} - the SIAM Kondo temperature, which does not depend on Vg, (c) T_{K1} - the half-width of the Kondo peak at $T = 0$, as estimated in [17] , and (d) T_{K2}- the half-width of the Kondo peak at a finite temperature as mentioned in [18]. In both cases (c) and (d) the Kondo temperature depends on Vg, and the scaling is satisfactory, resulting in the data collapse. For all curves, $V = T_K/100$. Other parameters are: $E_d = -0.25$, $\Gamma = 0.05$,$\tau = 0.006$ (measured in the lead bandwidth).

The equilibrium conductance described by the above formula in the limit $V = 0$ is plotted in Fig.3a as function of the gate potential V_g. Due to multilevel structure of the lateral dot the equilibrium conductance shows a sequence of resonances, their Fano character being obvious. The Fano *zeros* occur at integer values of V_g which is measured in units of charging energy of the side dot.

At nonequilibrium, the conductance as function of the bias V will present mainly the usual peak at $V = 0$, except for values of the V_g in vicinity of the Fano zero, as mentioned. An important property of the conductance in the Kondo regime is the scaling. The usual function for which a scaling is searched is $\Delta G(V,T) = G(0,T) - G(V,T)$ [19, 20, 21]. In Fig.3 we check weather $\Delta G(V,T)$ can be expressed in terms of scaled variables V/T_K and T/T_K; it turns out from Fig.3a that the scaling does not work if we choose as Kondo temperature that one of the isolated single dot (described by the single impurity Anderson model -SIAM, corresponding in our case to $\tau = 0$). Figs 3(c and d) shows however that the scaling works if a suitable definition of T_K is considered for the double dot system.

In [14], we have shown that the scaling properties are preserved even in the Fano-Kondo regime, if Vg is varied along a Fano line in the regions with peak-like behavior. Supplementary in this paper, we show that the good scaling properties maintain also when the gate is varied across multiple Fano resonances as seen in Fig.3.

Journal of Physics: Conference Series **338** (2012) 012001 doi:10.1088/1742-6596/338/1/012001

4. Conclusions

In summary, we investigate the nonliner conductance in two different complex dot systems :i) many-lead single dot, and ii) side-coupled double dot (similar to a mesoscopic transistor set up). In both cases the bias dependence of the differential conductance can show a peak-dip crossover around the zero bias. The crossover is controlled by external parameters as the asymmetry of the applied bias in the case i) ,and the temperature or gate potential in the case ii). For the double dot, $\Delta G(V,T) = G(0,T) - G(V,T)$ shows scaling properties provided a proper definition of the Kondo temperature is chosen. The scaling cannot be universal because of different behaviour (peak/dip-like) of the differential conductance in different ranges of the parameters.

Acknowledgments

We acknowledge support from the Core Programme under the contract No 45N/2009 and from PNCDI2-Research Programme under the grant No.515/2008.

References

[1] Goldhaber-Gordon D, Shtrikman H, Mahalu D, Abusch-Magder D, Meirav U and Kastner M A 1998 *Nature* **391** 156
[2] Lebanon E and Schiller A 2011 *Phys. Rev. B* **65** 035308
[3] Sun Q, Guo H 2001 *Phys. Rev. B* **64** 153306
[4] Shah N and Rosch A 2005 *Phys. Rev. B* **73** 081309
[5] Ţolea M, Dinu I V, Aldea A 2009 *Phys. Rev. B* **79** 033306
[6] Leturcq R, Schmid L, Ensslin K, Meir Y, Driscoll D C, and Gossard A C 2005 *Phys. Rev. Lett.* **95** 126603
[7] Leturcq R, Schmid L, Ensslin K, Driscoll D C, and Gossard A C 2006 *pss-b* **243** 3653
[8] Cornaglia P S and Grempel D R 2005 *Phys. Rev. B* **71** 075305
[9] Chung C H , Zarand G, and Wölfle P 2008 *Phys. Rev. B* **77** 035120 , Chung C H and Lee T H 2010 *Phys. Rev. B* **82** 085325
[10] Wu B H, Cao J C, and Ahn K H 2005 *Phys. Rev. B* **72** 165313
[11] Sasaki S, Tamura H, Akazaki T, and Fujisawa T 2009 *Phys. Rev. Lett.***103** 266806
[12] Tamura H, Sasaki S 2010 *Physica E* **42** 864
[13] Zitko R 2010 *Phys. Rev. B* **81** 115316
[14] Aldea A, Ţolea M, Dinu I V 2011 *Phys. Rev. B* **83** 245317
[15] Swirkowicz R, Barnas J, and Wilczynski W 2003 *Phys. Rev. B* **68** 195318
[16] Dinu I V, Ţolea M, Aldea A 2007 *Phys. Rev. B* **76** 113302
[17] Entin-Wohlman O, Aharony A, and Meir Y 2005 *Phys. Rev. B* **71** 035333
[18] T_{K2} is calculated solving numerically the equation $ImG(Ef + T_{K2}/2) = 1/2ImG(Ef)$, at $T = T_{K2}/100$, after the background is removed (by background we mean the value of the density of states at high temperatures, when the Kondo peak is absent).
[19] Ralph D C, Ludwig A W W, von Delft J, and Buhrman R A 1994 *Phys. Rev. Lett.* **72** 1064
[20] Schiller A and Herschfield S 1995 *Phys. Rev. B* **51** 12896
[21] Grobis M, Rau I G, Potok R M, Shtrikman H, and Goldhaber-Gordon D 2008 *Phys. Rev. Lett.***100** 246601

Journal of Physics: Conference Series **338** (2012) 012002 doi:10.1088/1742-6596/338/1/012002

Universal features in the thermodynamics and heat transport by particles of any statistics

Dragoş-Victor Anghel

Department of Theoretical Physics, National Institute of Physics and Nuclear Engineering, 30 Reactorului Street, P.O.BOX MG-6, RO-077125 Magurele, Jud. Ilfov, ROMANIA

E-mail: dragos@theory.nipne.ro

Abstract. I discuss in parallel two universal phenomena: the independence of statistics of the heat capacity and entropy of ideal gases of the same, constant, density of states, on one hand, and the independence of statistics of the heat and entropy transport through one-dimensional channels, on the other hand. I show that there is a close similarity between the microscopic explanations of each of these phenomena.

1. Introduction

The *thermodynamic equivalence* is a name coined by M. H. Lee [1] for the property of ideal Bose and Fermi gases confined in two-dimensional (2D) boxes to have the same heat capacity, C_V, and entropy, S. This implies that from the thermodynamic perspective, 2D Bose and Fermi gases under canonical conditions (i.e. has a constant volume and can exchange only heat with the environment) are identical.

This property was first proved by Auluck and Kothari in Ref. [2] and then rediscovered by May [3] and Viefers, Ravndal and Haugset [4]. Nevertheless, it did not receive much attention until Lee introduced a unitary description of ideal Bose and Fermi gases [1, 5, 6, 7] in terms of polylogarithmic functions [8, 9].

In Ref. [10], Haldane introduced an extension of the Pauli exclusion principle, which was latter called the *fractional exclusion statistics* (FES). In FES, the system is divided into species of particles, that we shall identify by an index, e.g. $i = 0, 1, \ldots$. In each species there are G_i available single-particle states and N_i particles. The FES is defined by the fact that if we change for example N_i by δN_i, then the dimension of species j, for any $j = 0, 1, \ldots$, changes by $\delta G_j = -\alpha_{ji}\delta N_i$, where α_{ij} are called the FES parameters. The diagonal parameters (α_{ii}) are called *direct parameters*, whereas the rest of them (α_{ij}, $i \neq j$) are called the *mutual parameters*. For example for a continuous (i.e. macroscopic) FES system the species of particles are defined by coarse-graining the phase space, each grain representing a species of particles [11, 12, 13].

The general properties of the FES parameters were deduced only relatively recently [14, 15, 16]. A quite general ansatz for the FES parameters that obey the properties introduced in [14] is [11, 17]

$$\alpha_{ij} = \alpha_{ij}^{(e)} + \alpha_i^{(s)}\delta_{ij}. \quad (1)$$

The parameters $\alpha_{ij}^{(e)}$ are "extensive", i.e. they are proportional to G_i – the dimension of the species on which they act,

$$\alpha_{ij}^{(e)} \equiv a_{ij} G_i, \tag{2}$$

whereas the parameters $\alpha_i^{(s)}$ are direct parameters and are not extensive.

Typically, in the literature we find exclusion statistics parameters of the (s) type (see e.g. [18, 19, 20, 21, 22, 23, 24, 25]), so in general $\alpha_{ij} = 0$ for any $i \neq j$. If, moreover, we impose $\alpha_{ii} \equiv \alpha$ for any i, the thermodynamic calculations simplify considerably. If $\alpha = 0$ we obtain the Bose statistics, whereas if $\alpha = 1$ we obtain the Fermi statistics.

If $\alpha_{ij} \equiv \alpha\delta_{ij}$, one obtains a very surprising result: in 2D systems C_V and S are independent of α and equal to the heat capacity and entropy of ideal Bose and Fermi gases. Therefore 2D systems of *any* statistics, i.e. Bose, Fermi or FES, are thermodynamically equivalent [24].

The property which leads to the thermodynamic equivalence of 2D ideal gases is the constant density of single-particle states (DOS). A system of ideal particles in a d-dimensional (dD) box has a DOS of the form

$$\sigma_d(\epsilon) = V \frac{d}{(2\pi)^{d/2}\Gamma\left(\frac{d}{2}+1\right)} \left(\frac{m}{\hbar^2}\right)^{d/2} \epsilon^{(d/2)-1} \equiv C\epsilon^{(d/2)-1}, \tag{3}$$

where ϵ is the single-particle energy, V si the dD volume, m is the mass of the particle and Γ is the gamma function. If $d = 2$, then $\sigma_{d=2}(\epsilon) = Vm/(\pi\hbar^2) \equiv C$ is a constant. We shall see in section 2 that the ideal systems of any statistics, of the same, constant DOS are thermodynamically equivalent.

The thermodynamic equivalence has a correspondent in the transport properties of ideal gases. At first, Rego and Kirczenow observed that in the low temperature limit, the heat conductivity of ideal particles confined in a 1D channel is independent of the dispersion relation [26] and, moreover, is independent also of their statistics [27]. The statistics independence of the heat conductivity implies the independence of statistics of the entropy current in 1D channels, as it was shown by Blencowe [28].

The independence of the low temperature heat conductivity of the dispersion relation have been confirmed experimentally in 1D phonons [29] and photons [30, 31] channels.

The low temperature results of Rego and Kirczenow have been extended in Ref. [32]. In this paper we pointed to another class of universality in systems of ideal particles, namely we showed that the low temperature statistics independence of the 1D channel heat conductivity can be extended to any temperature. In other words, the heat conductivity of ideal particles through a 1D channel is independent of the statistics of the particles at any temperature.

In principle all the universal physical properties should have also an universal (and hopefully very simple) microscopic physical explanation; there should be a key, basic feature, common to all these systems, that is leading always to similar results. For the thermodynamic equivalence the key feature is the spectrum of excitations, which is independent of statistics [24], whereas for the 1D universality the key feature is the current of excitations, which is also independent of statistics [32]. The two microscopic explanations are very closely related and, as we shall see in the next two sections, the quantities related to the 1D transport have mathematical expressions very similar to the ones that describe the equilibrium thermodynamics.

In section 2 we shall discuss the thermodynamics of systems of constant DOS and we shall present the main results and formulae. We shall also present the microscopic explanation of the thermodynamic equivalence.

In section 3 we shall present the calculations of the heat, particle and entropy transport in 1D channels and, by employing the formulae of section 2, we shall show that they are independent of statistics at any temperature. The microscopic interpretation of this phenomenon is given in the end of section 3 and is very similar to the microscopic interpretation of the thermodynamic equivalence.

Section 4 is reserved for conclusions.

2. Thermodynamic equivalence

Let us assume that we have a macroscopic system of ideal FES particles, with a DOS of the form $\sigma(\epsilon) = C\epsilon^s$. The FES parameters are $\alpha_{ij} \equiv \alpha\delta_{ij}$, for any i and j. We associate the same energy to all the particles in a species (ϵ_i, $i = 0, 1, \ldots$) and the same chemical potential, μ, to all the species. If we denote by G_i the number of states in the species i when there are no particles in it, then in the presence of N_i particles the number of available states becomes $G_i - \alpha N_i$ [10, 18, 21]. The number of particle arrangements that we can have in the species i is

$$W(G_i, N_i) = \frac{[G_i + (1-\alpha)N_i - 1]!}{N_i!\,[G_i - \alpha N_i - 1]!}. \tag{4}$$

Therefore the total number of microscopic configurations in the system, when the population of each species is fixed, is given by the product [18]

$$W(\{G_i, N_i\}) = \prod_i \frac{[G_i + (1-\alpha)N_i - 1]!}{N_i!\,[G_i - \alpha N_i - 1]!}. \tag{5}$$

The grandcanonical partition function of the system, at temperature T and chemical potential μ, is

$$\mathcal{Z} = \sum_{\{N_i\}} \mathcal{Z}(\{G_i, N_i\}) = \sum_{\{N_i\}} W(\{G_i, N_i\}) \exp\left[\sum_i \beta N_i(\mu - \epsilon_i)\right], \tag{6}$$

where $\beta = 1/(k_{\rm B}T)$ and $\mathcal{Z}(\{G_i, N_i\})$ is the "partition function" corresponding to a particular choice of the set $\{N_i\}$.

The standard procedure to find the particle populations of the species, denoted by $n_i \equiv N_i/G_i$, is to maximize $\mathcal{Z}(\{G_i, N_i\}) \equiv \mathcal{Z}(\{G_i, n_i\})$ with respect to the n_i's: $\partial\mathcal{Z}(\{G_i, n_i\})/\partial n_i = 0$. In this way it was obtained [18]

$$n(\epsilon) = \{w(\zeta_\epsilon) + \alpha\}^{-1}, \tag{7a}$$

where w is a function that satisfies the equation

$$w(\zeta_\epsilon)^\alpha[1 + w(\zeta_\epsilon)]^{1-\alpha} = \zeta_\epsilon^{-1} \equiv e^{\beta(\epsilon-\mu)} \tag{7b}$$

Now we apply the results above to our system of DOS $\sigma(\epsilon) = C\epsilon^s$. From (6) we write the expression for the grandcanonical thermodynamic potential,

$$-PV = \Omega = -k_{\rm B}T\log(\mathcal{Z}) = k_{\rm B}T\int_0^\infty d\epsilon\, C\epsilon^s \log\{(1-\alpha n)/[1 + (1-\alpha)n]\}. \tag{8a}$$

Integrating by parts and using the equations (7) we obtain

$$PV = \frac{1}{s+1}\int_0^\infty d\epsilon\, C\epsilon^{s+1} n(\epsilon) \equiv \frac{U}{s+1}, \tag{8b}$$

where U is the internal energy of the system.

Similarly, the total number of particles is

$$N = \int_0^\infty d\epsilon\, C\epsilon^s n(\epsilon). \tag{9}$$

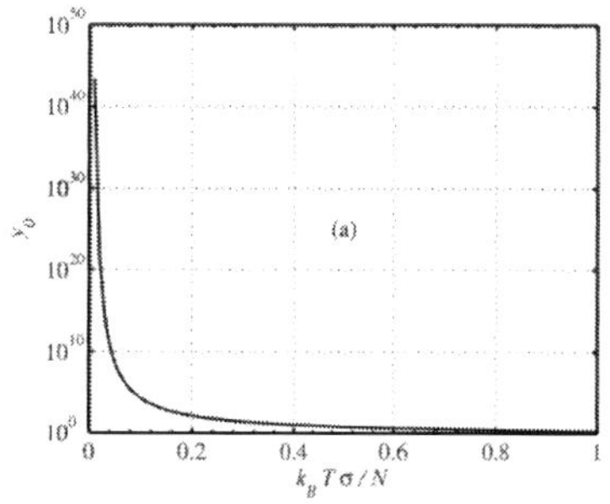

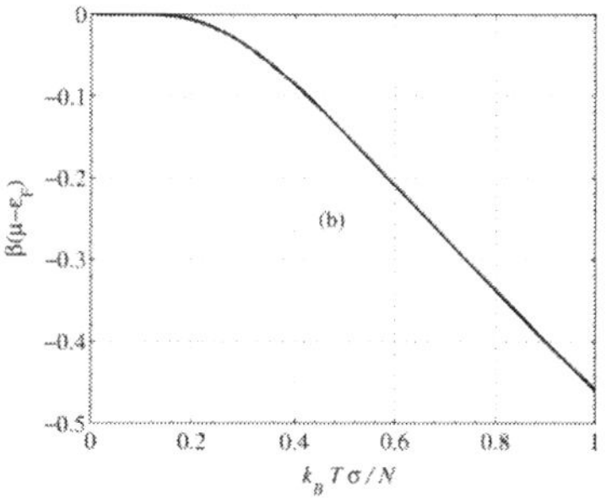

Figure 1. The statistics independent functions, y_0 (a) and $\beta(\mu - \epsilon_{\rm F})$ (b), vs the scaled temperature, $k_{\rm B}T\sigma/N$.

All these functions may be calculated for example by expressing ϵ in terms of w,

$$\beta\epsilon = \log\left[w^\alpha(1+w)^{1-\alpha}\right] + \beta\mu, \tag{10}$$

but the integrals cannot be performed analytically for general s and T.

For $s = 0$ ($\sigma \equiv C$), all the thermodynamic quantities can be expressed in terms of elementary or polylogarithmic functions [24]. I start with N, which, by simple algebra may be written as

$$N = k_{\rm B}T\sigma\log(1+y_0), \tag{11}$$

where $y_0 \equiv 1/w(\zeta_{\epsilon=0})$ satisfies the equation

$$(1+y_0)^{1-\alpha}/y_0 = \zeta_{\epsilon=0}^{-1} \equiv {\rm e}^{-\beta\mu}. \tag{12}$$

We observe from eq. (11) that y_0 is a function of the dimensionless variable $x \equiv k_{\rm B}T\sigma/N$, and does not depend on α. Moreover, as seen in figure 1a, $y_0(x)$ is a monotonically decreasing function, with $y_0(x\to\infty) = 0$ and $y_0(x \to 0) \to \infty$.

From the equations (11) and (12) follows

$$\exp\left[(\mu - \alpha N/\sigma)/k_{\rm B}T\right] = 1 - \exp\left[-N/(\sigma k_{\rm B}T)\right], \tag{13}$$

where we can identify the (generalized) Fermi energy as $\epsilon_{\rm F} \equiv \lim_{T\to 0}\mu = \alpha N/\sigma$ and observe that $\mu - \epsilon_{\rm F}$ is also independent of α (see figure 1b).

After some algebra we obtain

$$\begin{aligned} -\Omega = U &= -(k_{\rm B}T)^2\sigma\left[\frac{1-\alpha}{2}\log^2(1+y_0) + Li_2(-y_0)\right] \\ &= (k_{\rm B}T)^2\sigma\left[\frac{\alpha}{2}\log^2(1+y_0) + Li_2\left(\frac{y_0}{1+y_0}\right)\right] \end{aligned} \tag{14}$$

where $Li_2(x) = \sum_{k=1}^\infty x^k/k^2$ is Euler's dilogarithm [8]. The two expressions for U in (14) are connected by Landen's relation, $Li_2(x) + Li_2[-x/(1-x)] = -(1/2)\log^2(1-x)$, which is valid for any $x < 1$ [8].

Using (11) into (14) we obtain

$$-\frac{\sigma}{N^2}\Omega = \frac{\sigma}{N^2}U = -\left(\frac{k_{\rm B}T\sigma}{N}\right)^2 Li_2(-y_0) + \frac{\alpha}{2} - \frac{1}{2} \tag{15a}$$

$$= \left(\frac{k_{\rm B}T\sigma}{N}\right)^2 Li_2\left(\frac{y_0}{1+y_0}\right) + \frac{\alpha}{2} \tag{15b}$$

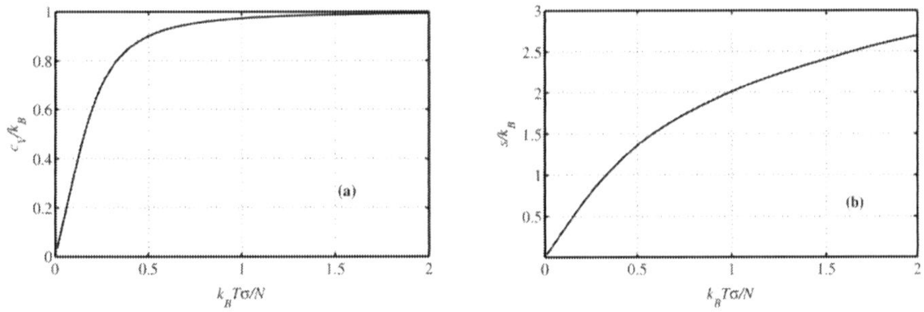

Figure 2. The specific heat (a), $c_V/k_{\rm B} \equiv C_V/(Nk_{\rm B})$, and entropy per particle (b), $s/k_{\rm B} \equiv S/(Nk_{\rm B})$, for a system of particles of constant DOS, σ, and any α.

In the expression (15a), $U_{\rm F} = -(k_{\rm B}T)^2\sigma Li_2(-y_0)$ is the internal energy of a Fermi gas ($U = U_{\rm F}$, if $\alpha = 1$), whereas in the expression (15b), $U_{\rm B} = (k_{\rm B}T)^2\sigma Li_2[y_0/(1+y_0)]$, represents the internal energy of a Bose gas ($U = U_{\rm B}$, if $\alpha = 0$).

Since y_0 does not depend on α, but just on $k_{\rm B}T\sigma/N$ and T, it is obvious that the differences between the thermodynamic potentials of gases with different α's come just from an additive constant. All the temperature dependence is the same.

To calculate C_V, we use the expression

$$C_V = \frac{\partial U}{\partial T} - \frac{\partial U}{\partial \mu}\frac{\partial N}{\partial T}\left(\frac{\partial N}{\partial \mu}\right)^{-1} \tag{16}$$

and obtain

$$C_V = -\frac{N^2}{T\sigma}\frac{1+y_0}{y_0} - 2k_{\rm B}^2 T\sigma Li_2(-y_0). \tag{17a}$$

To calculate the entropy we use $S = U/T + PV/T - \mu N/T$, together with equations (11), (12) and (15), to obtain

$$S = -k_{\rm B}^2 T\sigma[2Li_2(-y_0) + \log(1+y_0)\log y_0]. \tag{17b}$$

Equations (17a) and (17b) are independent of α and they express eventually in the most compact way the thermodynamic equivalence.

In figure 2 (a) and (b) we plotted $c_V/k_{\rm B} \equiv C_V/(Nk_{\rm B})$ and $s/k_{\rm B} \equiv S/(Nk_{\rm B})$, respectively.

Another way to calculate the entropy is by using the definition

$$TS = k_{\rm B}T\log W, \tag{18}$$

where W is given by (5). From the expression (5) we take only the most probable distribution, i.e. the one that satisfies the equations (7), and, after some algebra we obtain

$$\begin{aligned}\log W &= \sum_i G_i\{[1+(1-\alpha)n_i]\log[1+(1-\alpha)n_i] - n_i\log n_i - [1-\alpha n_i]\log[1-\alpha n_i]\} \\ &= \sigma\int_0^\infty d\epsilon\{[1+(1-\alpha)n(\epsilon)]\log[1+(1-\alpha)n(\epsilon)] - n(\epsilon)\log n(\epsilon) \\ &\quad -[1-\alpha n(\epsilon)]\log[1-\alpha n(\epsilon)]\}\end{aligned} \tag{19}$$

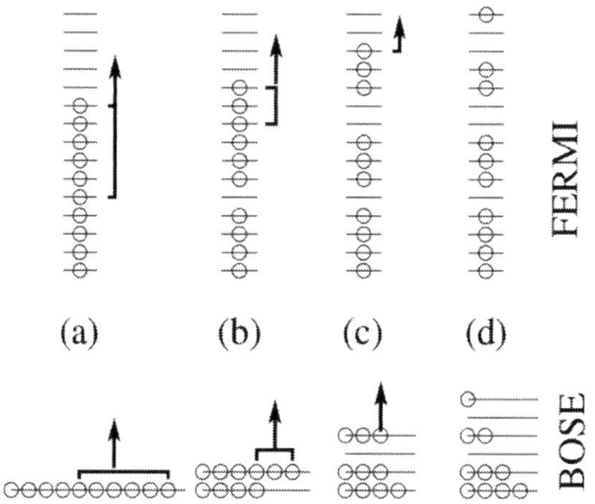

Figure 3. Excitations in a Bose and a Fermi gas, each of 10 particles and the same DOS. To each configuration of bosons it corresponds a configuration of fermions, with the same excitation energy. In (a) the two systems are in the ground state; in (b) six particles in each are excited on the first "level"; from the first level, 3 particles are excited two levels up (c), whereas in (d), one particle from the uppermost "level" is excited two states up.

where we have used the Stirling approximation, $\log(N!) \approx N\log(N/\mathrm{e})$, valid for $N \gg 1$, and neglected the terms of the order of $1/G_i$. Plugging (7) into (19) we obtain

$$\begin{aligned} TS &= k_{\rm B}T\sigma\int_0^\infty \frac{(w+1)\log(w+1)-w\log w}{w+\alpha}\mathrm{d}\epsilon \\ &= (k_{\rm B}T)^2\sigma\int_{y_0^{-1}}^\infty\left[\frac{\log(w+1)}{w}-\frac{\log w}{w+1}\right]\mathrm{d}\epsilon \\ &= (k_{\rm B}T)^2\sigma\int_0^{y_0}\left[\frac{\log(y+1)}{y}-\frac{\log y}{y+1}\right]\mathrm{d}\epsilon \\ &= -(k_{\rm B}T)^2\sigma[2Li_2(-y_0)+\log(1+y_0)\log y_0], \end{aligned} \tag{20}$$

which is equation (17b), as expected.

Since according to Eq. (11) $y_0 \to \infty$ when $T \to 0$, making use of the asymptotic behaviour of the dilogarithm, $Li_2(-y_0) \sim -[\pi^2/6 + \log^2(y_0)/2]$ one can show that in the limit of low temperatures

$$C_{\rm V} \sim (\pi^2/3)k_{\rm B}^2T\sigma\{1 - O[\log^2(1+y_0)/y_0]\}. \tag{21}$$

2.1. Microscopic interpretation

The microscopic interpretation of the remarkable identities expressed by Eqs. (17) was given in Ref. [24]. We showed in that paper that if we have two FES systems of the same, constant, DOS, but different α's, one can establish a one-to-one correspondence between configurations of particles in the two systems that have the same excitation energy. The excitation energy is the difference between the energy of the system in the given configuration and the energy of the ground state; the average of the excitation energy is $U_{\rm B}$, defined above.

In figure 3 we give an example of correspondence between configurations of fermions and configurations of bosons with the same excitation energies. In this figure we can see how the correspondence is established. Based on this method, in Ref. [33] we showed how we can transform a general system of fermions into a system of bosons.

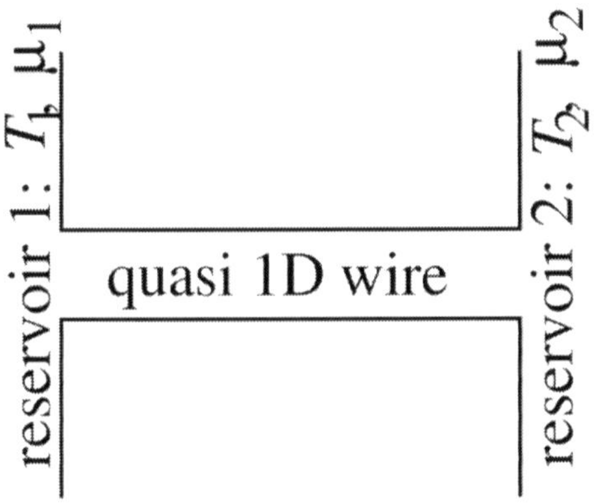

Figure 4. Schematic view of the heat and particle transport experiment. The system, which consists of two reservoirs connected by a quasi-1D wire, contains FES particles of parameter α. The two reservoirs are at the temperatures and chemical potentials (T_1, μ_1) and (T_2, μ_2), respectively.

3. Transport equivalence

In Ref. [27], Rego and Kirczenow proved theoretically that the low temperature limit of the 1D heat conductance if ideal particles is equal to $\kappa_0 \equiv \pi^2 k_{\rm B}{}^2 T/(3h)$ and is independent of the statistics of the particles. We extended this result in Ref. [32] and showed that the heat conductivity of 1D channels is independent of statistics at any temperature.

Let us now consider a two-terminal transport experiment in which the reservoirs 1 and 2 are connected by a quasi 1D wire, like in figure 4. The temperatures and chemical potentials in the two reservoirs will be denoted by T_r and μ_r, respectively $(r = 1, 2)$. The particles in the system obey FES of parameter α. If there are $\mathcal{N}_c$ transport channels through the wire, the Landauer formula for the particle and energy fluxes between the two reservoirs are [27, 28]

$$I = \sum_{p=1}^{\mathcal{N}_c} \int_0^\infty \frac{dk}{2\pi} v_p(k) \left[n_1(k) - n_2(k)\right] \zeta_p(k), \tag{22a}$$

$$\dot{U} = \sum_{p=1}^{\mathcal{N}_c} \int_0^\infty \frac{dk}{2\pi} \epsilon v_p(k) \left[n_1(k) - n_2(k)\right] \zeta_p(k), \tag{22b}$$

respectively, where $v_p(k)$ is the group velocity of the particles of momentum k and $n_r(k)$ denotes the thermal particle population in the reservoir r, i.e. $n_r(k) \equiv n[\epsilon_r(k), \mu_r, T_r]$ and satisfies Eqs. (7); finally, $\zeta_p(k)$ is the particle transmission coefficient through the wire.

Following [32] and using the thermodynamic relation for heat exchange, $Q = T\delta S = \delta U - \mu\delta N$ (see in section 2 the relation for S), where δS is the entropy variation in the process, we write the heat flux,

$$\dot{Q} = \dot{U} - \mu I \tag{23}$$

For massless particles, like phonons, $\dot{U} \equiv \dot{Q}$, since $\mu = 0$.

Since the particle group velocity is $v_p(k) = \hbar^{-1}(d\epsilon(k)/dk)$, and assuming further that $\zeta_p(k) \equiv \zeta_p$ is independent of k, Eqs. (22) get the simple form

$$I = \sum_{p=1}^{\mathcal{N}_c} \frac{\zeta_p}{h} \int_{\epsilon_p(0)}^\infty d\epsilon \left[n(\epsilon, \mu_1, T_1) - n(\epsilon, \mu_2, T_2)\right] \tag{24a}$$

$$\dot{U} = \sum_{p=1}^{\mathcal{N}_c} \frac{\zeta_p}{h} \int_{\epsilon_p(0)}^{\infty} d\epsilon\, \epsilon \left[n(\epsilon,\mu_1,T_1) - n(\epsilon,\mu_2,T_2)\right] \tag{24b}$$

where $\epsilon_p(0)$ is the lowest energy level in the channel p.

We are working in the linear regime, i.e. $|T_1 - T_2|/T_r \ll 1$ and $|\mu_1 - \mu_2| \ll k_B T_r$ $(r = 1, 2)$, so we shall denote the average temperature, $(T_1 + T_2)/2$, by T. We introduce the notations $I_1(\mu,T) = \sum_p I_{p,1}(\mu,T)$ and $\dot{U}_1(\mu,T) = \sum_p \dot{U}_{p,1}(\mu,T)$, with

$$I_{p,1}(\mu,T) \equiv \frac{\zeta_p}{h} \int_{\epsilon_p(0)}^{\infty} n_\alpha(\epsilon,\mu,T) d\epsilon, \tag{25a}$$

$$\dot{U}_{p,1}(\mu,T) \equiv \frac{\zeta_p}{h} \int_{\epsilon_p(0)}^{\infty} \epsilon n_\alpha(\epsilon,\mu,T) d\epsilon, \tag{25b}$$

and define I and $\dot{U}$ as

$$I \equiv \delta I_1 = \frac{\partial I_1}{\partial T}\delta T + \frac{\partial I_1}{\partial \mu}\delta\mu, \tag{26a}$$

$$\dot{U} \equiv \delta \dot{U}_1 = \frac{\partial \dot{U}_1}{\partial T}\delta T + \frac{\partial \dot{U}_1}{\partial \mu}\delta\mu. \tag{26b}$$

Rego and Kirczenow calculated the low temperature limit of $\kappa_{\dot{U}} \equiv \partial \dot{U}_1/\partial T$ – which we shall call here the energy conductance – and obtained $\kappa_{\dot{U}} \stackrel{T\to 0}{\sim} \mathcal{N}_c \kappa_0$ for all α's [27, 28], if $\epsilon_p(0)$ takes the same value for all p.

If we use the relations (23) and (26), together with the condition $I = 0$, we obtain [32] $\dot{Q} = \dot{U}$ and

$$\kappa \equiv \frac{\partial \dot{U}_1}{\partial T} + \frac{\partial \dot{U}_1}{\partial \mu}\left(\frac{d\mu}{dT}\right)_{I=0}. \tag{27}$$

We can observe now already that in general the expression (27) for the heat conductivity is similar to the expression for the heat capacity, if we replace $\dot{U}_1$ by U, i.e. the energy current by the internal energy.

Let us now set $\mathcal{N} = 1$ and focus only on one channel conduction. In such a case, $I_1(\mu,T) \equiv I_{p=1,1}(\mu,T)$ and we observe that if we set $\epsilon_{p=1}(0) \equiv 0$, then $I_1(\mu,T)$ and $\dot{U}_1(\mu,T)$ have the same expressions as the particle number, $N(\mu,T)$, and internal energy, $U(\mu,T)$, respectively, of a FES system of parameter α and constant DOS, $\sigma = \zeta_{p=1}/h$, in equilibrium at temperature T and chemical potential μ. Therefore we can apply here directly the results from the previous section.

Let us drop the subscript 1 from $I_1(\mu,T)$ and $\dot{U}_1(\mu,T)$ and rewrite the equilibrium thermodynamics equations of section 2, using the quantities corresponding to the stationary transport. We define again a quantity y_0, which satisfies

$$I(\mu,T) \equiv \frac{\zeta k_B T}{h}\log(1 + y_0), \tag{28a}$$

$$(1 + y_0)^{1-\alpha}/y_0 = e^{-\beta\mu}, \tag{28b}$$

$$\exp\left[(\mu - \alpha h I/\zeta)/k_B T\right] = 1 - \exp\left[-hI/(\zeta k_B T)\right], \tag{28c}$$

in analogy to eqs. (11)-(13)

From the eqs. (28) and observing that $\dot{U}(\mu,T)$ has an expression similar to that of $U(\mu,T)$, we obtain

$$\dot{U} = -\frac{1-\alpha}{2}\frac{hI^2}{\zeta} - (k_B T)^2 \frac{\zeta}{h} Li_2(-y_0) = \frac{\alpha h I^2}{2\zeta} + \frac{\zeta (k_B T)^2}{h} Li_2\left(\frac{y_0}{y_0+1}\right), \tag{28d}$$

where we used again the Landen's relations.

From Eqs. (28) and using

$$\left(\frac{\partial y_0}{\partial T}\right)_\mu = -\frac{\mu y_0(1+y_0)}{k_B T^2(1+\alpha y_0)} \tag{29a}$$

and

$$\left(\frac{\partial y_0}{\partial \mu}\right)_T = \frac{y_0(1+y_0)}{k_B T(1+\alpha y_0)} \tag{29b}$$

we calculate

$$\frac{\partial I}{\partial T} = \frac{\zeta k_B}{h}\log(1+y_0) - \frac{\zeta\mu}{hT}\frac{y_0}{1+\alpha y_0}, \tag{30a}$$

$$\frac{\partial I}{\partial \mu} = = \frac{\zeta y_0}{h(1+\alpha y_0)}. \tag{30b}$$

$$\frac{\partial \dot{U}}{\partial T} = -(1-\alpha)\frac{hI}{\zeta}\frac{\partial I}{\partial T} - 2k_B^2 T\frac{\zeta}{h}Li_2(-y_0) + \frac{\mu I(1+y_0)}{T(1+\alpha y_0)}, \tag{31a}$$

$$\frac{\partial \dot{U}}{\partial \mu} = -(1-\alpha)\frac{hI}{\zeta}\frac{\partial I}{\partial \mu} - 2k_B^2 T\frac{\zeta}{h}Li_2(-y_0) + \frac{I(1+y_0)}{1+\alpha y_0}, \tag{31b}$$

With eqs. (30) and (31) and using the identity

$$\left.\frac{d\mu}{dT}\right|_{I=0} = -\left.\frac{\partial I}{\partial T}\right|_\mu \left(\left.\frac{\partial I}{\partial \mu}\right|_T\right)^{-1} \tag{32}$$

we can calculate the heat conductivity. Plugging (32) into (27) we obtain

$$\kappa = \left.\frac{d\dot{U}}{dT}\right|_I = \frac{\partial \dot{U}}{\partial T} - \frac{\partial \dot{U}}{\partial \mu}\left.\frac{\partial I}{\partial T}\right|_\mu \left(\left.\frac{\partial I}{\partial \mu}\right|_T\right)^{-1} \tag{33}$$

which, if we replace I by N and $\dot{U}$ by U, becomes identical to the expression for the heat capacity, C_V (17a). Therefore we can transcribe Eq. (17a) into an equation for the heat conductivity as

$$\kappa = -\frac{I^2 h}{T\zeta}\frac{1+y_0}{y_0} - \frac{2k_B^2 T\zeta}{h}Li_2(-y_0), \tag{34}$$

which is independent of α at *any temperature*. In the low temperature limit Eq. (34) becomes the universal asymptotic expression, κ_0, which is equivalent to (21) from equilibrium thermodynamics, but with $\sigma \equiv \zeta/h$.

The heat flux is related to the entropy flux by the relation $d\dot{U}_{1,I=0} = Td\dot{S}_1$, from where we obtain an entropy flux

$$\dot{S}_1 = \int_0^T \frac{\kappa}{T'}\mathrm{d}T', \tag{35a}$$

which is consistent with the definition used in equilibrium thermodynamics,

$$S = \int_0^T \frac{C_V}{T'}\mathrm{d}T'. \tag{35b}$$

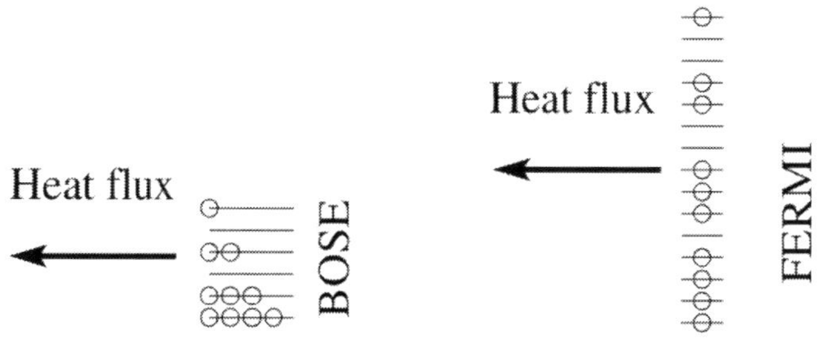

Figure 5. The microscopic analysis of the statistics independence of the heat conductivity: there is a one-to-one correspondence between micro-configurations of particles of different statistics which have the same excitation energy, $\dot{U}_{B,n,1}$, and therefore carry the same heat fluxes.

But since S of (35b) is consistent with the Boltzmann definition (18) and therefore with the integral of eq. (20), it follows that the definition (35a) is also consistent with

$$\begin{aligned} \dot{S} &= \frac{k_{\rm B}\zeta}{h}\int_0^\infty \{[1+(1-\alpha)n(\epsilon)]\log[1+(1-\alpha)n(\epsilon)] \\ &\quad -n(\epsilon)\log[n(\epsilon)] - [1-\alpha n(\epsilon)]\log[1-\alpha n(\epsilon)]\}d\epsilon \\ &= -\frac{k_{\rm B}^2 T\zeta}{h}[2Li_2(-y_0) + \log(1+y_0)\log y_0], \end{aligned} \tag{36}$$

which represents the flux of the number of configurations of particle populations, $\{n(\epsilon)\}$, from one reservoir to the other. But we notice again that eq. (36), like eq. (20), is independent of α at *any temperature*. Therefore the entropy flux is independent of α at any T.

3.1. Microscopic interpretation

From the r.h.s. of eq. (28d) we observe that the temperature dependent part of the energy flux $\dot{U}$ is the excitation energy flux, $\dot{U}_B \equiv \frac{\zeta(k_{\rm B}T)^2}{h} Li_2\left(\frac{y_0}{y_0+1}\right)$, which is independent of α. For $\alpha = 0$, $\dot{U} \equiv \dot{U}_B$. The microscopic interpretation of this statistics independence is similar to the one given in section 2.

As in Ref. [24], one can realize a one-to-one correspondence between micro-configurations of particles that transport the same excitation energies, although they have different exclusion statistics. An example of two micro-configurations like this, one of bosons and one of fermions, is given in figure 5. Since two such micro-configurations transport the same excitation energies and the same particle fluxes, then they transport the same heat fluxes. Moreover, since the entropy flux is the flux of the number of micro-configurations, the entropy flux should also be independent of the statistics [32].

4. Conclusions

I reviewed the thermodynamic equivalence of systems of the same, constant density of states (DOS) and the independence of statistics of the heat conductivity of one-dimensional (1D) transport channels. I showed that these two classes of apparently different physical phenomena have many similarities and practically the formulae that describe the equilibrium thermodynamics of constant DOS systems can be directly transcribed for the transport of 1D channels.

I also presented the microscopic explanation of these universal features of FES gases. I showed that the thermodynamic equivalence is caused by the independence of statistics of the particle excitation spectra, whereas the universality of the heat conductivity of 1D channels is due to the independence of statistics of particle excitation fluxes.

In conclusion, the similarity between the mathematical formulation of the equilibrium thermodynamics of constant DOS systems and the mathematical formulation of the transport properties of 1D channels is a reflection of the microscopic similarities between these two types of physical phenomena.

Acknowledgments
The work was supported by the Romanian National Authority for Scientific Research, CNCS-UEFISCDI project PN-II-ID-PCE-2011-3-0960 and project PN09370102/2009. The travel support from the Romania-JINR Dubna collaboration project Titeica-Markov and project N4063 are gratefully acknowledged.

References
[1] Lee M H 1997 *Phys. Rev. E* **55** 1518
[2] Auluck F C and Kothari D S 1946 *Proc. Cambridge Philos. Soc.* **42** 272
[3] May R M 1964 *Phys. Rev.* **135** A1515
[4] Viefers S, Ravndal F, and Haugset T 1995 *Am. J. Phys.* **63** 369
[5] Lee M H 1995 *J. of Math. Phys.* **36** 1217
[6] Lee M H and Kim J 2002 *Physica A* **304** 421
[7] Lee M H 2009 *Acta Physica Polonica B* **40** 1279
[8] Lewin L 1958 *Dilogarithms and Associated Functions.* (Macdonald, London)
[9] Lee M H 1997 *Phys. Rev. E* **56** 3909
[10] Haldane F D M 1991 *Phys. Rev. Lett.* **67** 937
[11] Anghel D V 2007 *J. Phys. A: Math. Theor.* **40** F1013
[12] Anghel D V 2008 *Phys. Lett. A* **372** 5745
[13] Anghel D V *Rom. J. Phys.* **54** 281
[14] Anghel D V 2009 *EPL* **87** 60009
[15] Anghel D V 2010 *Phys. Rev. Lett.* **104** 198901
[16] Wu Y-S 2010 *Phys. Rev. Lett.* **104** 198902
[17] Anghel D V 2010 *EPL* **90** 10006
[18] Wu Y-S 1994 *Phys. Rev. Lett.* **73** 922
[19] Murthy M V N and Shankar R 1994 *Phys. Rev. Lett.* **73** 3331
[20] Sen D and Bhaduri R K 1995 *Phys. Rev. Lett.* **74** 3912
[21] Murthy M V N and R. Shankar R 1999 *Phys. Rev. B* **60** 6517
[22] Bhaduri R K, Reimann S M, Viefers S, Choudhury A G, and Srivastava M K 2000 *J. Phys. B* **33** 3895
[23] Hansson T H, Leinaas J M, and Viefers S 2001 *Phys. Rev. Lett.* **86** 2930
[24] Anghel D V 2002 *J. Phys. A: Math. Gen.* **35** 7255
[25] Anghel D V 2007 *Rom. Rep. Phys.* **59** 235
[26] Rego L G C and Kirczenow G 1998 *Phys. Rev. Lett.* **81** 232
[27] Rego L G C and Kirczenow G 1999 *Phys. Rev. B* **59** 13080
[28] Blencowe M 2004 *Phys. Rep.* **395** 159
[29] Schwab K, Henriksen E A, Worlock J M, and Roukes M L 2000 *Nature* **404** 974
[30] Meschke M, Guichard W, and Pekola J P 2006 *Nature* **444** 187
[31] Timofeev A V, Helle M, Meschke M, Möttönenm M, and Pekola J P 2009 *Phys. Rev. Lett.* **102** 200801
[32] Anghel D V 2011 *EPL* **94** 60004
[33] Anghel D V 2008 *Rom. J. Phys.* **53** 689

Absorption Spectra of Defect-Free Semiconductor Quantum Dots

T O Cheche[1, 3], **R Vidican**[2], **E Barna**[1]

[1] Faculty of Physics, University of Bucharest, Bucharest 077125, Romania
[2] Department of Mathematics, Polytechnic University of Bucharest, Bucharest, Romania

E-mail: tiberiuscheche@yahoo.com

Abstract. The presence of accompanying resonances to the longitudinal optical phonon satellites in the optical spectra of semiconductor quantum dots is confirmed theoretically by a non-adiabatic approach of the optical absorption. The theory is applied to simulate features of the optical spectra of spherical quantum dots.

1. Introduction

The optical spectra of polar semiconductor quantum dots (QDs) present several particularities whose understanding is of crucial importance for the technical applications of such nanostructures. Thus, in both emission and absorption spectra longitudinal optical (LO) phonon replicas and accompanying resonances have been observed [1-3]. Also, a red shift with temperature of the 0LO phonon line in the photoluminescence spectra of QDs is reported in experiment [4]. The presence of the strong phonon replicas in the photoluminescence (PL) spectra of QDs of weakly polar III-V compounds is a striking result since no such strong phonon replicas are usually observed in the luminescence of bulk III-V compounds, and not always in the PL spectra of QDs. On the other hand, some theoretical analyses within an adiabatic approach for defect-free QDs fail in calculation of the Huang-Rhys (HR) factors (a measure for the coupling strength of a particular electronic (excitonic) state); the HR factors predicted are smaller from one to four orders of magnitude than by those obtained from experiment [3, 5].

To offer an explanation to the mentioned experimental and theoretical observations, in this paper we consider a non-adiabatic approach of the optical absorption. Non-adiabatic treatments, necessary when the inter-level energy of the electron-hole pair (EHP) induced by irradiation is comparable to the LO phonon energy, have already been proposed [6, 7]. The present work, comparatively to [6, 7] predicts appearance of accompanying resonances in the optical spectra of QDs in addition to the usual LO phonon satellites by an intrinsic mechanism (defect-free QDs in absence of Auger processes and surface optical transitions). Technically, instead of evaluating the lines intensity from the oscillator strength using the eigenstates of the EHP-phonon Hamiltonian with a Fröhlich coupling at temperature $T = 0\mathrm{K}$, as in [7], the temperature effect in the absorption spectra is obtained by averaging over the equilibrium phonon states. Completing the work from [8], where only two electronic levels involved in transition are considered, the present treatment is able to take into account several electronic (excitonic) levels.

[3] To whom any correspondence should be addressed.

The paper is organized as follows. In section 2, a general expression of the *linear absorption coefficient* is derived by applying a semi-classical theory of the radiation-matter interaction to the EHP-LO phonons system. In section 3, within the effective mass approximation, we describe the *energy structure* of spherical QDs. The EHP-LO phonon interaction is modelled by Fröhlich interaction and the electron-hole charge separation appears as result of the confinement potentials of the QD heterostructure. The theory is applied to CdSe/ZnS and GaAs/AlAs QDs. In section 4, the EHP-LO phonon coupling matrix elements (CMEs) are calculated, and the absorption coefficient is simulated for spherical CdSe/ZnS and GaAs/AlAs QDs.

2. The Optical Absorption Coefficient

In the non-adiabatic approach the different electronic states are coupled by phonons. For QDs this happens in the medium to weak confinement regime for electronic inter-level energies close to the LO optical phonon energy. The Hamiltonian of the EHP-LO phonon system is described by an extension of the Huang-Rhys model of F centers,

$$H=\sum_{f}E_{f}B_{f}^{+}B_{f}+\sum_{\mathbf{q}}\hbar\omega_{0}b_{\mathbf{q}}^{+}b_{\mathbf{q}}+\sum_{\mathbf{q},f,f'}M_{\mathbf{q}}^{ff'}B_{f}^{+}B_{f'}(b_{\mathbf{q}}+b_{-\mathbf{q}}^{+}),\tag{1}$$

where the first term is the EHP Hamiltonian, the second term is the LO phonon Hamiltonian, and the third is the EHP-LO phonon interaction Hamiltonian. B_f^+ (B_f), $b_{\mathbf{q}}^+$ ($b_{\mathbf{q}}$) are creation (annihilation) operators of the EHP, and phonons, respectively, E_f ($|f\rangle$) are the EHP eigenvalues (eigenstates), $M_{\mathbf{q}}^{ff'}\equiv\langle f|M_{\mathbf{q}}|f'\rangle$ is the CME between the EHP states $|f\rangle$ and $|f'\rangle$, ω_0 is the frequency of the dispersionless LO phonon modes, and $\mathbf{q}$ is the LO phonon wave vector. The EHPs are considered as bosons (the EHP spin is an integer). This statement, generally valid in the dilute limit of exciton complexes is exact for the case we discuss here, where only one exciton in the QD is considered. On the other hand, in [2] one finds that though the exciton-LO phonon is usually enhanced in QDs, the Fermi Golden Rule is applicable at least for QDs with low relaxation rates. In what follow the optical transition probabilities are computed with the Fermi Golden Rule.

The absorption coefficient for a single QD is given by [9]

$$\alpha(\omega)=\frac{2\pi\hbar\omega}{ncE_0^2V_0}R_{abs}\tag{2}$$

where ω is the frequency, E_0 amplitude of the monochromatic radiation field, n is the refractive index of the environment, V_0 is the absorptive volume, and R_{abs} is the radiation absorption rate. R_{abs} is given by

$$R_{abs}=\frac{2\pi}{\hbar}\mathrm{Av}\sum_{\mathscr{G},\mathscr{F}}|W_{\mathscr{G},\mathscr{F}}|^2\,\delta(\hbar\omega+E_{\mathscr{G}}-E_{\mathscr{F}})\tag{3}$$

The average Av involved by equation (3) means a quantum average over the finite number of the exciton states in the QD and a statistical average over the phonon modes at thermal equilibrium; $E_{\mathscr{G}}$ is the total energy of the system in the ground state $|\mathscr{G}\rangle=|0\rangle|\gamma\rangle$ ($|\gamma\rangle$ is the phonon state), $E_{\mathscr{F}}$ is the total energy of the system in an exciton+phonons state $|\mathscr{F}\rangle=|f;\varphi\rangle$, and $W_{\mathscr{G},\mathscr{F}}=\langle\mathscr{G}|W|\mathscr{F}\rangle$ is the transition probability between the initial state $|\mathscr{G}\rangle$ and the final state $|\mathscr{F}\rangle$, with W the radiation-QD Hamiltonian. Greek characters are used for phonon states, Latin characters for exciton states, and capital handwriting characters for exciton+phonon system. Equation (3) can explicitly be written as follows

$$R_{abs}=\frac{2\pi}{\hbar}Tr\left\{\rho\sum_{\mathscr{G},\mathscr{F}}|W_{\mathscr{G}\mathscr{F}}|^2\,\delta(\hbar\omega+E_{\mathscr{G}}-E_{\mathscr{F}})\right\}=\frac{2\pi}{\hbar}Tr\left\{\rho\sum_{\mathscr{G},\mathscr{F}}|W_{\mathscr{G}\mathscr{F}}|^2\frac{1}{2\pi\hbar}\int_{-\infty}^{\infty}dte^{i\omega t}e^{(E_{\mathscr{G}}-E_{\mathscr{F}})t/\hbar}\right\}$$

$$
\begin{aligned}
&= \frac{1}{\hbar^2}\int_{-\infty}^{\infty} dt e^{i\omega t}\, Tr\left\{\rho \sum_{\mathscr{G}} \langle \mathscr{G}| e^{itH_{ph}/\hbar} W e^{-itH/\hbar}\left(\sum_{\mathscr{F}} |\mathscr{F}\rangle\langle \mathscr{F}| + |\mathscr{G}\rangle\langle\mathscr{G}|\right) W|\mathscr{G}\rangle\right\}\\
&= \frac{1}{\hbar^2}\int_{-\infty}^{\infty} dt e^{i\omega t}\, Tr\left\{\rho \sum_{\mathscr{G}} \langle \mathscr{G}| e^{itH_{ph}/\hbar} W e^{-itH/\hbar} W|\mathscr{G}\rangle\right\}\\
&= \frac{1}{\hbar^2}\int_{-\infty}^{\infty} dt e^{i\omega t} \sum_{\mu}\langle\mu|\sum_{\nu}\left(|\nu\rangle\rho_\nu\langle\nu|\right)|\mu\rangle\langle\mu|\langle 0|\widetilde{W}(t)W|0\rangle|\mu\rangle\\
&= \frac{1}{\hbar^2}\int_{-\infty}^{\infty} dt e^{i\omega t}\, Tr\left\{\rho\langle 0|\widetilde{W}(t)W|0\rangle\right\} \equiv \frac{1}{\hbar^2}\int_{-\infty}^{\infty} dt e^{i\omega t}\left\langle\langle 0|\widetilde{W}(t)W|0\rangle\right\rangle_0
\end{aligned}
\tag{4}
$$

where $\rho = \sum_{\nu}|\nu\rangle\rho_\nu\langle\nu|$ is the density matrix of the phonons, with $\rho_\nu = e^{-\beta E_\nu}/Tr\left(e^{-\beta H_{ph}}\right)$ the probability of the phonon state $|\nu\rangle$ in the equilibrium statistical ensemble of the phonons, $Tr\{A\} = \sum_\nu \langle\nu|A|\nu\rangle = \sum_\nu A_{\nu\nu}$, and $\langle A\rangle_0 \equiv Tr\{\rho A\} = \sum_\nu \rho_\nu\langle\nu|A|\nu\rangle = \sum_\nu \rho_\nu A_{\nu\nu}$. The closure relation

$$\sum_{\mathscr{F}}|\mathscr{F}\rangle\langle\mathscr{F}| + |\mathscr{G}\rangle\langle\mathscr{G}| = \mathbf{1} \tag{5}$$

was used in the third equality of equation (4), where $\sum_{\mathscr{G}}|\mathscr{G}\rangle\langle\mathscr{G}| = \sum_\mu |0\rangle|\mu\rangle\langle\mu|\langle 0|$ that has no effect on the matrix element was inserted. Using the adiabatic picture the state $|\mathscr{F}\rangle$ is written as a product of states, $|\mathscr{F}\rangle = |f\rangle|\varphi\rangle$, and the significance of the closure relation is more transparent:

$$\sum_{\mathscr{F}}|\mathscr{F}\rangle\langle\mathscr{F}| + |\mathscr{G}\rangle\langle\mathscr{G}| = \sum_{f,\nu}|f\rangle|\nu\rangle\langle\nu|\langle f| + \sum_\mu|0\rangle|\mu\rangle\langle\mu|\langle 0| = \left(\sum_f |f\rangle\langle f| + |0\rangle\langle 0|\right)\sum_\nu|\nu\rangle\langle\nu| = \mathbf{1}.$$

In equation (4), $\widetilde{W}(t) = e^{itH_{ph}/\hbar}We^{-itH/\hbar}$. With equations (2-5) one obtains

$$\alpha(\omega) = \frac{2\pi \mathrm{e}^2}{ncm_0^2\hbar\omega V_0}\int_{-\infty}^{\infty} dt e^{i\omega t}\left\langle\langle 0|\widetilde{W}(t)W|0\rangle\right\rangle_0 \tag{6}$$

By using the bosonic commutation rules for creation and annihilation of EHP and phonons, and the operator relation $e^{A+B} = e^A e^B e^{-[A,B]/2}$, one obtains

$$
\begin{aligned}
\left\langle\langle 0|\widetilde{W}(t)W|0\rangle\right\rangle_0 &= \sum_{f,f'\neq 0} Tr\left\{\rho\langle 0|e^{itH_{ph}/\hbar}B_f e^{-itH/\hbar}B^+_{f'}|0\rangle\right\} = \sum_{f,f'\neq 0}\langle 0|B_f e^{-itH_{ex}/\hbar}Tr\left\{\rho\, e^{it(H_{ph}+H_{ex})/\hbar}e^{-itH/\hbar}\right\}B^+_{f'}|0\rangle\\
&= \sum_{f,f'\neq 0} e^{-it\omega_f}\langle 0|B_f\langle U(t)\rangle_0 B^+_{f'}|0\rangle
\end{aligned},
$$

where $\langle U(t)\rangle_0 \equiv Tr\left\{\rho e^{itH_0/\hbar}e^{-itH/\hbar}\right\}$ is the statistical average over the phonon system at thermal equilibrium of the time evolution operator, and $H_0 = H_{ph} + H_{EHP}$. By expanding the time evolution operator, equation (6) is transformed to the form

$$\alpha(\omega) = \frac{2\pi e^2}{ncm_0^2\hbar\omega V_0}\sum_{f,f'\neq 0}\left[P_{0f}P_{f'0}\int_{-\infty}^{\infty} dt \exp[i(\omega-\omega_f)t]\langle 0|B_f\left\langle \hat{T}\exp\left[-\frac{i}{\hbar}\int_0^t dt_1\widetilde{V}(t_1)\right]\right\rangle_0 B^+_{f'}|0\rangle\right] \tag{7}$$

where $\hat{T}$ is the time-ordered operator, $\widetilde{V}(t) = \exp(itH_0/\hbar)H_{EHP-ph}\exp(-itH_0/\hbar)$, $P_{0f} \equiv \langle 0|(\boldsymbol{\varepsilon}\cdot\mathbf{P})|f\rangle$, and $\mathbf{P} \equiv \sum_i \mathbf{p}_i$ is the total electronic momentum (with $\mathbf{p}_i$ the electron momentum).

For the linear phononic coupling from equation (7), the cumulant expansion method gives the result that all cumulants higher than the second order vanish. Thus, by using the bosonic commutation rules one obtains

$$\langle U(t)\rangle_0 = \exp\left[-\frac{1}{\hbar^2}\sum_{\mathbf{q},i_1,j_1,i_2,j_2} M_{\mathbf{q}}^{i_1 j_1} M_{-\mathbf{q}}^{i_2 j_2} B_{i_1}^{+} B_{j_1} B_{i_2}^{+} B_{j_2} \times \int_0^t dt_1 \int_0^{t_1} dt_2 \exp(it_1\omega_{i_1 j_1})\exp(it_2\omega_{i_2 j_2}) D^0(t_1-t_2)\right] \equiv \exp(S) \tag{8}$$

where $D^0(t_1-t_2) \equiv \left[\bar{N}\exp\left(i\omega_0|t_1-t_2|\right) + (\bar{N}+1)\exp\left(-i\omega_0|t_1-t_2|\right)\right]$, $\omega_{ij} = \left(E_i - E_j\right)/\hbar$, and $\bar{N} = 1/\left(e^{\beta\hbar\omega_0}-1\right)$ is the phonon occupation number. Using the series expansion of the exponentials in equation (8), we find

$$\langle 0|B_f S^n B_{f'}^{+}|0\rangle = \sum_{i_3,i_5,\ldots,i_{2n-1}} \sum_{i_2}(fi_2i_2i_3)\sum_{i_4}(i_3i_4i_4i_5)\ \ldots \sum_{i_{2n}}(i_{2n-1}i_{2n}i_{2n}f') \xrightarrow{\text{cutting}} \left[\sum_i (fiif')\right]^n \tag{9}$$

where $\sum_{k'}\left(i_k i_{k'} i_{k'} i_{p'}\right) \equiv \sum_{\mathbf{q},k'}\left[M_{\mathbf{q}}^{kk'} M_{-\mathbf{q}}^{k'p'} \int_0^t dt_1 \int_0^{t_1} dt_2 \exp(it_1\omega_{kk'})\exp(it_2\omega_{k'p'}) D^0(t_1-t_2)\right] \equiv \sum_{\mathbf{q},k'}\left[M_{\mathbf{q}}^{kk'} M_{-\mathbf{q}}^{k'p'} I(t,k,k',k',p')\right]$. A partial resummation (PR) of $\langle 0|B_f S^n B_{f'}^{+}|0\rangle$ is obtained by the "cutting" procedure suggested in equation (9). We use the notations $g_{kk'pp'} \equiv \sum_{\mathbf{q}}\left[M_{\mathbf{q}}^{kk'} M_{-\mathbf{q}}^{pp'}/(\hbar\omega_0)^2\right]$ ($g_{pppp} \equiv g_p$ is the HR factor) and $G_{piis} \equiv g_{piis}\omega_0^2 I(t,p,i,i,s)$. The PR yields

$$\alpha(\omega) = \frac{2\pi e^2}{ncm_0^2\hbar\omega V_0}\sum_{p,s\neq 0}\left[P_{0p}P_{s0}\int_{-\infty}^{\infty} dt \exp\left[i(\omega-\omega_p)t\right]\exp\left(-\sum_{i\neq 0} G_{piis}\right)\right]. \tag{10}$$

Comparatively to the case presented in [6], the CMEs do not have to respect the restriction, $\max\left|M_{\mathbf{q}}^{ff'}\right| << 1$, which may be difficult to fulfill for large $\left|g_{kk'pp'}\right|$, as long as few LO phononic modes are strongly coupled with the EHPs (see, e.g., [8]). If the off-diagonal coupling terms in equation (1) are disregarded then equation (8) is exact and it *recovers* the adiabatic limit (the Franck-Condon progression):

$$\alpha^{ad}(\omega) = \frac{4\pi^2 e^2}{ncm_0^2\hbar\omega V_0}\sum_{f\neq 0}\left|P_{0f}\right|^2 \exp\left[-g_f(2\bar{N}+1)\right] \times \sum_{n=-\infty}^{\infty} I_n\left(2g_f\sqrt{\bar{N}(\bar{N}+1)}\right)\exp\left(n\beta\hbar\omega_0/2\right)\delta\left(\omega-\omega_f+\Delta_f^{ad}-n\omega_0\right) \tag{11}$$

where I_n are the modified Bessel functions, and $\Delta_f^{ad} = \omega_0\sum_{\mathbf{q}}\left(\left|M_{\mathbf{q}}^{ff}\right|^2\hbar^{-2}\omega_0^{-2}\right) \equiv \omega_0 g_f$ is the self-energy. The relative intensity of the absorption lines is given by the coefficients of the Dirac delta functions.

3. Spherical Quantum Dot Model

We consider a spherical QD in the case of size-quantized energies of QD (or equivalently, QD with dimensions smaller than its corresponding exciton Bohr radius so that the exciton effect, Coulombic interaction is neglected) [10]. For spherical QDs the one-band model by which conduction and valence states are computed from single-particle Schrödinger equations in the effective mass approximation is a good approximation for type I heterostructures [11], and we situate our analysis within such an approach.

In order to describe the energy structure, we consider the confinement potential energy $V_e(r_e)=0$ for $r_e \in [0, R_0]$, and $V_e(r_e)=V_{0e}$ for $r_e > R_0$ (similar equation is written for holes by replacing r_e by r_h); R_0 is the QD radius. The single-particle wave function is written as the product $\varphi_{nlm}(\mathbf{r}) = R_{nl}(r)Y_{lm}(\theta,\varphi)$, where $R_{nl}(r)$ is the radial function and $Y_{lm}(\theta,\varphi)$ is the spherical harmonics function. By using the second quantization language, and disregarding the spin dependence, the EHP state may be written as $|f\rangle \to |\varphi_{ab}\rangle = \int d\mathbf{r}_e\, d\mathbf{r}_h\, \varphi_a(\mathbf{r}_e)\,\varphi_b(\mathbf{r}_h) a_{\boldsymbol{e}}^{c+} a_{\boldsymbol{h}}^{v}|0\rangle$, where $a_{\boldsymbol{e}}^{c+}(a_{\boldsymbol{h}}^{v})$ are the creation (annihilation) fermionic operator of an electron in the conduction band at $\mathbf{r}_e$ (valence band at $\mathbf{r}_h$) and a (b) holds for the set of quantum numbers n_e, l_e, m_e (n_h, l_h, m_h) of electrons (holes). The model takes into account the difference in the effective masses between the nano-sphere and its surrounding and considers finite confinement potentials for both electrons and holes. These factors cause a *non-vanishing* EHP-LO phonon interaction. Following [12, 13], the expression of orthonormalized $R_{nl}(r)$ and the secular equation of energy are as follows

$$R_{nl}(r) = \sqrt{\frac{2}{R_0^3}}\left[j_l^2(x)k_{l-1}(y)k_{l+1}(y) - k_l^2(y)j_{l-1}(x)j_{l+1}(x)\right]\begin{cases} k_l(y)j_l(xr/R_0), & r < R_0 \\ j_l(x)k_l(yr/R_0), & r > R_0 \end{cases} \tag{12}$$

$$\mu_2 x k_l(y) j_l'(x) = \mu_1 y j_l(x) k_l'(y) \tag{13}$$

where $x = R_0\sqrt{(2\mu_1 E_{n,l})/\hbar^2}$, $y = R_0\sqrt{(2\mu_2(V_{0c} - E_{n,l}))/\hbar^2}$, k_l the modified spherical Bessel functions, μ_1 (μ_2) the effective mass in the dot (surrounding medium), V_{0c} the band offset of the carriers (c becomes e for electron or h for hole), and n, $l \to n_e$, l_e or n_h, l_h for electron and hole, respectively.

The optically active EHPs are obtained with the optical selection rules which dictate the allowed transitions. In the linear response theory and long wave approximation the particle-radiation interaction Hamiltonian for a carrier of charge Q and mass M reads

$$H_{Q-R} = -Q(Mc)^{-1}\mathbf{A}\cdot\mathbf{P}, \tag{14}$$

where c is the speed of light, $\mathbf{A}$ is the vector potential, and $\mathbf{P}$ is the carrier momentum. With Maxwell's equation $\mathbf{E} = -c^{-1}\partial\mathbf{A}/\partial t$, within Coulomb gauge, and with the space-independent electric field $\mathbf{E} = \boldsymbol{\varepsilon}E_0\cos\omega t$ (monochromatic field of frequency ω, amplitude E_0, and direction of oscillation along the unit polarization vector $\boldsymbol{\varepsilon}$), the corresponding vector potential reads $\mathbf{A} = -\boldsymbol{\varepsilon}A_0\sin\omega t = -\boldsymbol{\varepsilon}E_0 c\omega^{-1}\sin\omega t$. Translated to the EHP problem the momentum operator reads

$$\mathbf{P} = \sum_f \left[|0\rangle\langle 0|\mathbf{P}|f\rangle\langle f| + |f\rangle\langle f|\mathbf{P}|0\rangle\langle 0|\right] \equiv \sum_f \left[\langle 0|\mathbf{P}|f\rangle B_f + \langle f|\mathbf{P}|0\rangle B_f^+\right] \tag{15}$$

where $|0\rangle$ stands for the EHP vacuum state, and $\mathbf{P} \equiv \sum_i \mathbf{p}_i$ the total electronic momentum (with $\mathbf{p}_i$ the electron momentum). With equations (14, 15) the semi-classical EHP-radiation interaction reads

$$H_{EHP-R} = -\mathrm{e}E_0(m_0\omega)^{-1}\boldsymbol{\varepsilon}\cdot\sum_{f\neq 0}\left[\langle 0|\mathbf{P}|f\rangle B_f + h.c.\right]\sin\omega t \equiv W\sin\omega t \tag{16}$$

With the EHP eigenstates and the appropriate definition of the momentum [14], $\mathbf{P} = \mathbf{p}_{cv}^0\int d\mathbf{R}\, a_{\mathbf{R}}^{c+}a_{\mathbf{R}}^{v} + h.c.$, where $\mathbf{p}_{cv}^0$ is the momentum matrix element between the valence-band and the conduction-band at the Γ point and $\mathbf{R}$ integrates over unit cells, one obtains the optical matrix element

$$\langle\varphi_{ab}|\mathbf{P}|0\rangle = \mathbf{p}_{cv}^0\delta_{l_e l_h}\delta_{m_e m_h}\int_0^\infty dr\, r^2 R_{n_e l_e}(r)R_{n_h l_h}(r) \equiv \mathbf{p}_{cv}^0\delta_{m_e m_h}A_{n_e n_h l}, \tag{17}$$

with $l_e = l_h = l$. Thus, according to the optical selection rules $l_e = l_h$ for an optically active state.

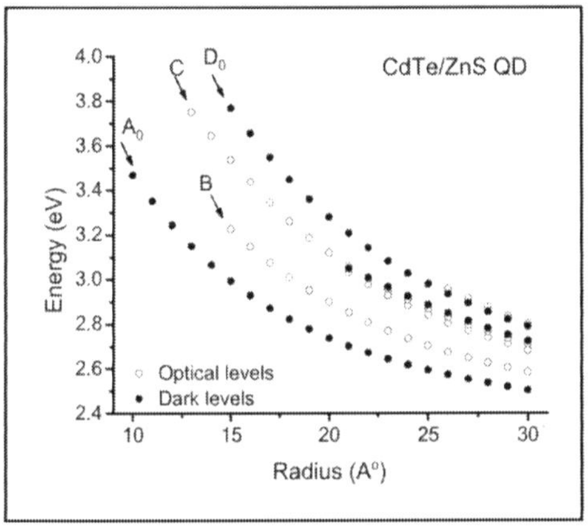

Figure 1. The energy spectrum of spherical CdSe/ZnS QDs. The capital characters indicate dark (not indexed) and optically active (indexed by 0) EHP levels.

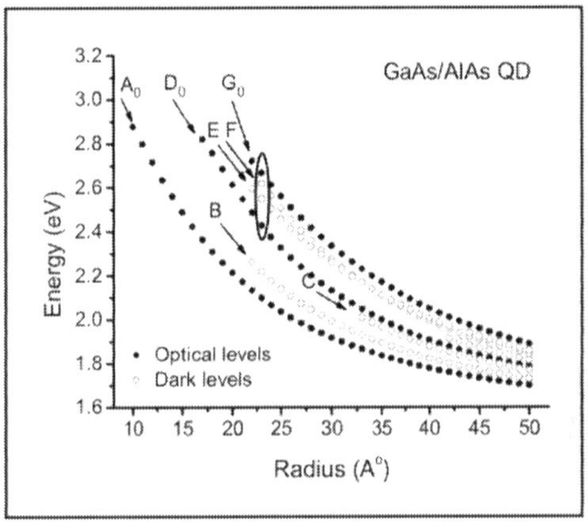

Figure 2. The energy spectrum of spherical GaAs/AlAs QDs. The capital characters indicate dark (not indexed) and optically active (indexed by 0) EHP levels.

Next, we apply the model to concrete heterostructures. First, we consider CdSe microcrystallites embedded in a ZnS matrix (II-VI heterostructures) and simulate the energy structure within an adiabatic approach. The necessary material parameters found in the literature are as follows: bulk CdSe energy gap $E_g = 1.8$ eV, CdSe (ZnS) electron effective mass $\mu_e/m_0 = 0.13$ ($\mu_e/m_0 = 0.34$), hole effective mass $\mu_h/m_0 = 0.45$ ($\mu_h/m_0 = 0.23$), conduction band offset $V_{0e} = 1.2$ eV, and the valence band offset $V_{0h} = 0.5$ eV; m_0 is the electron mass. The energy spectrum is obtained from equations (12, 13), and the EHP energy $E_{n_e,l_e;n_h,l_h} = E_g + E_{n_e,l_e} + E_{n_h,l_h}$ is computed as a function of the QD radius. To obtain values of the absorption lines as reported by experiment [15] (see next section), we increased the bulk CdSe energy gap by 0.43eV. This is in accordance with the mismatch strain effect which increases the energy gap in QD, comparatively to bulk. In figure 1 are shown the first several energy levels that are interesting for the analysis. Some particular levels are labelled by the set of quantum numbers, $(n_e,l_e,m_e;n_h,l_h,m_h)$ as follows: $\mathrm{A}_0 \to (1,0,0;1,0,0)$, $\mathrm{B} \to (1,0,0;1,1,m_h)$ - dark level, $\mathrm{C} \to (1,0,0;1,2,m_h)$ - dark level, $\mathrm{D}_0 \to (1,0,0;2,0,0)$. According to the energy levels distribution from figure 1, at $R_0 = 20$ Å the phonon coupling is highly improbable at least at low temperatures: $(E_{\mathrm{B}} - E_{\mathrm{A}_0})/\hbar\omega_0 \approx 5.9$ (CdSe bulk LO phonon energy is $\hbar\omega_0 = 27.3$meV), and the optical spectra are properly discussed in an adiabatic framework.

For GaAs microcrystallites embedded in AlAs matrix (III-V heterostructures), we use the material parameters from [16]: GaAs energy gap $E_g = 1.5177$ eV, GaAs (AlAs) electron effective mass $\mu_e/m_0 = 0.0665$ ($\mu_e/m_0 = 0.124$), hole effective mass $\mu_h/m_0 = 0.45$ ($\mu_h/m_0 = 0.5$), conduction band offset $V_{0e} = 0.968$ eV, and the valence band offset $V_{0h} = 0.6543$ eV. GaAs/AlAs is a nearly lattice-matched heterostructure, and the weak strain effect makes the bulk GaAs energy gap to be used in calculation. The energy spectrum obtained from equations (12, 13) is shown in figure 2. Some particular levels are labeled by the set of quantum numbers, $(n_e,l_e,m_e;n_h,l_h,m_h)$ as follows: $\mathrm{A}_0 \to (1,0,0;1,0,0)$, $\mathrm{B} \to (1,0,0;1,1,m_h)$-dark level, $\mathrm{C} \to (1,0,0;1,2,m_h)$-dark level, $\mathrm{D}_0 \to (1,0,0,2,0,0)$, $\mathrm{E} \to (1,1,m_e;1,0,0)$-dark level, $\mathrm{F} \to (1,0,0;2,1,m_h)$-dark level, $\mathrm{G}_0 \to (1,1,m_e;1,1,m_h)$. For B, for example, one considers $m_h = 0$, the only value for which CME does not vanish. Based on the distribution of energy levels and taking into account the exciton Bohr radius (larger than 100Å), we consider $R_0 = 50$ Å as a reasonable upper-limit for neglecting the Coulombic interaction. On the other hand, possible phonon mixing effect could manifest starting with $R_0 \approx 23$ Å (see the ellipse mark at figure 2), between the optically active level G_0 and the dark level F. But, the phonon-assisted transition between G_0 and D_0 is improbable (at least in the low temperature limit) since $\left(E_{\mathsf{E}} - E_{\mathsf{D}_0}\right)/\hbar\omega_0 = 3.37$ (GaAs LO phonon energy is $\hbar\omega_0 = 36.2\text{meV}$). For the first two optically active levels, the adiabatic treatment is safe for $R_0 < 22$ Å and may be accepted as satisfactory for $R_0 < 32$ Å, beyond which the optically active level C appears.

The quantity $P_{0f}P_{f'0} = \left|\mathbf{p}_{cv}^0\right|^2 3^{-1} A_{n_e n_h l} A_{n_e' n_h' l'} \delta_{m_e m_h} \delta_{m_e' m_h'}$ in equation (7) is obtained by averaging over all space polarization directions. The Fröhlich coupling is written for dispersionless bulk LO phonons (for a spherical QD the interface modes are weak and do not couple with the exciton states, see [17]).

Within the pure-EHP approximation the EHP-phonon interaction reads [18, 19]

$$M_{\mathbf{q}}^{ff'} \to M_{\mathbf{q}}^{ab;a'b'} = V_0^{-1/2} f_0 q^{-1} \int d\mathbf{r}_e\, d\mathbf{r}_h\, \varphi_a^*(\mathbf{r}_e)\varphi_b^*(\mathbf{r}_h)\varphi_{a'}(\mathbf{r}_e)\varphi_{b'}(\mathbf{r}_h)\left[\exp(i\mathbf{q}\mathbf{r}_e) - \exp(i\mathbf{q}\mathbf{r}_h)\right] \tag{18}$$

which for spherical QDs becomes

$$M_{\mathbf{q}}^{ff'} \to M_{\mathbf{q}}^{ab;a'b'} = V_0^{-1/2} f_0 q^{-1} \times\left[\delta_{bb'}\int_0^\infty dr\, r^2 R_{n_e l_e}(r) R_{n_e' l_e'}(r) Q_{l_e' m_e'}^{l_e m_e}(q,r) - \delta_{aa'}\int_0^\infty dr\, r^2 R_{n_h l_h}(r) R_{n_h' l_h'}(r) Q_{l_h' m_h'}^{l_h m_h}(q,r)\right] \tag{19}$$

where

$$Q_{l'm'}^{lm}(q,r) \equiv \int d\Omega\, Y_{lm}^*(\Omega) Y_{l'm'}(\Omega)\exp(i\mathbf{q}\mathbf{r}) = \sqrt{\frac{2l'+1}{2l+1}}\delta_{mm'} \sum_{\substack{l''=|l-l'| \\ l''+l'+l \to even}}^{l+l'} i^{l''}(2l''+1)\langle l'' l' 00 | l0\rangle\langle l'' l' 0m' | lm\rangle j_{l''}(qr) \tag{20}$$

with $j_l(x)$ the spherical Bessel functions, $\langle l'' l' m'' m' | lm\rangle$ the Clebsh-Gordan coefficients, and $f_0 = \sqrt{2\pi\hbar\omega_0 e^2\left(\varepsilon_\infty^{-1} - \varepsilon_0^{-1}\right)\kappa^{-1}}$ the Fröhlich coupling constant. Equation (20) is obtained by using the expansion of plane waves, $\exp(i\mathbf{q}\mathbf{r}) = \sum_{l'=0}^{\infty} i^{l'}\sqrt{4\pi(2l'+1)}\, j_{l'}(qr) Y_{l'0}(\Omega)$, and the integration of three spherical harmonics over the solid angle. Inspection of equation (19) shows the states φ_{ab}, $\varphi_{a'b'}$ are not coupled by phonons if both $a \neq a'$ and $b \neq b'$.

3. Application

With the theory developed in section 2, we simulate the absorption coefficient for the semiconductor heterostructures introduced in section 3.

First, we consider CdSe/ZnS QD, and evaluate the absorption coefficient for a QD radius $R = 20$ Å, case which is appropriate for an adiabatic approach. For the ground EHP state we obtain $g = 0.094$, a value close of the value reported in [19], namely 0.07 for a radius of 2nm. In figure 3, we show the absorption coefficient with the typical LO resonances obtained in the optical spectra. The spectrum which is generated by the ground EHP state is obtained with equation (11). The 0LO phonon line is in accordance with [15]; for $R_0 = 23$ Å the model generates a value of 2.69eV approximately equal to the value of ZnS overcoated CdSe QD of the same radius. One can see the effect of

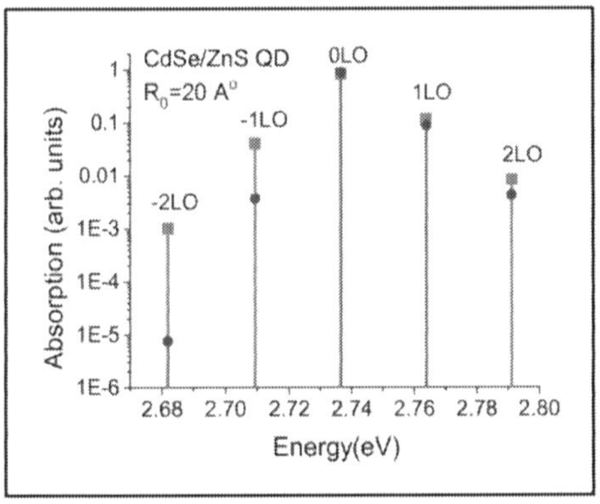

Figure 3. The simulated absorption spectra of CdSe/ZnS nanocrystal QDs for different temperatures (see text).

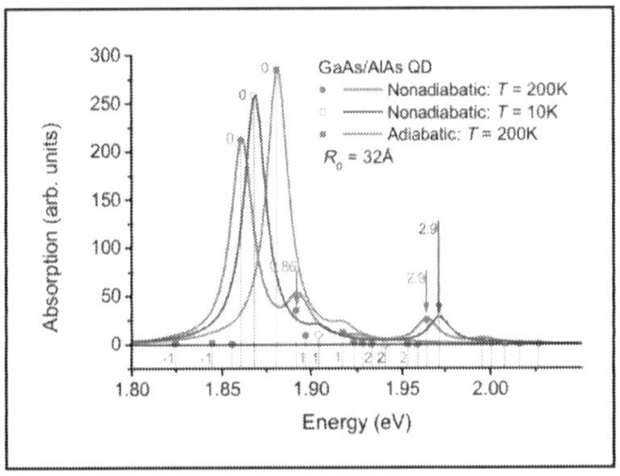

Figure 4. The simulated absorption spectra of GaAs/AlAs nanocrystal QDs for different temperatures (see text).

temperature is of increasing intensity of LO resonances and decreasing intensity of the 0LO line. The right phonon satellites in figure 2 are more intense for higher temperature, where LO phonon population is higher, and consequently probability of LO phonon absorption is increased.

Second, we consider GaAs/AlAs QD, and evaluate the absorption coefficient in more detail. The large values of $|g_{kk'pp'}|$ obtained for such nanostructures less polar than CdSe, e.g., are the result of charge separation induced by the finite confinement potential.

For the only two optical levels which appear at $R_0 = 20$ (see figure 2), with an inter-level energy of approximately $11\hbar\omega_0$ (the energy of LO phonons is considered as $\hbar\omega_0 = 36.2\text{meV}$), the absorption coefficient from equation (7) reads

$$\alpha(\omega) = \frac{2\pi e^2 |p_{cv}^0|^2}{3ncm_0^2\hbar\omega V_0} \sum_{i=1,2} \int_{-\infty}^{\infty} dt \exp[i(\omega-\omega_i)t] \times \left[A^2_{n_e^{(i)} n_h^{(i)} l^{(i)}} \langle i|\langle U(t)\rangle_0|i\rangle + A_{n_e^{(1)} n_h^{(1)} l^{(1)}} A_{n_e^{(2)} n_h^{(2)} l^{(2)}} \sum_{f=1,2} \langle i|\langle U(t)\rangle_0|f\rangle \right] \tag{21}$$

The first term in the integrand of equation (21) can be approximated by using the PR procedure, and the contribution to the absorption coefficient is evaluated by equation (10). The second term of the integrand, responsible for the phonon mixing of different optically active states can be neglected [20]. Thus, equation (21) written for the lines centered on the line A_0 reads

$$\begin{aligned} \alpha_1(\omega) = {} & \frac{2\pi e^2 |\mathbf{p}_{cv}^0|^2}{3ncm_0^2\hbar\omega V_0} A_{110}^2 \exp(-\Lambda_1) \sum_{p=-\infty}^{\infty} \sum_{k,r=0}^{\infty} \Big[I_p\left(2g_1\sqrt{\overline{N}(\overline{N}+1)}\right) \\ & \times \left(\frac{\overline{N} g_{1221}\beta^2}{k!}\right)^k \left(\frac{(\overline{N}+1) g_{1221}\gamma^2}{r!}\right)^r \exp\left(\frac{-p\hbar\omega_0}{2k_B T}\right) \\ & \times \delta[\omega - \omega_1 + \Delta_1 + p\omega_0 - k(\omega_{21}-\omega_0) - r(\omega_{21}+\omega_0)] \Big] \end{aligned} \tag{22}$$

where $\Lambda_1 = g_1(2\overline{N}+1) + g_{1221}\left[\overline{N}\beta^2 + (\overline{N}+1)\gamma^2\right]$, $\Delta_1 = \omega_0\left[g_1 + g_{1221}(\overline{N}\beta + (\overline{N}+1)\gamma)\right]$ is the *self-energy*, $\beta \equiv \omega_0/(\omega_{21}-\omega_0)$, and $\gamma \equiv \omega_0/(\omega_{21}+\omega_0)$. Within the QD model we work with one obtains $g_1 = 0.117$, $g_{1221} = 0.314$. The non-adiabatic corrections are, as expected from the large inter-level energy, not significant and the non-adiabatic spectrum plotted with equation (22) is practically identical with the adiabatic spectrum obtained with equation (11). The spectra, not plotted here show the ordinary phonon satellites.

Situation is different for $R_0 = 32$ Å, where the the dark level B is located between two optical levels A_0 and D_0 (see figure 2). The effect of the dark level B is roughly estimated by considering the PR procedure shown in equation (9). Thus, the contribution of the optical and dark levels to the absorption centered on line A_0 is described by an equation similar to equation (22), but written to account for the dark level too (see details in [20]). The non-adiabaticity effect is shown in figure 4, where the absorption spectrum at different temperatures is plotted (we dressed the lines by Lorentzians with a finite width of 15meV to simulate the EHP-acoustic phonons interaction). The adiabatic spectrum obtained with equation (11) has no temperature-induced shift and its maxima are not significantly changed with temperature. The following quantities obtained within the adopted QD model have been used: $E_1 = 1.8822\text{eV}$, $E_2 = 2.0738\text{eV}$, $E_3 = 1.9496\text{eV}$, $g_1 = 0.039$, $g_{1221} = 0.234$, and $g_{1331} = 0.904$. The stronger accompanying resonances are marked by arrows. The energy of some resonances are indicated by factors which multiply the LO phonon energy; they are placed to the left of the lines or arrows. The temperature dependence of the spectra, weak in the case of adiabatic treatment, becomes important now. Thus, decrease of intensity (by 37%) and red shift (from 1.87eV to 1.85eV) of the 0PL lines are obtained when temperature increases from 10K to 300K. This agrees with the behavior observed experimentally for CdTe QDs [21]. On the other hand, the *simulated* HR factors reach values larger by two orders of magnitude than those of the bulk phase (0.0079 in [19]). A similar behavior is reported for small self-assembled InAs/GaAs QDs in [22]. Thus, by the non-adiabatic activated channel at +0.86LO, the predicted HR factor obtained as the ratio of the line intensities for this accompanying resonance increases from 0.084 at $T = 10\text{K}$ to 0.23 at $T = 200\text{K}$. On the other hand, the non-adiabaticity effect manifests by strong resonances at 2.9LO (see figure 4), close to the third LO phonon replica as reported by some experiments (see, e.g., [2]). The usual Franck-Condon progression is obtained by the adiabatic treatment (see the gray line in figure 4).

Concluding, the non-adiabatic treatment presented, in accordance with the experimental observation, predicts: (i) accompanying resonances to the LO phonon satellites in the optical spectra of QDs; (ii) a red shift of the 0LO phonon lines and increased intensities of the accompanying resonances with temperature in the absorption spectra of QDs.

Acknowledgments
The work was supported by the strategic grant POSDRU/89/1.5/S/58852, Project "Postdoctoral programme for training scientific researcher" co-financed by the European Social Found within the Sectorial Operational Program Human Resources Development 2007-2013.

References

[1] See, e.g., Sarkar D, van der Meulen H P, Calleja J M, Becker J M, Haug R J, and Pierz K 2005 *Phys. Rev.* B **71** 081302(R)
[2] Heitz R, Veit M, Ledentsov N N, Hoffmann A, Bimberg D, Ustinov V M, Kop'ev P S, and Alferov Zh I 1997 *Phys. Rev.* B **56** 435
[3] Bissiri M, Baldassarri Höger von Högersthal G, Bhatti A S, Capizzi M, Frova A, Frigeri P, and Franchi S 2000 *Phys. Rev.* B **62** 4642
[4] Besombes L, Kheng K, Marsal L, and Mariette H 2001 *Phys. Rev* B **63** 155307
[5] Krauss T D and Wise F W 1997 *Phys. Rev.* B **55** 9860
[6] Fomin V M, V. N. Gladilin V N, Devreese J T, Pokatilov E P, Balaban S N, and Klimin S N 1998 *Phys. Rev.* B **57** 2415
[7] Verzelen O, Ferreira R, and Bastard G 2002 *Phys. Rev. Lett.* **88** 146803
[8] Vasilevskiy M I, Anda E V, and Makle S S 2004 *Phys. Rev.* B **70** 035318

[9] Mittin V V, Kochelap V A, Stroscio M A 1999 Quantum Heterostructures. Microelectronics and Optoelectronics (Cambridge: University Press)
[10] Hanamura E 1988 *Phys. Rev.* B **37** 1273
[11] Sercel P C and Vahala K J 1990 *Phys. Rev.* B **42** 3690
[12] Vahala K J 1988 *IEEE J. Quantum Electron.* **24** 523
[13] Chamberlain M P, Trallero-Giner C and Cardona M 1995 *Phys. Rev.* B **51** 1680
[14] Takagahara T 1993 *Phys. Rev.* B **47** 4569
[15] Dabbousi B O, Rodriguez-Viejo J, Mikulec F V, Heine J R, Mattoussi H,Obe r R, Jensen K F, and Bawendi M G *J. Phys. Chem. B* 1997 **101** 9463
[16] Menéndez E, Trallero-Giner C and Cardona M 1997 *Phys. Status Solidi* (b) **199** 81
[17] Melnikov D V and Fowler W B 2002 *Phys. Rev.* B **64** 245320
[18] Voigt J, Spielgelberg F, and Senoner M 1979 *Phys. Status Solidi* (b) **91** 189
[19] Nomura S and Kobayashi T 1992 *Phys. Rev.* B **45** 1305
[20] Cheche T O and Chang M C 2005 *Chem. Phys Lett.* **406** 479
[21] Besombes L, Kheng K, Marsal L, and Mariette H 2001 Phys. Rev B **63** 155307
[22] García-Cristóbal A, Minnaert W E, Fomin V M, Devreese J T, Silov A Yu, Haverkort J E M, and Wolter J H 1999 *Phys. Status Solidi* (b) **215** 331

Account of non-diagonal coupling between *d* electrons at describing the transition-metal pair potentials

N Dubinin[1,2,*]
[1]Institute of Metallurgy of the Ural Branch of Russian Academy of Science, 101 Amundsen st., Ekaterinburg, 620016, Russia
[2]Ural Federal University, 19 Mira st., Ekaterinburg, 620006, Russia

E-mail: ned67@mail.ru

Abstract. In the framework of the Wills-Harrison (WH) model the correction due to the non-diagonal coupling between *d* electrons in different atoms is suggested. Effect of this correction on the WH effective pair interaction is studied for liquid Fe, Co, and Ni. It is found that our approach gives a shift of the WH pair potential to the right-hand side for all metals under consideration (in most degree for Fe and in less degree for Ni). As a rule, such tendency leads to significant changes in calculating thermodynamic properties.

1. Introduction

The Wills-Harrison (WH) model [1] is widely used in thermodynamics calculations for liquid transition metals and their alloys [2-5]. However, the small argument expansion of the *d*-state radial function was used to obtain the effective pair potential in Ref. [1]. Only diagonal *d-d* coupling between different atoms arises in the framework of that approximation. At the same time, an influence of the non-diagonal coupling to the transition-metal potential energy can be significant. The difficulty is that the deviation from the diagonal coupling can not be expressed in the exact form.

Here, we introduce the probability that not only the diagonal coupling is possible between *d* electrons centred at the neighbouring atoms, and study how the WH potential corrected by such a way depends on the magnitude of this probability. In fact, this correction gives an additional magnetic contribution to the pair interaction due to the indirect orbital interaction between *d* electrons of different atoms. The WH effective pair potentials are investigated in pure liquid Fe, Co, and Ni near their melting temperatures.

2. Theory

Wills and Harrison [1] presented the effective pair potential for transition metals as a sum of two terms. The first one arises due to *d* electrons, while the second is taken from the nearly-free-electron (NFE) theory (with the effective number of valence *s* electrons per atom - z_s):

[*]To whom any correspondence should be addressed.

$$V_{\rm WH}(r) = V_d(r) + V_{\rm NFE}(r), \tag{1}$$

Following to Refs. [4, 5], we use the Bretonnet-Silbert (BS) local model pseudopotential [6] to calculate the contribution $V_{\rm NFE}(r)$. The exchange-correlation function is taken in the Vashishta-Singwi form [7].

The d-electron-dependent part of the WH potential consists of two terms – the contribution $V_b(r)$ due to the d-band width, and the contribution $V_c(r)$ due to the shift of the d-band heavy centre with respect to the d level of the single atom (hereafter, in the atomic units (a.u.)):

$$V_d(r) = V_b(r) + V_c(r), \tag{2}$$

$$V_b(r) = -K_b\left(\frac{10-z_d}{10}\right)\left(\frac{12}{a}\right)^{1/2}\frac{z_d r_d^3}{r^5}, \tag{3}$$

$$V_c(r) = K_c\frac{z_d r_d^6}{r^8}, \tag{4}$$

where z_d is the effective number of valence d electrons per atom, r_d is the d-state radius, a is the coordination number (taken for each Fe, Co, and Ni equal to 12), K_b and K_c are some coefficients.

Wills and Harrison used the small argument expansion of the d-state radial function, which gives only diagonal coupling between d-electron states of different atoms with respect to the magnet quantum number, m (i.e. the coupling between d electrons with the same m only is possible). In this case, for the orbital quantum number $l = 2$,

$$K_b = \left(\sum_{m=-2}^{2}\frac{y_m^2}{5}\right)^{\frac{1}{2}}, \tag{5}$$

$$K_c = -2\sum_{m=-2}^{2}\frac{y_m x_m}{5}, \tag{6}$$

where y_m and x_m are combinatoric coefficients:

$$y_m = -\frac{(-1)^m 180}{\pi(2+m)!(2-m)!}, \tag{7}$$

$$x_m = -\frac{1}{8}\left(1+\frac{4m^2-1}{9}\right)y_m. \tag{8}$$

From Eqs. (5) – (8)

$$K_b = \frac{28.06}{\pi}, \tag{9}$$

$$K_c = \frac{225}{\pi^2} \,. \tag{10}$$

In the case when only diagonal coupling is possible, there are only 5 equiprobable couplings between two atoms, and the multiplier 1/5 in Eqs. (5), (6) corresponds to this case.

On the contrary, when a free change of m is allowed for the d-state coupling, there are 25 equiprobable couplings between two atoms and instead of Eqs. (5), (6) we have the following equations (denote $y_m = y_{|m|}$ and $x_m = x_{|m|}$):

$$K_b = \left(\frac{y_0^2}{25} + \frac{4y_1^2}{25} + \frac{4y_2^2}{25} + \frac{4y_0 y_1}{25} + \frac{4y_0 y_2}{25} + \frac{8y_1 y_2}{25} \right)^{\frac{1}{2}}, \tag{11}$$

$$K_c = -2\left(\frac{y_0 x_0}{25} + \frac{2y_0 x_1}{25} + \frac{2y_0 x_2}{25} + \frac{2x_0 y_1}{25} + \frac{2x_0 y_2}{25} + \frac{4y_1 x_1}{25} + \frac{4y_1 x_2}{25} + \frac{4y_2 x_1}{25} + \frac{4y_2 x_2}{25} \right), \tag{12}$$

Let us now denote the probability that not only diagonal coupling is possible as p. Then, the probability that only diagonal coupling is possible will be $(1-p)$, and the resulting expressions for K_b and K_c became as follows:

$$K_b = \left[\frac{1}{5}\left(\left(1-\frac{4p}{5}\right) y_0^2 + \left(2-\frac{6p}{5}\right)(y_2^2 + y_1^2) + \frac{4p}{5} y_0 (y_1 + y_2) + \frac{8p}{5} y_1 y_2 \right) \right]^{\frac{1}{2}}, \tag{13}$$

$$K_c = -\frac{2}{5}\left[\left(1-\frac{4p}{5}\right) y_0 x_0 + \left(2-\frac{6p}{5}\right)(y_1 x_1 + y_2 x_2) + \frac{2p}{5}(y_0(x_1 + x_2) + x_0(y_1 + y_2)) + \frac{4p}{5}(y_1 x_2 + y_2 x_1) \right] \tag{14}$$

At $p = 0$, Eqs. (13) and (14) lead to Eqs. (9) and (10), and at $p = 1$ to Eqs. (11) and (12), respectively.

3. Results and Discussion

The effective pair potentials of pure liquid Fe, Co, and Ni near the melting points are considered. Needed values of the mean atomic volume, Ω, for each metal are taken from experiment [8] and are listed in Table 1. The BS parameters, R_c and A, and the WH parameter, r_d, are taken from Refs. [9] and [1], respectively (Table 1). z_d is defined as $(z - z_s)$, where z is the total number of valence electrons per atom. For all three metals z_s equal to 1.4 following to Ref. [9].

Pair potentials near their first minimums at different p are presented in Figs.1-3. It is found that with increase of the p magnitude the position of the first minimum of the pair potential is being shifted to larger r and its depth becomes smaller in all cases independently on the kind of the metal. Both above tendencies are most sharp for Fe, intermediate for Co, and lowest for Ni.

Table 1. Input data for calculations

	Fe	Co	Ni
r_d (a.u.)	1.512	1.437	1.342
z_d	6.6	7.6	8.6
R_c (a.u.)	1.540	1.641	1.030
A (a.u.)	0.363	0.393	0.207
Ω (a.u.)	89.29	85.85	85.24

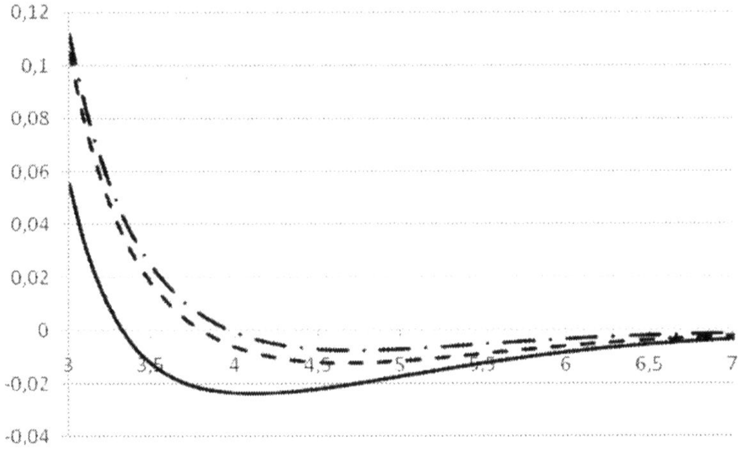

Figure 1. $V_{WH}(r)$ at $p = 0$ (Fe – solid line; Co – dashed line; Ni – dashed-dotted line)

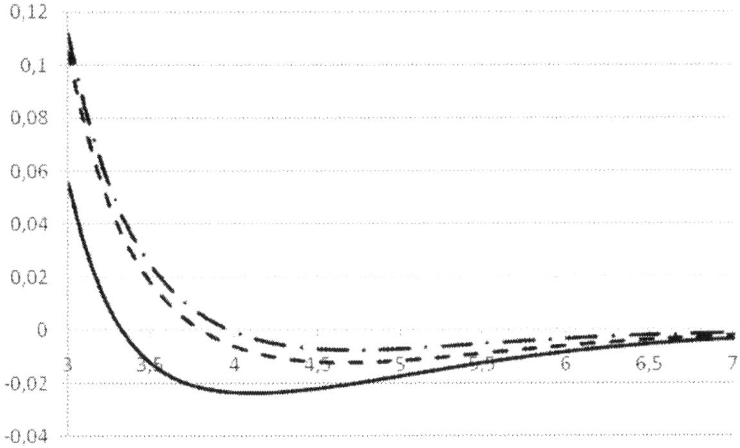

Figure 2. $V_{WH}(r)$ at $p = 0.5$ (Fe – solid line; Co – dashed line; Ni – dashed-dotted line)

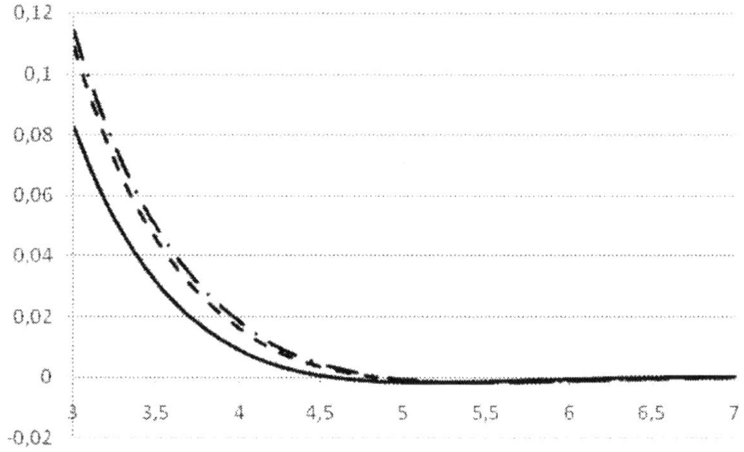

Figure 3. $V_{WH}(r)$ at $p = 1$ (Fe – solid line; Co – dashed line; Ni – dashed-dotted line)

4. Conclusion

Results obtained show that the suggested correction of the WH model leads to significant changes in the corresponding pair interaction. Since the behaviour of the effective pare potential influences on thermodynamic properties, it is possible to conclude that this correction will give more considerable changes for liquid Fe-Co alloy and less considerable for Co-Ni one in future thermodynamics calculations.

Acknowledgments

This work is supported by the Ural Branch of the Russian Academy of Sciences, and Russian Ministry of Science and Education (contract 02.740.11.0641).

References

[1] Wills J M and Harrison W A 1983 *Phys. Rev. B* **28** 4363
[2] Hausleitner C and Hafner J 1988 *J. Phys.F:Metal Phys.* **18** 1025
[3] Bretonnet J L and Derouiche A 1991 *Phys. Rev. B* **43** 8924
[4] Dubinin N E, Son L D and Vatolin N A 2008 *J. Phys.: Condens. Matter* **20** 114111
[5] Dubinin N E 2009 *J. Phys.: Conf. Ser.* **144** 012115
[6] Bretonnet J L and Silbert M 1992 *Phys. Chem. Liq.* **24** 169
[7] Vashishta P and Singwi K 1972 *Phys. Rev. B* **6** 875
[8] Waseda Y1981 *The Structure of Non-crystalline Materials – Liquids and Amorphous Solids* (New York: McGraw-Hill)
[9] Jakse N and Bretonnet J L 1995 *J. Phys.: Condens. Matter* **7** 3803

Advanced many-body and statistical methods in mesoscopic systems IOP Publishing
Journal of Physics: Conference Series **338** (2012) 012005 doi:10.1088/1742-6596/338/1/012005

Cooperative Resonance Interaction Between One- and Two-Photon Super-fluorescences Trough the Vacuum Field

Nicolae A. Enaki

Academiei str.5, Institute of Applied Physics of Academy of Sciences of Moldova, Chisinau, MD 2028, Republic of Moldova

E-mail: `enakinicolae@yahoo.com`

Abstract. The theoretical approach takes in consideration the cooperative phenomena which appear between three particle in two-photon resonance in the process of single- and two- photon decay. This type of single- and two-quanta cooperative effect between three subsystems of radiators are described by master equation which takes in to account three-particle cooperative resonance in the system. The resonance between the spontaneous emissions by two- and single photon transitions of three inverted radiators from the ensemble is proposed in order to accelerate the collective decay rate of the entangled photon pairs generated by the system. This effect is accompanied with the interferences between one-photon and two-quantum collective transitions of three inverted radiators from the ensemble. The three particle collective decay rate is defined in the description of three atomic correlation functions.

1. Introduction

Recently much attention is devoted to the problem of coherence which appears not only between the "individual" photons but also between groups of quanta. The generation of non-classical coherent electromagnetic field (EMF) in multi-photon emission and the interaction of coherent radiation with matter (nuclei, atoms and solids) have been the subjects of several theoretical and experimental studies in recent years [1]-[4].Examples include the higher-order coherence in multi-photon generation of light the two-photon micro-maser emission [4], two-photon lasers the parametric down conversion, four-wave mixing and other effects in optical diapason [2], and the possibility of coherent generation of photons in x and $gamma$ -spectral regions [3].

This article aims to investigate the cooperative emission of the inverted system of radiators taking into account the resonance between one- and two-photon cooperative transitions of three atomic subsystems. Since the two-photon cooperative phenomenon has a small two-quantum cooperative emission rate[5], we propose to extend our attention to the new type of cooperative resonance interaction between three radiators, in which the single-photon transitions of two radiators enter the two-quantum resonance with dipole forbidden transition of third atom. This cooperative effect between three particles occurs through the vacuum fluctuations of EMF and can amplify or diminish the spontaneous emission rates of each atomic subsystem as a function of the distances between them. To obtain more powerful pulses of entangled photons, it is proposed the cooperative interaction of three radiator subsystems in which one of them are inverted relatively the dipole forbidden transition of Hydrogen-like or Helium-like atoms (ions)

[6, 7]. The big number of classical experimental [7] and theoretical [6] descriptions of two-photon decay relatively the transitions $2^2S_{1/2} \to 1^2S_{1/2}$ and $2^1S_0 \to 1^1S_0$ give us the larger possibilities in the experimental realization of such resonances [8, 9]. Taking into account the elementary acts of two photon resonance between radiators, we have demonstrated the increasing of two-photon emission rate in one of radiator subsystem comparison with traditional two-photon super-fluorescence [5, 10]. The mutual influence of two single-photon super-fluorescence processes on the two-photon cooperative emission of the inverted atomic subsystem relatively dipole forbidden transition is examined in this paper. The elementary acts of new cooperative phenomenon consider the three-particle mutual interactions through the vacuum of EMF field, in which the product of vacuum polarization of two atoms enters in resonance with two-photon polarization of dipole forbidden transition.

In the second section of this article it is proposed the effective method of elimination of EMF operators from generalized equation of atomic subsystems in two-photon resonance. Note, that the master equation of such a system of radiators was obtained for the sample of atoms with the dimension commensurable with the radiation wavelengths. The quantum approach of elimination of bi-boson operators from master equation of atomic subsystem is proposed. The method gives us the possibility to take into account the quantum fluctuations among three radiators in two-quantum resonance through the vacuum field. The generalized equation of such three particle interaction is given in the next section.

In the third part of the paper we studied the spontaneous emission of three radiators j-th, l-th and m-th, situated at distance $r = r_{ln} = r_{ln}$ relatively the dipole forbidden atom D. As the one-photon resonance between atoms S and R is absent , the distance r_{jl} is considered as an arbitrary basis of isosceles triangle SRD. In this paper it is studied the mutual influence between the spontaneous emissions of three atomic subsystems localized in the isosceles triangle vertices SRD, considering that the exchange integral around the subsystem D consists from two domain.

2. Interaction Hamiltonian and Master Equation

Let us consider the interaction of three subsystems of radiators R, S, and D through the vacuum of EMF. The first two groups of radiators, R and S, prepared in excited state $|e_r\rangle \otimes |e_s\rangle$ can pass into Decke super-radiance [11] , [12] relatively the dipole active transitions $e_r \to g_r$ and $e_s \to g_s$ at frequencies ω_r and ω_s. Preparing in the excited state relative the dipole forbidden transition $2_d \to 1_d$, the D atomic subsystem passes in the ground state $|1_d\rangle$ simultaneously generation two quanta under the influences of first two atomic subsystems and vacuum fluctuations of EMF. This transition takes place through the virtual levels, which we replaced by the state $|3_d\rangle$ with opposite symmetry relative the states $|1_d\rangle$ and $|2_d\rangle$. Considering that the conservation energy law is established between the transitions of three atoms, $\hbar(\omega_r + \omega_s) = 2\hbar\omega_0$, one can propose the following Hamiltonian of the system

$$\hat{H} = \hat{H}_0 + \lambda \hat{H}_I, \tag{1}$$

where

$$\hat{H}_0 = \sum_k \hbar\omega_k \hat{a}_k^\dagger \hat{a}_k + \sum_{j=1}^{N_r} \hbar\omega_r \hat{R}_{zj} + \sum_{l=1}^{N_s} \hbar\omega_s \hat{S}_{zl} + 2\sum_{m=1}^{N} \hbar\omega_0 \hat{D}_{zm}$$

is the free and

$$\lambda \hat{H}_I = -\sum_k \sum_{j=1}^{N_a} (\mathbf{d}_r, \mathbf{g}_k)\{\hat{R}_j^+ \hat{a}_k \exp[i(\mathbf{k}, \mathbf{r}_j)] + \hat{R}_j^- \hat{a}_k^\dagger \exp[-i(\mathbf{k}, \mathbf{r}_j)]\}$$

$$
\begin{aligned}
&- \sum_k \sum_{l=1}^{N_b} (\mathbf{d}_s, \mathbf{g}_k)\{\hat{S}_l^+ \hat{a}_k \exp[i(\mathbf{k}, \mathbf{r}_l)] + \hat{S}_l^- \hat{a}_k^\dagger \exp[-i(\mathbf{k}, \mathbf{r}_l)]\} \\
&- \sum_{k_1,k_2} \sum_{m=1}^{N_b} (\mathbf{n}_{31}, \mathbf{e}_{\lambda_1})(\mathbf{n}_{32}, \mathbf{e}_{\lambda_2}) q(\omega_1, \omega_2)\{\hat{D}_m^+ \hat{a}_{k_2} \hat{a}_{k_1} \exp[i(\mathbf{k}_1 + \mathbf{k}_2, \mathbf{r}_m)] \\
&+ \hat{D}_m^- \hat{a}_{k_1}^\dagger \hat{a}_{k_2}^\dagger \exp[-i(\mathbf{k}_1 + \mathbf{k}_2, \mathbf{r}_l)]\}
\end{aligned}
$$

is the interaction parts of Hamiltonian (1). Here we have:

$$
q(\omega_1, \omega_2) = \frac{d_{23} d_{31} g_{k_2} g_{k_1}}{2\hbar} \left\{ \frac{1}{\omega_{32} + \omega_{k_1}} + \frac{1}{\omega_{31} - \omega_{k_2}} \right\}, \quad \mathbf{g}_k = \sqrt{\frac{2\pi\hbar\omega_k}{V}} \epsilon_\lambda,
$$

$\hat{a}_k$ and $\hat{a}_k^\dagger$ are annihilation and creation operators of EMF photons with wave vector $\mathbf{k}$, polarization ϵ_λ and frequency ω_k; $\mathbf{d}_r$ and $\mathbf{d}_s$ are dipole momentum transition between the ground and excited states for R and S atomic subsystems; d_{32i} and d_{31} are dipole momentum transitions in the three level system of atomic group D. The operators of R, S, and D atomic subsystems satisfy the commutation relations for $SU(2)$ algebra $[\hat{J}^+, \hat{J}^-] = 2\hat{J}_z$; $[\hat{J}_z\ , \hat{J}^\pm] = \pm\hat{J}^\pm$, where $\hat{J}^\pm$ is equivalent with $R^\pm$, $S^\pm$ and $D^\pm$. The inversion operator $\hat{J}_z$ is one of $\hat{R}_z$, $\hat{S}_z$ and $\hat{D}_z$ operators respectively. The operators of EMF satisfy the commutation relation $[\hat{a}_k, \hat{a}_{k'}^\dagger] = \delta_{k,k'}$; $[\hat{a}_k^\dagger, \hat{a}_{k'}^\dagger] = 0$, where $k = (\mathbf{k},\lambda)$ is the wave vector and polarization of the photon.

According to the Hamiltonian (1), let us represent the solution of Haisenberg equation for EMF operators $\hat{a}_k(t)$ and $\hat{a}_k^\dagger(t)$ through the sources and free operator parts

$$
\hat{a}_k(t) = \hat{a}_k(0)\exp[-i\omega_k t] + \hat{a}_k^s(t), \ \hat{a}_k^\dagger(t) = [\hat{a}_k(t)]^\dagger, \tag{2}
$$

where $\hat{a}_k^v(t) = \hat{a}_k(0)\exp[-i\omega_k t]$ and $\hat{a}_k^s(t)$ are the vacuum and source parts respectively.Here the source part is represented through the atomic polarization operators of the system

$$
\begin{aligned}
\hat{a}_k^s(t) &= 2i \sum_{n=1}^{N} \sum_{k_1} \frac{(\mathbf{n}_{31}, \mathbf{e}_{\lambda_1})(\mathbf{n}_{32}, \mathbf{e}_\lambda) q(\omega_1, \omega)}{\hbar} \exp[-i(\mathbf{k}_1 + \mathbf{k}, \mathbf{r}_n)] \int_0^t d\tau \exp[-i\omega_k \tau] \\
&\times \hat{D}_n^-(t-\tau)\hat{a}_{k_1}^\dagger(t-\tau) + \sum_{l=1}^{N_r} \frac{i(\mathbf{d}_r, \mathbf{g}_k)}{\hbar} \exp[-i(\mathbf{k}, \mathbf{r}_l)] \int_0^t d\tau \exp[-i\omega_k\tau]\hat{R}_l^-(t-\tau)] \\
&+ \frac{i(\mathbf{d}_s, \mathbf{g}_k)}{\hbar} \sum_{j=1}^{N_s} \exp[-i(\mathbf{k}, \mathbf{r}_j)] \int_0^t d\tau \exp[-i\omega_k\tau] S_j^-(t-\tau), \ \hat{a}_k^{\dagger,s}(t) = [\hat{a}_k^s(t)]^+
\end{aligned}
$$

Let us now consider the Heisenberg equation for the mean value of the arbitrary atomic operator $\langle 0, \Psi_a(0)|\hat{O}(t)|0, \Psi_a(0)\rangle$, where $|0, \Psi_a(0)\rangle = |0\rangle|\Psi_a(0)\rangle$ is the initial state of EMF and atomic subsystems. Taking into account, the action of vacuum parts of operators $\hat{a}_k(t)$ and $\hat{a}_k^\dagger(t)$ on the vacuum state of EMF $\hat{a}_k(0)|0\rangle_{ph} = \langle 0|_{ph}\hat{a}_k^\dagger(0) = 0$, we can partially eliminate the free parts of EMF field operators from the equation for atomic operators. Indeed considering that at initial time $t = 0$ the EMF is in the vacuum state, we can represent the Heisenberg equation for the mean value of the arbitrary atomic operator $\langle \hat{O}(t)\rangle$ through the normal product of EMF operators

$$
\frac{d}{dt}\langle \hat{O}(t)\rangle = -i \sum_k \sum_{j=1}^{N_r} \frac{(\mathbf{d}_r, \mathbf{g}_k)}{\hbar} \langle [\hat{R}_j^+(t), \hat{O}(t)]\hat{a}_{ks}(t)\rangle \exp[i(\mathbf{k}, \mathbf{r}_j)]
$$

$$-i\sum_{k}\sum_{l=1}^{N_s}\frac{(\mathbf{d}_s,\mathbf{g}_k)}{\hbar}\langle[\hat{S}_l^+(t),\hat{O}(t)]\hat{a}_{ks}(t)\rangle\exp[i(\mathbf{k},\mathbf{r}_l)]$$
$$-i\sum_{k_1,k}\sum_{m=1}^{N}\frac{(\mathbf{n}_{31},\mathbf{e}_{\lambda_1}(k_1))(\mathbf{n}_{23},\mathbf{e}_{\lambda}(k))q(\omega_{k_1},\omega_k)}{\hbar}\langle[\hat{D}_m^+(t),\hat{O}(t)]\hat{a}_{k_1}(t)\hat{a}_{ks}(t)\rangle\exp[i(\mathbf{k}_1+\mathbf{k},\mathbf{r}_m)]$$
$$+H.c.(\hat{O}^+\to\hat{O}). \quad (3)$$

As follows from the equation (3), the free part of operators (2) are partially eliminated. To obtain the master equation for the atomic subsystem, we must totally eliminate the vacuum parts of EMF operators from the expression (3). According to equation (3) it is impossible to move the free part $\hat{a}^v_{k_1}(t)$ of operator $\hat{a}_{k_1}(t)$ in the correlation $\langle[\hat{D}_m^+(t),\hat{O}(t)]a_{k_1}(t)\hat{a}_{ks}(t)\rangle$, because it don't commute with the operator $\hat{a}^s_k(t)$.

Let us further eliminate the EMF field operators from generalized equation (3). Looking carefully in the right hand site of this equation, we observe that the fool EMF operator $\hat{a}_{k_1}$ don't commute with source part $[\hat{a}_{k_1}(t),\hat{a}^s_k(t)]\neq 0$. Of course we can represent the correlation $\langle\hat{A}(t)\hat{a}_{k_1}(t)\hat{a}^s_k(t)\rangle$ of equation(3) in opposite case when the free part of operator $\hat{a}_{k_1}$ is eliminated $\langle\hat{A}(t)\hat{a}_k(t)\hat{a}_{k_1}(t)\hat{a}^s_k(t)\rangle$ bat this representation don't solve the problem of simultaneous elimination of vacuum parts of both operators $\hat{a}_{k_1}$ and $\hat{a}_k(t)$ (here $\hat{A}(t)=[\hat{D}_m^+(t),\hat{O}(t)]$). Observing that the source part of EMF operator $\hat{a}^s_k(t)$ is expressed through the linear combination of retardation atomic operator $\hat{S}_j^-(t-\tau)$ and $\hat{R}_l^-(t-\tau)$), for elimination of vacuum part of second operator $\hat{a}_k(t)$ from above correlation we propose the following lemma

Lemma 1 *If the Bose $\hat{a}_k(t)$ and $\hat{a}_k^+(t)$ operators lie between two operators of atomic subsystem $\hat{A}(t_1)$ and $\hat{B}(t_2)$, which belong to other time moments t_1 and t_2 , the elimination of the free part of these operators yields to the following expression for the correlation*

$$\left\langle\hat{A}(t_1)\hat{a}_k(t)\hat{B}(t_2)\right\rangle=\left\langle\hat{A}(t_1)\hat{a}^s_k(t)\hat{B}(t_2)\right\rangle-e^{-i\omega_3(t-t_2)}\left\langle\hat{A}(t_1)[\hat{a}^s_k(t_2),\hat{B}(t_2)]\right\rangle,$$
$$\left\langle\hat{A}(t_1)\hat{a}_k^+(t)\hat{B}(t_2)\right\rangle=\left\langle\hat{A}(t_1)\hat{a}_k^{\dagger,s}(t)(t)(t)\hat{B}(t_2)\right\rangle-e^{i\omega_3(t-t_1)}\left\langle[\hat{A}(t_1),\hat{a}_k^{\dagger,s}(t_1)]\hat{B}(t_2)\right\rangle \quad . \quad (4)$$

Proof. The commutations in (4) play the highest role in the two-photon spontaneous emission and such commutations bring the main contribution to the two-photon resonances between three particles. The problem is reduced to the elimination of vacuum part, which lies between the operators $\hat{A}(t_1)$ and $\hat{B}(t_2)$

$$<\hat{A}(t_1)a_k(t)\hat{B}(t_2)> \;=\; <A(t_1)(\hat{a}^v_k(t)+\hat{a}^s_k(t))\hat{B}(t_2)>.$$

Since $\hat{a}_k(t)=\hat{a}^v_k(t)+\hat{a}_{ks}(t)$, we will represent the vacuum part $\hat{a}^v_k(t)=\hat{a}_k(0)\exp[-i\omega_k t]$ through the vacuum operator at the time moment t_1. Taking into account the identity (2), we can represent the vacuum part in the following form $\hat{a}^v_k(t)=\hat{a}^v_k(t_2)e^{-i\omega_k(t-t_2)}=\{\hat{a}_k(t_2)-\hat{a}_{ks}(t_2)\}e^{-i\omega_k(t-t_2)}$. After the substitution of $\hat{a}^v_k(t)$ into correlation it is obtained the expression

$$\langle\hat{A}(t_1)\hat{a}_k(t)\hat{B}(t_2)\rangle=\langle\hat{A}(t_1)\hat{a}^s_k(t)\hat{B}(t_2)\rangle+e^{-i\omega_k(t-t_2)}\langle\hat{A}(t_1)\{\hat{a}_k(t_2)-\hat{a}^s_k(t_2)\}\hat{B}(t_2)\rangle.$$

We observe that operator $\hat{a}_k(t_2)$ commutes with the operator $\hat{B}(t_2)$. Consequently taking into account, that $\hat{a}_k(t_2)\ \hat{B}(t_2)|0>=\ \hat{B}(t_2)\hat{a}_{ks}(t_2)|0\rangle$, it is easily obtained that

$$\langle\hat{A}(t_1)\{\hat{a}_k(t_2)-\hat{a}^s_k(t_2)\}\ \hat{B}(t_2)\rangle=-\langle\hat{A}(t_1)[\hat{a}^s_k(t_2),\hat{B}(t_2)]\rangle.$$

This identity proofs the Lemma (4). ■

This lemma (4) can be used in the last term of generalized equation (3) for correlation functions like $\langle[\hat{D}_j^+(t),\hat{O}(t)]\hat{a}_k(t)S_n^-(t-\tau)\rangle$, $\langle[\hat{D}_j^+(t),\hat{O}(t)]a_k(t)\hat{R}_l^-(t-\tau)\rangle$ and $\langle[\hat{R}_j^+(t),\hat{O}(t)]\hat{D}_n^-(t-\tau)\hat{a}_{k_1}^\dagger(t-\tau)\rangle$, $\langle[\hat{S}_l^+(t),\hat{O}(t)]\hat{D}_n^-(t-\tau)\hat{a}_{k_1}^\dagger(t-\tau)\rangle$. Taking into account the lemma (4), we observe that the above correlation functions can be simply represented through atomic operators

$$\begin{aligned}\langle[\hat{D}_j^+(t),\hat{O}(t)]a_k(t)\hat{S}_n^-(t-\tau)\rangle &= \left\langle[\hat{D}_j^+(t),\hat{O}(t)]\hat{a}_{ks}(t)\hat{S}_n^-(t-\tau)\right\rangle \\ &- e^{-i\omega_k\tau}\left\langle[\hat{D}_j^+(t),\hat{O}(t)][\hat{a}_{ks}(t-\tau),S_n^-(t-\tau)]\right\rangle,\end{aligned} \tag{5}$$

$$\begin{aligned}\langle[\hat{R}_j^+(t),\hat{O}(t)]\hat{a}_k^\dagger(t-\tau)\hat{D}_n^-(t-\tau)\rangle &= \langle[\hat{R}_j^+(t),\hat{O}(t)]a_{ks}^\dagger(t-\tau)\hat{D}_n^-(t-\tau)\rangle \\ &- e^{-i\omega_k\tau}\langle[[\hat{R}_j^+(t),\hat{O}(t)],\hat{a}_{ks}^\dagger(t)]\hat{D}_n^-(t-\tau)\rangle.\end{aligned} \tag{6}$$

The interaction between the electromagnetic field and atomic subsystems can be found in the third order of interaction constants $(\mathbf{d}_r,\mathbf{g}_k)(\mathbf{d}_s,\mathbf{g}_k)q(\omega_1,\omega_2)$ respectively. According to this condition, the smooth correlation functions can be obtained only for the following terms of expressions (5) and (6) : $\left\langle R_l^-(t-\tau')[D_j^+(t),O(t)]S_n^-(t-\tau)\right\rangle$ and $\langle R_l^+(t-\tau')[R_j^+(t),O(t)]D_n^-(t-\tau)\rangle$. Other terms of expressions (5) and (6) give the contribution in the higher order of decomposition on the small parameter λ (see the interaction Hamiltonian (1)).

The lemma (4) is non-applicable for correlation functions in which it is used simultaneously the creation and annihilations Boson operators belonging to different time moments : $\langle[D_j^+(t),O(t)]a_k(t)D_n^-(t-\tau)a_{k_1}^\dagger(t-\tau)\rangle$ and its hermit conjugate part $\langle a_{k_1}(t-\tau)D_n^+(t-\tau)a_k^\dagger(t)[O(t),D_j^-(t)]\rangle$. To eliminate the vacuum part of operators $a_k(t)$ and $a_k^\dagger(t')$ let us formulate the following rule.

Lemma 2 *If the arbitrary operator $\hat{A}(t_1)$ contains the creation operators of EMF and $\hat{B}(t_2)$ contains the annihilation operators of EMF, the elimination of vacuum part of annihilation $\hat{a}_k(t)$ or creation $\hat{a}_k^\dagger(t)$ operators situated between these operators $\hat{A}(t_1)$ and $\hat{B}(t_2)$ takes place according with Lemma 1. In opposite situation, we can represent the operator $A(t_1)$ through the product of atomic operator $\mathcal{A}(t_1)$ and annihilation field operators $A(t_1)=\mathcal{A}(t_1)a_{k_1}(t_1)a_{k_2}(t_1)...a_{k_n}(t_1)$ and the operator $\hat{B}(t_2)$ through the product of creation field operators and atomic operator $\mathcal{B}(t_2)$ so that $B(t_2)=\mathcal{B}(t_2)a_{k_1}^\dagger(t_2)a_{k_2}^\dagger(t_2)...a_{k_m}^\dagger(t_2)$. Here the elimination of vacuum part of the operators $a_k(t)$ and $a_k^\dagger(t)$ can be represented through the following identities*

$$\begin{aligned}\langle A(t_1)a_k(t)B(t_2)\rangle &= \langle A(t_1)a_k^s(t)B(t_2)\rangle - \exp[-i\omega_k(t-t_2)]\{\langle A(t_1)[a_k^s(t_2),B(t_2)]\rangle \\ &- \delta_{k,k_1}\left\langle A(t_1)\mathcal{B}(t_2)a_{k_2}^\dagger(t_2)...a_{k_m}^\dagger(t_2)\right\rangle \\ &- ... - \delta_{k,k_m}\left\langle A(t_1)\mathcal{B}(t_2)a_{k_1}^\dagger(t_2)a_{k_2}^\dagger(t_2)...a_{k_{m-1}}^\dagger(t_2)\right\rangle\},\end{aligned} \tag{7}$$

$$\begin{aligned}\left\langle A(t_1)a_k^\dagger(t)B(t_2)\right\rangle &= \left\langle A(t_1)a_k^{\dagger,s}(t)B(t_2)\right\rangle - \exp[i\omega_k(t-t_2)]\{\left\langle[A(t_1),a_k^{\dagger,s}(t_1)]B(t_2)\right\rangle \\ &- \delta_{k,k_1}\langle\mathcal{A}(t_1)a_{k_2}(t_1)...a_{k_n}(t_1)B(t_2)\rangle \\ &- ... - \delta_{k,k_n}\langle\mathcal{A}(t_1)a_{k_1}(t_1)a_{k_2}(t_1)...a_{k_{n-1}}(t_1)B(t_2)\rangle\}.\end{aligned} \tag{8}$$

Proof. Taking in-to account the lemma (4), we can represent the third correlation function $\langle A(t_1)a_k(t)B(t_2)\rangle$ of expression (7) in the following form

$$\begin{aligned}\langle A(t_1)a_k(t)B(t_2)\rangle &= \langle A(t_1)a_k^s(t)B(t_2)\rangle - \exp[-i\omega_k(t-t_2)]\langle A(t_1)[a_k^s(t_2),B(t_2)]\rangle \\ &+ \exp[-i\omega_k(t-t_2)]\langle A(t_1)[a_k(t_2)B(t_2)-B(t_2)a_k^s(t_2)]\rangle.\end{aligned} \tag{9}$$

According to explicit expression of operator, $\hat{B}(t_2) = \mathcal{B}t_2)a^{\dagger}_{k_1}(t_2)a^{\dagger}_{k_2}(t_2)...a^{\dagger}_{k_m}(t_2)$, let us introduced it in the second term of right hand site of expression (9). Following the commutation rules of the boson operators of EMF, the operator $\hat{a}_k(t_2)$ can be permuted in the right hand site of the correlation function. Taking into consideration that $(\hat{a}^s_k(t_2) + \hat{a}_{kv}(t_2))\,|0\rangle = \hat{a}^s_k(t_2)\,|0\rangle$, this term becomes

$$\left\langle A(t_1)a_k(t_2)\mathcal{B}(t_2)a^{\dagger}_{k_1}(t_2)a^{\dagger}_{k_2}(t_2)...a^{\dagger}_{k_m}(t_2)\right\rangle = \delta_{k,k_1}\left\langle A(t_1)\mathcal{B}(t_2)a^{\dagger}_{k_2}(t_2)...a^{\dagger}_{k_m}(t_2)\right\rangle$$
$$+... + \delta_{k,k_m}\left\langle A(t_1)\mathcal{B}(t_2)a^{\dagger}_{k_1}(t_2)a^{\dagger}_{k_2}(t_2)...a^{\dagger}_{k_{m-1}}(t_2)\right\rangle$$
$$+\left\langle A(t_1)\mathcal{B}(t_2)a^{\dagger}_{k_1}(t_2)a^{\dagger}_{k_2}(t_2)...a^{\dagger}_{k_m}(t_2)a_{ks}(t_2)\right\rangle. \quad (10)$$

Introducing this relation in expression (10) , it is not difficult to observe that the new expression for correlation $\left\langle \hat{A}(t_1)\hat{a}_k(t)\hat{B}(t_2)\right\rangle$ coincides with expression (7) from the lemma. The similar procedure of permutation of vacuum part of creation operator $a^{\dagger}_k(t)$ demonstrates the identities (8) of lemma 2. ■

After the substitution of source part of the EMF operators (2), the next step consists in the elimination of operator $\hat{a}^{\dagger}_k(t-\tau')$ from the correlation $\langle[\hat{R}^+_j(t),\hat{O}(t)]a^{\dagger}_k(t-\tau')\hat{D}^-_m(t-\tau')\rangle$ of first term of equation (3). According to lemma (4) and taking into account only the resonance terms in the right hand site of equation (3), it is not difficult to observe, that in the third order of the interaction constant λ the term proportional to $(\mathbf{d}_s,\mathbf{g}_k)q(\omega_{k_1},\omega_k)$ describes the tho-photon resonance between the atomic subsystems D and S,R. Neglecting the terms proportional to the λ^4, we obtain the following resonance between S, R and D subsystems

$$\langle[R^+_j(t),O(t)]a^{\dagger}_k(t-\tau')D^-_m(t-\tau')\rangle = \frac{i(\mathbf{d}_s,\mathbf{g}_k)}{\hbar}\sum_{l=1}^{N_s}\int_0^t d\tau\exp[i\omega_k(\tau-\tau')]$$
$$\times\langle([S^+_l(t-\tau)[R^+_j(t),O(t)]] + \theta(\tau-\tau')[R^+_j(t),O(t)]S^+_l(t-\tau))D^-_m(t-\tau')\rangle. \quad (11)$$

Replacing in expression (11) the operator of S subsystem with operators of R atoms, it is obtained the similar correlation (11) from second term of equation (3).

As follows from the representation (2) and generalized equation (3), the procedure of elimination of EMF field operators must continue in the last term of expression (3) too. For correlation $\langle[\hat{D}^+_m(t),\hat{O}(t)]\hat{a}_{k_1}(t)\hat{R}_j(t-\tau)\rangle$ it is obtained the same resonances in the third order of small interaction constant

$$\langle[D^+_m(t),O(t)]a_{k_1}(t)R_j(t-\tau)\rangle = \frac{i(\mathbf{d}_s,\mathbf{g}_k)}{\hbar}\sum_{l=1}^{N_s}\int_0^t d\tau'\exp[-i\omega_{k_1}\tau']$$
$$\times\langle[D^+_m(t),O(t)](S^+_l(t-\tau')R_j(t-\tau) - \theta(\tau'-\tau)[R_j(t-\tau),S^+_l(t-\tau')])\rangle. \quad (12)$$

According to the lemma (7) from the last term of equation (3), we obtain the following contribution to two-photon cooperative decay diagrams of D subsystem.

$$\langle[D^+_j(t),O(t)]a_k(t)D^-_n(t-\tau)a^{\dagger}_{k_1}(t-\tau)\rangle = -\{\langle[D^+_j(t),O(t)][a_{ks}(t-\tau),D^-_n(t-\tau)a^{\dagger}_{k_1}(t-\tau)]\rangle$$
$$-\delta_{k,k_1}\langle[D^+_j(t),O(t)]D^-_n(t-\tau)\rangle\}\exp[-i\omega_k\tau] + \langle[D^+_j(t),O(t)]a_{ks}(t)D^-_n(t-\tau)a^{\dagger}_{k_1}(t-\tau)\rangle. \quad (13)$$

Neglecting the non-resonance terms in equation (3) according with expressions (11), (12) and (13), we obtain the following master equation for arbitrary operator $\hat{O}(t)$ in thread approximation of interaction constant λ

$$\begin{aligned}
\frac{d\langle O(t)\rangle}{dt} &= \sum_k \sum_{l,j=1}^{N_r} \frac{(\mathbf{d}_r,\mathbf{g}_k)^2}{\hbar^2} \int_0^t d\tau \exp[-i\omega_k\tau + i(\mathbf{k},\mathbf{r}_j-\mathbf{r}_l)] \langle [R_j^+(t),O(t)]R_l^-(t-\tau)\rangle \\
&+ \sum_k \sum_{l,l=1}^{N_s} \frac{(\mathbf{d}_s,\mathbf{g}_k)^2}{\hbar^2} \int_0^t d\tau \exp[-i\omega_k\tau + i(\mathbf{k},\mathbf{r}_j-\mathbf{r}_l)] \langle [S_j^+(t),O(t)]S_l^-(t-\tau)\rangle \\
&+ \sum_{k_1,k_2} \sum_{l,j=1}^{N} \frac{(\mathbf{n}_{31},\mathbf{e}_{\lambda_1})^2(\mathbf{n}_{23},\mathbf{e}_{\lambda_2})^2 q^2(\omega_1,\omega_2)}{\hbar^2} \int_0^t d\tau \langle [D_j^+(t),O(t)]D_l^-(t-\tau)\rangle \\
&\times \exp[-i(2\omega_0-\omega_{k_1}-\omega_{k_2})\tau]\exp[i(\mathbf{k}_1+\mathbf{k}_2,\mathbf{r}_j-\mathbf{r}_l)] \\
&+ i\sum_{k,k'}\sum_{n=1}^{N}\sum_{j=1}\sum_{l=1} \frac{(\mathbf{n}_{31},\mathbf{e}_\lambda)(\mathbf{n}_{23},\mathbf{e}_{\lambda'})q(\omega_k,\omega_{k'})}{\hbar^3} d\tau' \exp[i(\mathbf{k},\mathbf{r}_n-\mathbf{r}_j)+i(\mathbf{k}',\mathbf{r}_n-\mathbf{r}_l)] \\
&\times \int_0^t d\tau \int_0^t \exp[-i\omega_k\tau - i\omega_{k'}\tau'][(\mathbf{d}_s,\mathbf{g}_{k'})(\mathbf{d}_r,\mathbf{g}_k)\langle [D_n^+(t),O(t)]R_j^-(t-\tau)S_l^-(t-\tau')\rangle \\
&+ (\mathbf{d}_r,\mathbf{g}_{k'})(\mathbf{d}_s,\mathbf{g}_k)\langle [D_n^+(t),O(t)]S_j^-(t-\tau)R_l^-(t-\tau')\rangle] \\
&- i\sum_{k,k'}\sum_{m=1}^{N}\sum_{j=1}\sum_{l=1} \frac{(\mathbf{n}_{31},\mathbf{e}_\lambda)(\mathbf{n}_{23},\mathbf{e}_{\lambda'})q(\omega_k,\omega_{k'})}{\hbar^3} \exp[-i(\mathbf{k},\mathbf{r}_m-\mathbf{r}_l)+i(\mathbf{k}',\mathbf{r}_j-\mathbf{r}_m)] \\
&\times \int_0^t d\tau' \int_0^t d\tau \exp[i\omega_k\tau - i\omega_k\tau' - i\omega_{k'}\tau'][(\mathbf{d}_s,\mathbf{g}_k)(\mathbf{d}_r,\mathbf{g}_{k'})\langle\{[S_l^+(t-\tau)[R_j^+(t),O(t)]] \\
&+ \theta(\tau-\tau')[R_j^+(t),O(t)]S_l^+(t-\tau)\}D_m^-(t-\tau')\rangle + (\mathbf{d}_s,\mathbf{g}_{k'})(\mathbf{d}_r,\mathbf{g}_k) \\
&\times \langle\{[R_l^+(t-\tau)[S_j^+(t),O(t)]] + \theta(\tau-\tau')[S_j^+(t),O(t)]R_l^+(t-\tau)\}D_m^-(t-\tau')\rangle] \\
&+ H.c.(O^+\to O).
\end{aligned} \tag{14}$$

This method give as the possibility to eliminate the vacuum parts of bi-boson operators $\hat{a}_{k_1}(t)\hat{a}_k(t)$. The master equation (14) for three atomic subsystem in interaction through the vacuum of EMF is obtained, taking into account the normal ordering of annihilation and creation operators of EMF in correlation functions.

To avoid these divergences it is proposed the integration procedure, which takes into account the causality appearing between the radiators [13]. Following this method the right hand site of master equation (14) takes the form

$$\begin{aligned}
\frac{d}{dt}\langle O(t)\rangle &= \frac{1}{2\tau_r}\sum_{l,j=1}^{N_r}\chi_r(j,l)\langle [R_j^+(t),O(t)]R_l^-(t)\rangle + \frac{1}{2\tau_s}\sum_{l,l=0}^{N_s}\chi_s(j,l)\langle [S_j^+(t),O(t)]S_l^-(t)\rangle \\
&+ \frac{1}{2\tau_d}\sum_{l,j=1}^{N}\chi_d(j,l)\langle [D_j^+(t),O(t)]D_l^-(t)\rangle \\
&+ \frac{i}{2\tau_{srd}^0}\sum_{m=1}^{N}\sum_{l=1}^{N_r}\sum_{l=0}^{N_s} U(j,l,m)\langle [D_m^+(t),O(t)]R_j^-(t)S_l^-(t)\rangle \\
&- \frac{i}{4\tau_{srd}^0}\sum_{m=1}^{N}\sum_{j=1}^{N_r}\sum_{l=0}^{N_s} V(j,l.m)[\langle S_l^+(t)[R_j^+(t),O(t)]D_m^-(t)\rangle \\
&+ \langle R_j^+(t)[S_l^+(t),O(t)]D_m^-(t)\rangle] + H.C.(O^+\to O).
\end{aligned} \tag{15}$$

Here $\chi_p(j,l) = [\exp[i\omega_p r_{jl}/c] - 1]/[i\omega_p r_{jl}/c], p \equiv S, \ R$ is the intrinsic exchange integrals between atoms of subsystems S or R. The two-photon exchange integral between j-th and l-th radiators according to the theory of super-radiance [5] is described by $\Re\chi_d(j,l) = \sin[(2k_0 - k)r_{jl}]/[(2k_0 - k)r_{jl}] \times \sin[kr_{jl}]/(kr_{jl})$ respectively, where k_i is the wave vector of emitted photons.The spontaneous emission times, τ_s and τ_r, for S and R dipole active atomic subsystems are defined in the literature $\tau_\alpha^{-1} = 4d^2\omega_\alpha^3/(3\hbar c^3$; $\alpha = r, s$.

We observe, that in the right hand part of the equation (15) the third order terms contain resonances between the single-photon radiators S , R and two-photon radiator D, described by the correlation functions $\langle S_l^+(t)[R_j^+(t), O(t)]D_m^-(t)\rangle$, $\langle[\hat{D}_n^+(t), \hat{O}(t)]\hat{R}_j^-(t)\hat{S}_l^-(t)\rangle$ and $\langle\hat{R}_l^+(t)[\hat{S}_j^+(t), \hat{O}(t)]\hat{D}_m^-(t)\rangle$. Indeed, representing the exchange integral between three atoms in the similar form as in the paper [5], we can find cooperative emission rate of three atoms belong to S, R and D subsystems situated at small distance $r \ll \lambda_0$

$$\frac{1}{\tau_{srd}^0} = \left(\frac{4}{3}\right)^2 \frac{d_s d_r d_{23} d_{31} \omega_s^3 (\omega_r)^3}{4\pi c^6 \hbar^2} \left\{\frac{1}{\omega_{32} + \omega_s} + \frac{1}{\omega_{31} - \omega_r}\right\}, \qquad (16)$$

The resonance interaction of a dipole forbidden atom m and two dipole active radiators l and j are described by the interference function

$$U(j,l,m) = -\frac{c^2[\exp[i\omega_r r_{mj}/c] - 1][\exp[i\omega_s r_{ml}/c] - 1]}{\omega_r \omega_s r_{mj} r_{ml}};$$
$$V(j,l,m) = \exp[-i\omega_r r_{mj}/c]U(j,l,m). \qquad (17)$$

The master equation (15) can be used for description of cooperative interaction between the dipole forbidden and dipole active subsystems of radiators.

3. Kinetic equations for correlation functions

In this section it is proposed to simplify the problem of interaction between three subsystems of radiators, taking into account the generalized equation (15). The closed system of equations for correlation functions can be found studding the chain of equations for correlations between dipole forbidden transitions of D subsystem and dipole active transitions of S and R subsystems of radiators. Below it is studied two situations: A. the small number of atoms in which it is possible two photon resonance, B. the large number of atoms in each S, R and D subsystems for which the semiclassical approach is possible.

A. Considering, that the numbers of atoms in the subsystems are relatively small, the generalized equation (15) can be reduced to closed systems of equations. Let us annualize the simple situation for three particle system which contains one atom in each subsystems S , R and D. According to the master equation (15), we can introduce the mean value of exited numbers of atoms, $\langle\hat{N}_\alpha\rangle = \langle\hat{J}_{z\alpha}(t)\rangle + 0.5$, $\alpha = s,\ r,\ d$ of type S, R, and D. Here operator $\hat{J}_{z\alpha}$ is considered equivalent to the inversions of each atom S_z, R_z, D_z. Let us considered the situation, when two atoms R and S are situated at equal distance r relatively the radiator D. This distance may have the same magnitude as the emission wavelength of photons from the system. When emission frequencies of the one-photon radiators coincides $\omega_s \simeq \omega_s$, this dependence of exchange integral $V(j,l,m)$ becomes real defined function $[\exp[2\pi i\omega_s r/c]-1][\exp[-2\pi i\omega_r r/c]-1 \simeq 2[1-\cos(\omega_r r/c)]$. In this case the mutual interference between S, R and D subsystems are described by more simple exchange integrals

$$V_{srd} = \frac{2(1-\cos(2\pi r/\lambda_s)}{(2\pi r/\lambda_s)^2}, \quad U_{srd} = exp[i\omega_s r/c]V_{srd}, \qquad (18)$$

Here r is the distance between R, S and D subsystems, $\lambda_s \simeq \omega_s$; . Introducing the correlation function between the polarizations of three atoms $\langle F\rangle = i[\langle\hat{D}^+(t)\hat{S}^-(t)\hat{R}^-(t)\rangle -$

$\langle\hat{S}^+(t)\hat{R}^+(t)\hat{D}^-(t)\rangle]$, we obtain the closed system of equations. According to this procedure the closed system of equations is obtained from master equation (15)

$$\begin{aligned}
\frac{d}{dt}\langle\hat{N}_s(t)\rangle &= -\frac{\langle\hat{N}_s\rangle}{\tau_s}-\frac{1}{2\tau_{srd}}\langle\hat{F}\rangle,\quad \frac{d}{dt}\langle\hat{N}_r(t)\rangle = -\frac{\langle\hat{N}_r\rangle}{\tau_r}-\frac{1}{2\tau_{srd}}\langle F\rangle,\\
\frac{d}{dt}\langle\hat{N}_d(t)\rangle &= -\frac{\langle\hat{N}_d\rangle}{\tau_d}-\frac{\cos(x)}{\tau_{srd}}\langle\hat{F}\rangle,\quad \frac{d}{dt}\langle\hat{N}_s\hat{N}_r\rangle = -B\langle\hat{N}_s\hat{N}_r\rangle,\\
\frac{d}{dt}\langle\hat{N}_s\hat{N}_d\rangle &= -C\langle\hat{N}_s\hat{N}_r\rangle,\quad \frac{d}{dt}\langle\hat{N}_s\hat{N}_d\rangle = -D\langle\hat{N}_s\hat{N}_r\rangle,\\
\frac{d}{dt}\langle\hat{F}(t)\rangle &= -\frac{A}{2}\langle F(t)\rangle+\frac{1}{\tau_{srd}}[2(\cos(x)+1)\langle\hat{N}_s\hat{N}_r\hat{N}_d\rangle-\cos(x)\langle\hat{N}_s\hat{N}_r\rangle-\langle\hat{N}_s\hat{N}_d\rangle/2\\
&- \langle\hat{N}_r\hat{N}_d\rangle/2],\quad \frac{d}{dt}\langle\hat{N}_s\hat{N}_r\hat{N}_d\rangle = -A\langle\hat{N}_s\hat{N}_r\hat{N}_d\rangle. \qquad (19)
\end{aligned}$$

Here the following correlation functions between the atomic excitations was introduced: $\langle\hat{N}_s\hat{N}_r\hat{N}_d\rangle$, $\langle\hat{N}_s\hat{N}_r\rangle$, $\langle\hat{N}_s\hat{N}_d\rangle$, and $\langle\hat{N}_r\hat{N}_d\rangle$. The new three particle cooperative time between the atoms situated at relative distance x is introduced $\tau_{srd}(x)=\tau^0_{srd}/V_{srd}(x)$, $x=\omega_s r/c$. This system of equation is exactly solvable. The solution is

$$\begin{aligned}
\langle\hat{N}_s\hat{N}_r\hat{N}_d\rangle &= \exp(-At),\ \langle N_sN_r\rangle=\exp(-Bt),\ \langle\hat{N}_s\hat{N}_d\rangle=\exp(-Ct),\ \langle N_dN_r\rangle=\exp(-Dt),\\
\langle\hat{F}(t)\rangle &= [4(\cos(x)+1)\frac{1-\exp[-At/2]}{A}-2\cos(x)\frac{1-\exp[-(B-\tau_d^{-1})t/2]}{B-\tau_b^{-1}}\\
&- \frac{1-\exp[-(C-\tau_r^{-1})t/2]}{C-\tau_r^{-1}}-\frac{1-\exp[-(C-\tau_s^{-1})t/2]}{C-\tau_s^{-1}}]\frac{\exp(-At/2)}{\tau_{srd}},\\
\langle\hat{N}_d\rangle &= \exp[-t/\tau_d]-\frac{\cos x}{\tau_{srd}}\int_0^t\exp[-(t-t')/\tau_d]\langle F(t')\rangle,\quad \langle\hat{N}_i\hat{N}_j\rangle=\exp[-(\frac{1}{\tau_i}+\frac{1}{\tau_j})t]\\
\langle\hat{N}_i\rangle &= \exp[-t/\tau_i]-\frac{1}{2\tau_{srd}}\int_0^t\exp[-(t-t')/\tau_i]\langle\hat{F}(t')\rangle, i,j\to s,\ r.
\end{aligned}$$

where the collective rates are defined $A=1/\tau_r+1/\tau_d+1/\tau_s$; $B=1/\tau_r+1/\tau_s$; $C=1/\tau_d+1/\tau_s$; $D=1/\tau_d+1/\tau_s$. The solution of this system of equation is plotted in Figs. 1.The dependence of the atomic inversion $\langle D_z(t,r)\rangle$ as functions of relative time t/τ_d and relative distance r/λ_s between S , R and D subsystems for the parameters of the system $\tau_d/\tau_r=\tau_d/\tau_s=5$, $\tau_d/\tau^0_{srd}=2$ is plotted in the Figs.1. The significant influence of single-photon transition on the two-quantum transitions is observed at small distance r with the magnitude comparable with emission wavelength of the photons. This dependence follows from analytical expression (20). According to two-dimension surface (t,r) represented in Figs.1 and 1, we can observe that with decreasing of distance between the radiators the relative decay rate increases τ_s/τ_{srd}. This process is accompanied with the substantially increasing of two-photon emission intensity of subsystem D.

As follows from Fig.1 the cooperative exchanges between the radiator accelerate the two-photon decay processes so that the the emission rate $V=-\frac{dD_z}{dt}$ achieves the maximal value. After same time this rate decreases till zero value as in the typical super-radiant process.

B. Let us now describe the time behavior of big number of radiators in subsystems. Neglecting the quantum fluctuations of inversion operators $\langle R_{z_j}\rangle$, $\langle S_{zl}\rangle$, $\langle D_{zn}\rangle$, we obtain the following closed system of nonlinear equations

$$\frac{d}{dt}R_z(t) = -\frac{1}{\tau_r}\{N_r(N_r+2)/4-R_z^2+R_z\}-\frac{1}{2\tau_{srd}}F(t),$$

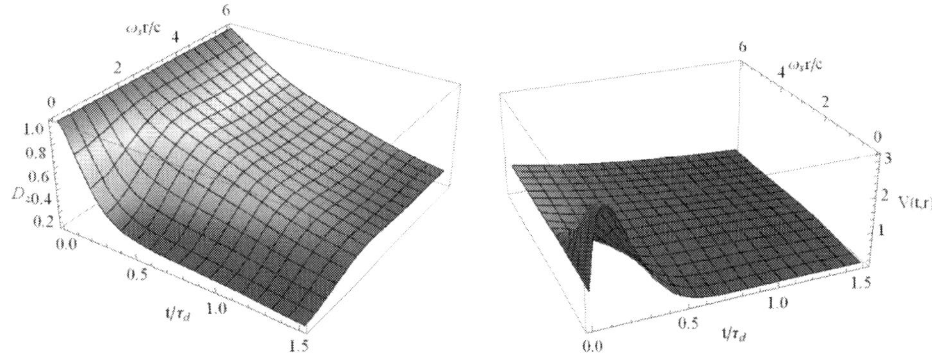

Figure 1. The decay process of inversion $D_z(t)$ and the two-photon decay rate $V = -dD_z(t)/dt$ stimulated by single photon processes for the parameters of the system relative decay rates of S- and R- atoms $\tau_d/\tau_r = \tau_d/\tau_s = 5$, $\tau_d/\tau^0_{srd} = 2$.

$$\begin{aligned}
\frac{d}{dt}S_z(t) &= -\frac{1}{\tau_s}\{N_s(N_s+2)/4 - S_z^2 + S_z\} - \frac{1}{2\tau_{srd}}F(t),\\
\frac{d}{dt}D_z(t) &= -\frac{1}{\tau_d}\{N(N+2)/4 - D_z^2 + D_z\} - \frac{\cos(x)}{\tau_{srd}}F(t),\\
\frac{d}{dt}F(t) &= \{\frac{1}{\tau_s}[S_z(t)-1] + \frac{1}{\tau_r}[R_z(t)-1] + \frac{1}{\tau_d}[D_z(t)-1]\}F\\
&+ \frac{1}{\tau_{srd}}[2\cos(x)D_z\{N_s^2/4 - S_z^2\}\{N_r^2/4 - R_z^2\}\\
&+ R_z\{N_s^2/4 - S_z^2\}\{N^2/4 - D_z^2\} + S_z\{N_r^2/4 - R_z^2\}\{N^2/4 - D_z^2\}\\
&+ 2(1+\cos x)N_sN_rN\exp[-A*t] - \cos(x)N_s*N_r*\exp[-Bt]\\
&- 0,5N_sN\exp[-Ct] - 0.5N_rN\exp[-Dt]]. \qquad (20)
\end{aligned}$$

the solution of which for three atomic system pass into (19). Here $R_z(t) = \sum_{j=0}^{N_r}\langle R_{zj}(t)\rangle$, $S_z(t) = \sum_{l=0}^{N_s}\langle S_{zl}(t)\rangle$ and $D_z(t) = \sum_{m=0}^{N}\langle D_{zm}(t)\rangle$ are the atomic inversions of R S and D subsystems respectively; $F(t) = i[\langle D^+(t)S^-(t)R^-(t)\rangle - \langle S^+(t)R^+(t)D^-(t)\rangle]$ describes three particle correlation function in collective interaction between subsystems, $D^+(t) = \sum_{m=0}^{N} D_m^+(t)$, $R^+(t) = \sum_{j=0}^{N_r} R_j^+(t)$ and $S^+(t) = \sum_{l=0}^{N_s} S_l^+$ describe the collective atomic excitation operators for subsystems D, R and S with the dimension less than emission wavelength.

In comparison with three atoms cooperative interaction, described by the closed system of quantum equations (19), the system of equations (20) is obtained in the semi-classical approximation and contains the quantum aspect of interaction between three atoms of subsystems S, R and D. In other words when the number of atoms in each subsystems is equal to one, $N_s{=}N_r{=}N{=}1$, the additional terms in the system of equations 20) give us the possibility to reduce this equations to the following system (19).

Taking into account the relative expressions of the decay rates described in Figs. 2, we obtain the numerical simulation of the inversion D_z and its derivative. As follows from the Figs. 2 for relative parameter $\tau_d/\tau^0_{drs} > 0.2$ the decay rate of of two-photon emission passes from a monotonic time function to the oscillatory time dependence. The influence of single photon cooperative emission to two-quantum decay process increases with decreases of the relative distance $r/\lambda_s < 1$ between the subsystems. In Figs. 2 is plotted the time dependence of the inversion $D_z(t)$ of dipole forbidden radiators as function of the relative time t/τ_d and relative

Journal of Physics: Conference Series **338** (2012) 012005 doi:10.1088/1742-6596/338/1/012005

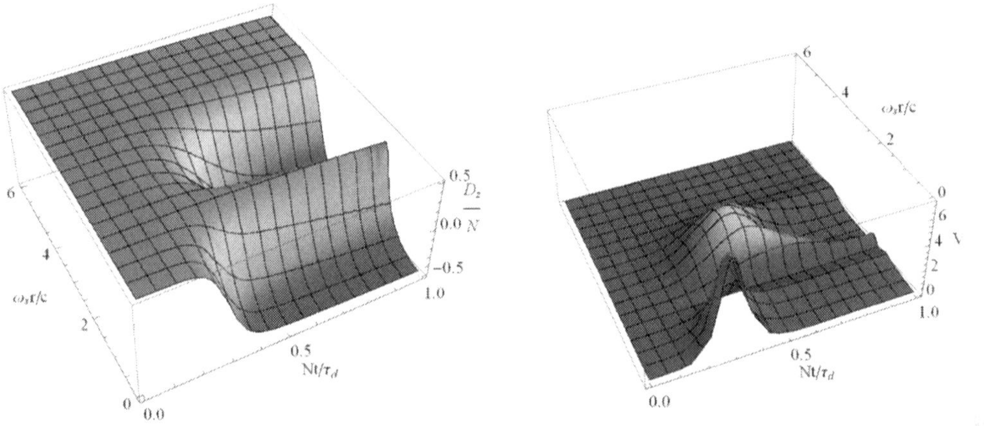

Figure 2. The time dependence of inversion $D_z(t)$ Relative emission rate $V = -\frac{\tau_d}{N^2}\frac{dD_z(t)}{dt}$ of dipole forbidden radiator subsystem for the values of the parameters number of atoms in the subsystems S, R and D is $N_s = N_r = N = 50$ respectively; the relative decay times of the subsystems are $\tau_b/\tau_s = \tau_b/\tau_r = 6$ and $\tau_d/\tau^0_{drs} = 1/6$; The distance r between the subsystems S, R and D is changed between 0 and 2π.

distance r/λ_s. The behavior of decay rate of D-atomic subsystem is plotted in Figs. 2 for different values of cooperative rate $1/\tau_{srd}(x)$ and distance r, where $\tau_d/\tau^0_{srd} = 0.2$.

The strong mutual influences between single and two-photon super-radiance processes in three particle interaction are characteristic for these numerical representations. It is observed again two regions $0 \le r \le \lambda_s/2$ and $\lambda_s/2 \le r \le \lambda_s$ in which the exchange integral U_{srd} changes the sign. In the point $r = \lambda_s/2$ we observe non-significant influence of subsystems$S$ and R in two-photon decay process. The oscillations in the two and single photon rates are observed in all subsystems for the parameter $\tau_d/\tau^0_{srd} > 0.2$. This effect plays an important role in the collective decay process of the systems of radiators with the dimension smaller then wavelength.

4. Conclusions

It finds the effective interaction between three subsystems of radiators in two-photon resonance, using the method of elimination of vacuum field operators. The new cooperative interaction between one dipole-forbidden and two-dipole active subsystems of radiators was proposed. For the description of cooperative emission of three radiators (19) in two-quantum resonance interaction through the vacuum of EMF, the new master equation (15) is obtained. The exact analytical solution of three radiators in interaction with vacuum field is represented in expressions (20) and in Figs. 1. This solution takes into account quantum interference between two dipole active and one dipole forbidden radiators situated at distance r. The exchange integrals between such subsystems can be described by the expressions (18)and describe the acceleration of decay rates of the radiators. The magnitude of this cooperative emission depends on $1/\tau_{srd}$ as a function of the distances between the radiators. The exchange integral U_{srd} pass from positive to negative sign in the regions, $0 \le r \le \lambda_s/2$ and $\lambda_s/2 \le r \le \lambda_s$ around the D-atom. If the atoms S and R are situated on the circle with radius $r = \lambda_s/2$ around the radiator D, the influence of single photon radiators on two-photon decay of D radiator can be neglected. From numerical results Figs. 1 follows that the mutual influence between the radiators depends only on the module of interaction integral (18).

The semi-classical approach is proposed for numerical studies of cooperative resonance between single-photon and two-quantum transitions described by correlation function $\langle S_l^+(t)R_j^+(t)D_m^-(t)\rangle$. Using the master equation (15), and neglecting the quantum fluctuation of the inversion, the closed system of kinetic equation (20) for three atomic subsystems have been obtained. The big number of radiators drastically changes the decay rate as a function of time and distance r. For small relative rate $\tau_d/\tau_{sdr}^0 = 0.2$ the energy exchanges between the atoms is plotted on the Fig. 2. Studying the time behavior of the cooperative rate of the atomic subsystems we observe two regions in the decay rates as in the case of three radiator system. From the behavior of inversion of D subsystem and emission rate of bi-photons plotted in Figs. 2 respectively we observe, that the nutation process stimulated by mutual influences between the subsystems of radiators substantially increases with increasing the number of radiators. For equal numbers of radiators in all subsystems, $N = 50$, it is established the point $\tau_d/\tau_{rsd}^0 = 0.2$ in which the inversion of subsystem D passes into nutation process.

References

[1] Nikolaus B , Zhang D, and Toschek P 1981 *Phys. Rev. Lett.* **47** 171

[2] Gauthier D J, Wu Q, Morin S E and Mossberg T W 1987 *Phys.Rev. Lett* **68** 464

[3] Rivlin L A 1997 *Hiperfine Interaction* **107**, 57

[4] M. Bruno, J. M. Raimond, P. Goy 1987 Phys. Rev. Lett.**59** 1899

[5] N.A.Enaki 1988 Zh. Eksp. Teor.Fiz.**94** 135; N.A.Enaki 1990 Zh. Eksp. Teor.Fiz., **98** 783

[6] Goeppert-Mayer M 1931 *Ann. Phys.* **9** 273 ; Breit G and Teller E 1940 *Astrophys. J.* **91** 215; Spitzer L. and Greenstain J.L. 1940 *Astrophys. J.* **144** 215

[7] Pearel A S 1970 *Phys. Rev. Lett.* **24** 703; Van Dyck R S Jr, Jonson C E, Shugart H A 1971 *Phys.Rev.Lett.* **25**, 1403; Marrus R and Schmieder R 1972 *Phys. Rev. A* **5** 1160

[8] Alexander S G and Eszaros P M 1991 *Astrophysics J.* **327**, 554; Chluba J. and Sunyaev R. A. 2006 Astronomy and Astrophysics, **446** 39

[9] Jentschura U.D. 2008 *J.Phys.A*, **41** 155307; U.D. Jentschura and A. Surzhykov 2008 *Phys. Rev. A* **77** 042507

[10] Kalinkin A A , Kalachev A A , and Samartsev V.V. 2004 *Optics and Spectroscopy* **96** 803

[11] Dicke R H 1954 *Phys. Rev.* **93**, 99

[12] Andreev A V, Emelyanov V I, and Ilinskii Yu A 1992 *Cooperative Effects in Optics* IOP Publishing, Bristol

[13] Enaki N A 1996 *Soviet Phys. JETP* **82** 607; Enaki N A and Galeamov E 2008 *International Journal of Theoretical Physics* **47** 911

Journal of Physics: Conference Series **338** (2012) 012006 doi:10.1088/1742-6596/338/1/012006

Nonlinear effects in the theory of superconductivity

Nicolae Enaki and Sergiu Colun

Abstract. It is proposed the non-linear cooperative interaction between the electrons through the vibration states of thermostat (the large system with big degree of freedom). As an example is studied the cooperative interaction between the cooper pairs through the non-linear lattice vibration in single and two-phonon exchange processes between the electrons. In the case of superconductivity phase transition, the cooperative interaction between the carriers in two and single phonon exchange processes depends on the temperature. This temperature dependence of exchange integral between the quasi-particles drastically changes the second order diagram as this is demonstrated in paper [1, 2]. Unlike the approach proposed in papers [1, 2], in this report it is takes into account the non-linear vibration of thermostat modes (phonon, optical modes etc.). It is observed that this effect influence the temperature dependence of the order parameter in the second order phase transition like superconductivity, super-radiance, ferromagnetism, etc.

Institute of Applied Physics of Academy of Sciences of Moldova, Academiei str. 5, Chisinau, MD-2028, Republic of Moldova

E-mail: enache@asm.md

1. Introduction

The collective processes in condensed matter have many properties with analogies in cooperative radiation effects in quantum optics [3]. The Bardeen-Cooper-Schrieffer (BCS) [4], theory of single phonon phase transition in superconductivity was in the center of attention of many authors [5]-[8]. Following the ideas developed in these papers, the new cooperative emission phenomenon for dipole-forbidden transitions of an inverted system of radiators can be observed in processes of two-photon spontaneous emission [1, 10]. Following the ideas developed in these papers, the new theory of nonlinear superconductivity in which it is take in to account the single and two-phonon exchanges between the band electrons is proposed. Using the method of elimination of boson operators it is obtain the temperature dependence of two-quanta exchange integral between Cooper-electrons is obtained. This integral increase with increasing the temperature due to the fact that thermal field of vibration lattice stimulates the two-quantum exchanges between the electrons. This is one of the main differences between the single and two phonon superconductivity theories. The non-linear (quadratic) electron-phonon coupling dominates in $Bi_2\ Sr_2CaCu_2O_8$ [6]. At the microscopic level any non-linear electron-phonon interaction can be described in terms of the coupling of electrons to a few phonon-complexes [6]. Such coupling significantly influences the critical temperature [7]

In this paper it is proposed the new method of interaction of electronic subsystem with vibration modes of lattice. Firstly we introduced the new creation and annihilation operators for vibration modes which takes into account nonlinear (quadratic) effect in the lattice vibration modes with potential energy $V(x) = m\omega^2\hat{x}^2/2 + K\hat{x}^4$. After that we have studied the collective nonlinear effects in interaction of electronic subsystem with nonlinear thermal bath of lattice oscillators.

The interference between the single- and two-phonon exchange mechanisms in superconductivity is studied using projection operator method. It is observed the situation for which the temperature dependence part of exchange integral between the electrons can be large than constant part.

In the third part of this paper it is demonstrated that for positive nonharmonic constant K the order parameter firstly increase with temperature achieving the maximal value. After that it decreases as in traditional phase transition effects. In this case the correlation effect between first and second order electron-phonon interaction takes place through nonlinear effects. In opposite case $K < 0$ the superconductor state is dumped by temperature.

2. Effective Hamiltonian of Nonlinear Interaction of Electrons with vibration Lattice

The collective processes in condense matter have many analogical proprieties with cooperative radiation effects in quantum optics [3]. The new cooperative emission phenomenon for dipole-forbidden transitions of inverted system of radiators can be observed in the processes of the two-photon spontaneous emission [10]- [11].

Let us study the electronic transitions between the states $|k\rangle$ and $|k'\rangle$ in the low band of a two band superconductor with absorption or radiation of phonons. In order to neglect the one phonon transitions between these states, let us consider that the second band is situated at the distance energetically larger than k_B (k_B is Boltzmann constant, T is the temperature) on the Fermi level of the first band. The interaction two-band electronic system, which interacts with the vibration field of the lattice, is proposed for theoretical discussion. For the simplicity of the problem the Columbian interaction between electrons is not taken into consideration. Therefore the two band electron gas in interaction with nonlinear vibration lattice is described by the Hamiltonian

$$H = H_{el} + H_{ph} + H_{int}, \tag{1}$$

Here the first part of the Hamiltonian (1)

$$H_{el} = \sum_{m=1}^{2} \sum_{\mathbf{k}} (\varepsilon_m(\mathbf{k}) - \mu)\hat{a}^{\dagger}_{m,\mathbf{k}}\hat{a}_{m,\mathbf{k}}$$

describes two-bands electron gaze in which it is introduced the creation $\hat{a}^{\dagger}_{m,\mathbf{k}}$ and $\hat{a}_{m,\mathbf{k}}$ annihilation operators in the $m-$band which satisfy the anticommutation relation $[\hat{a}^{\dagger}_{m,k}, \hat{a}_{m',k'}]_{+} = \delta_{m,m'}\delta_{k,k'}$; $[\hat{a}_{m,\mathbf{k}}, \hat{a}_{m',\mathbf{k}'}]_{+} = 0$. The transition matrix element takes in-to account the wave vector and spin of electrons $k = (\mathbf{k},\sigma)$ from the bands n and m, $\varepsilon_m(\mathbf{k}) - \mu$ is the energy of the electron, which is calculated from the position of the chemical potential μ. In order to study the nonlinear Hamiltonian of lattice vibration let us introduced the model hamiltonian of bath of nonlinear oscillators described by the Hamiltonian

$$H_{ph} = \sum_{\mathbf{q}} [\hbar\omega_q \hat{b}^{\dagger}_{\mathbf{q}}\hat{b}_{\mathbf{q}} + \chi_{\mathbf{q}}\hat{b}^{\dagger}_{\mathbf{q}}\hat{b}^{\dagger}_{\mathbf{q}}\hat{b}_{\mathbf{q}}\hat{b}_{\mathbf{q}}], \tag{2}$$

Here $\hbar\omega_q$ is the phonon energy of harmonic part of the phonon Hamiltonian (2), the nonlinear part of this Hamiltonian is described by nonharmonic parameter $\chi_{\mathbf{q}}$, which for nonlinear oscillators describes the quadratic term in the potential energy $V(x) = m\omega^2\hat{x}^2/2 + K\hat{x}^4$. The phonon operators obey the following commutation relations: $\left[\hat{b}_q; \hat{b}^{\dagger}_{q_1}\right]_{-} = \delta_{q,q_1}$; $\left[\hat{b}_q, \hat{b}_{q_1}\right]_{-} = 0$. As the function of the sign of interaction constant $\chi_{\mathbf{q}}$ (or K) we can introduce the new operators in the Hamiltonian (2)

$$H_{ph} = \sum_{\mathbf{q}} |\chi_{\mathbf{q}}| \hat{K}^{+}_{\mathbf{q}}\hat{K}^{-}_{\mathbf{q}}, \tag{3}$$

where for positive value of nonharmonic parameter $\chi_{\mathbf{q}}$ we can can represent the $\hat{K}$−operators through the $\hat{I}$-nonlinear operators belonging to $SU(1,1)$ algebra: $\hat{I}^{+}_{\mathbf{q}} = \hat{b}^{\dagger}_{\mathbf{q}}\sqrt{\frac{\hbar\omega_q}{|\chi_{\mathbf{q}}|} + \hat{b}^{\dagger}_{\mathbf{q}}\hat{b}_{\mathbf{q}}}$ and $\rightarrow \hat{I}^{-}_{\mathbf{q}} = \sqrt{\frac{\hbar\omega_q}{|\chi_{\mathbf{q}}|} + \hat{b}^{\dagger}_{\mathbf{q}}\hat{b}_{\mathbf{q}}}\ \hat{b}_{\mathbf{q}}$ which satisfy the commutation relations

$$[I^{+}_{\mathbf{q}}, I^{-}_{\mathbf{q}'}] = -2\delta_{q,q'}I^{z}_{q}; \quad [I^{z}_{q}, I^{\pm}_{\mathbf{q}'}] = \pm\delta_{q,q'}I^{\pm}_{\mathbf{q}}. \tag{4}$$

In this case we have the conservation quasi vector $I^2 = (I^z_q)^2 - (I^x_q)^2 - (I^y_q)^2$ with the components $I^x_q = (I^{+}_{\mathbf{q}} + I^{-}_{\mathbf{q}})/2$; $I^y_q = (I^{+}_{\mathbf{q}} - I^{-}_{\mathbf{q}})/2i$; $I^z_q = \frac{\hbar\omega_q}{2|\chi_{\mathbf{q}}|} + b^{\dagger}_q b_q$. In opposite case, $\chi_{\mathbf{q}} < 0$, we can introduce the new operators in lattice Hamiltonian (3) $K^{+}_{\mathbf{q}} \rightarrow J^{+}_{\mathbf{q}} = b^{\dagger}_{\mathbf{q}}\sqrt{\frac{\hbar\omega_q}{|\chi_{\mathbf{q}}|} - b^{\dagger}_{\mathbf{q}}b_{\mathbf{q}}}$; $K^{-}_{\mathbf{q}} \rightarrow J^{-}_{\mathbf{q}} = \sqrt{\frac{\hbar\omega_q}{|\chi_{\mathbf{q}}|} - b^{\dagger}_{\mathbf{q}}b_{\mathbf{q}}}\ b_{\mathbf{q}}$ and $K^z_q \rightarrow J^z_{\mathbf{q}} = -\frac{\hbar\omega_q}{2|\chi_{\mathbf{q}}|} + b^{\dagger}_q b_q$, which belong to other symmetry

$$[J^{+}_{\mathbf{q}}, J^{-}_{\mathbf{q}'}] = 2\delta_{q,q'}J^{z}_{q}; \quad [J^{z}_{q}, J^{\pm}_{\mathbf{q}'}] = \pm\delta_{q,q'}J^{\pm}_{\mathbf{q}}, \tag{5}$$

the conservation law $J^2 = (J^z_q)^2 + (J^x_q)^2 + (J^y_q)^2$ of which shows that these operators are similar to angular momentum generators in quantum mechanics and belong to $su(2)$ algebra. These new operators take into account the sign of nonlinearity in vibration lattice subsystem and pass in boson operators for small value of $|\chi_{\mathbf{q}}|$: $\hat{K}^{+}_{\mathbf{q}} \simeq \hat{b}^{\dagger}_{\mathbf{q}}\sqrt{\frac{\hbar\omega_q}{|\chi_{\mathbf{q}}|}}$; $\hat{K}^{-}_{\mathbf{q}} \simeq \hat{b}_{\mathbf{q}}\sqrt{\frac{\hbar\omega_q}{|\chi_{\mathbf{q}}|}}$. Considering that the nonlinear displacement of the nuclei from equilibrium positions is proportional to the value of new variables $\hat{K}^{+}_{\mathbf{q}}$ and $K^{-}_{\mathbf{q}}$, we can follow the method of deformation potential developed in paper [9] and represent the interaction Hamiltonian between the electron and nonlinear vibrations of the lattice in the following form

$$\begin{aligned} H_{int} &= -\frac{1}{\sqrt{V}}\sum_{n,m}\sum_{\mathbf{k},\mathbf{q}}\{q_{m,n}a^{\dagger}_{m,k+q}a_{n,k}K^{-}_{\mathbf{q}} \\ &\quad + q^{*}_{m,n}a^{\dagger}_{n,\mathbf{k}}a_{m,\mathbf{k+q}}K^{\dagger}_{\mathbf{q}}\}, \end{aligned} \tag{6}$$

The matrix elements of the electron-phonon interaction are known [8]. It is not difficult to shown that for for electron in the first $n = m = 1$ we obtain the following matrix element band

$$q_{11}(\mathbf{q}) = -i\sqrt{\frac{|\chi_{\mathbf{q}}|}{2\rho\omega_q^2}}E_{11}q,$$

where $E_{11} \approx \frac{\hbar^2}{m_e}\int_{\tau_0} d^3\mathbf{r}\{(\mathbf{e_q},\nabla\mathbf{u}^{*}_{1,\mathbf{k}})(\mathbf{e_q},\nabla\mathbf{u}_{1,\mathbf{k}}) \sim \mu$. In a similar way we obtain the interaction constant for inter-band interaction

$$\begin{aligned} q_{21}(k,q) &= -\sqrt{\frac{|\chi_{\mathbf{q}}|}{2\rho\omega_q^2}}(\mathbf{e}_q\int_{\tau_0} d^3\mathbf{r}\mathbf{u}^{*}_{2,\mathbf{k+q}}\{\varepsilon_2(\mathbf{k}+\mathbf{q})-\varepsilon_1(\mathbf{k}) \\ &+ \frac{\hbar^2}{2m_e}(\mathbf{k}^2 - (\mathbf{k}+\mathbf{q})^2)\}, \nabla u_{1,\mathbf{k}}) \simeq -\sqrt{\frac{|\chi_{\mathbf{q}}|}{2\rho}}\frac{\varepsilon_{21}p_{21}}{\omega_q\hbar}. \end{aligned}$$

Here $\varepsilon_{21} = \varepsilon_2(k+q) - \varepsilon_1(k)$ is the energetic distance between bands one and two, $p_{21} = \hbar(\mathbf{e}_q, \int_{\tau_0} d^3ru^{*}_{2,\mathbf{k}+q}\nabla u_{1,\mathbf{k}})$, $u_{1,\mathbf{k}}$ is the periodical part of Bloch wave function.

In the case when the one phonon transition between the first and the second bands $q_{21}(q)$ is larger than the matrix element of the one phonon transitions in the intrinsic first band $q_{11}(q)$, the two-phonon transition between the states $|k\rangle$ and $|k'\rangle$ become larger than the single phonon transitions. This becomes possible, when the first and the second bands are formed from the local atomic levels with different symmetries. For example, if the first band provided by the local atomic level of S- symmetry crystal and the second band comes from the level of P- symmetry, it is clear that in the case when $k \sim k'$ the intrinsic matrix element $q_{11}(q)$ is less than the matrix element between the first and the second bands $q_{21}(q)$. In this case the two-phonon transitions between the states $|k\rangle$ and $|k'\rangle$ of the first band is similar to two-photon transitions between the forbidden states $|1S\rangle$ and $|2S\rangle$ of Hydrogen-like atom. The second band in such two-phonon transitions plays the role of a virtual state $|P\rangle$.

In this chapter the two-phonon cooperative processes between the electrons of the lowest band through the virtual state of the second band are proposed. For this it is necessary firstly to eliminate the operators of the second band from Hamiltonian (1). If the second band is situated at the energetic distance larger than $\kappa_b T$, we can eliminate the electronic operators of this band from interaction Hamilton (1). In order to solve this problem let us consider the operator $A(t)$ which does not depend on the operators of the second band. The Heisenberg equation for the mean value of this operator $\hat{A}(t)$ is

$$\langle\frac{d\hat{A}(t)}{dt}\rangle = \frac{i}{\hbar}\langle\left[\hat{H},\hat{A}(t)\right]\rangle$$

Using the Hamiltonian (1) one can obtain the following equation for mean value of operator $\hat{A}(t)$

$$\begin{aligned}
\langle\frac{d\hat{A}(t)}{dt}\rangle &= \frac{i}{\hbar}\sum_k \varepsilon_1(k)\langle[a^\dagger_{1k}(t)\hat{a}_{1k}(t),\hat{A}(t)]\rangle \\
&+ \frac{i}{\hbar\sqrt{V}}\sum_{k,\mathbf{q}}\{q_{1,1}(\mathbf{q})\langle[a^\dagger_{1,k+\mathbf{q}}(t)a_{1,k}(t)\hat{K}^-_{\mathbf{q}}(t),\hat{A}(t)]\rangle \\
&- H.c.t.(\hat{A}^\dagger(t)\rightarrow\hat{A}(t))\} \\
&+ \frac{i}{\hbar\sqrt{V}}\sum_{k,\mathbf{q}}\{q_{2,1}(\mathbf{q})\langle a^\dagger_{2,k+\mathbf{q}}(t)[\hat{a}_{1,k}(t)\hat{K}^-_{\mathbf{q}}(t),\hat{A}(t)]\rangle \\
&+ q_{1,2}(\mathbf{q})\langle[a^\dagger_{1,k+\mathbf{q}}(t)K^-_{\mathbf{q}}(t),A(t)]a_{2,k}(t)\rangle \\
&- H.c.t.(\hat{A}^\dagger(t)\rightarrow\hat{A}(t))\}. \qquad (7)
\end{aligned}$$

As the second band plays the role of virtual states for the transitions between the electron states in the first band, it is possible to eliminate the electronic operators of the second band. In this approximation, taking in to account the Heisenberg equation for electronic operators from the second band

$$\begin{aligned}
\frac{d\hat{a}_{2,k}(t)}{dt} &= -i\varepsilon_2(k)\hat{a}_{2,k}(t)/\hbar \\
&- \frac{i}{\hbar\sqrt{V}}\sum_{n=1}^{2}\sum_{\mathbf{q}_1}\{q_{2n}(\mathbf{q}_1)\hat{a}_{n,k-\mathbf{q}_1}(t)\hat{K}^-_{\mathbf{q}_1}(t)+q^*_{n2}(\mathbf{q}_1)\hat{K}^\dagger_{\mathbf{q}_1}(t)\hat{a}_{n,k+\mathbf{q}_1}(t)\}
\end{aligned}$$

we obtain the solutions

$$\hat{a}_{2,k}(t) = \hat{a}_{2,k}(t_0)\exp\left(-\frac{i}{\hbar}\varepsilon_2(k)(t-t_0)\right)$$

$$
\begin{aligned}
& -\frac{i}{\hbar\sqrt{V}}\sum_{n=1}^{2}\sum_{\mathbf{q}_1}\int_{t_0}^{t} d\tau \exp\left(-\frac{i}{\hbar}\varepsilon_2(k)(t-t_0-\tau)\right) \\
& \times\{q_{2n}(\mathbf{q}_1)\hat{a}_{n,k-\mathbf{q}_1}(\tau)\hat{K}^-{}_{\mathbf{q}_1}(\tau) + q^*_{n2}(\mathbf{q}_1)\hat{K}^\dagger_{\mathbf{q}_1}(\tau)\hat{a}_{n,k+\mathbf{q}_1}(\tau)\}; \\
\hat{a}^\dagger_{2,k}(t) \quad &= \quad [\hat{a}_{2,k}(t)]^\dagger. \qquad (8)
\end{aligned}
$$

In the right hand site of this operators we have the nonlinear phonon operators $K^-{}_{\mathbf{q}_1}(\tau)$ and $K^\dagger_{\mathbf{q}_1}(\tau)$. The Hasenberg equation for such operators are

$$
\frac{d\hat{K}^-{}_{\mathbf{q}}(\tau)}{d\tau} = \mp 2i|\chi_{\mathbf{q}}|\hat{K}^z{}_{\mathbf{q}}(\tau)\hat{K}^-{}_{\mathbf{q}}(\tau) \pm \frac{2i}{\hbar\sqrt{V}}\sum_{n,m}\sum_{k} q^*_{m,n}a^\dagger_{n,\mathbf{k}}(\tau)\hat{K}^z_q(\tau)a_{m,\mathbf{k+q}}(\tau)
$$

where the sings "+" or "−" corresponds to negative or positive value of enharmonically coefficient $\chi_{\mathbf{q}}$ respectively. Neglecting the interaction with electronic subsystem we can represent the solution of equation (9) in the following T-product $K^-{}_{\mathbf{q}}(\tau) = T\exp[\mp 2i|\chi_{\mathbf{q}}|\int_0^\tau K^z{}_{\mathbf{q}}(\tau')d\tau']$ $\hat{K}^-{}_{\mathbf{q}}(0)$. Considering that for short retardation time the operator $K^z{}_{\mathbf{q}}(\tau')$ can be approximated with the expression $\left\langle \hat{K}^z{}_{\mathbf{q}}\right\rangle \simeq \frac{\hbar\omega_q}{2\chi_{\mathbf{q}}} + \langle\hat{b}^\dagger_q\hat{b}_q\rangle$, we can represent the solution of operator $\hat{K}^-{}_{\mathbf{q}}(\tau)$ in expression (8) thorough the smooth amplitude $\hat{K}^-_{\mathbf{q}}(\tau) = \tilde{K}^-_{\mathbf{q}}(\tau)\exp[-i2i|\chi_{\mathbf{q}}|\langle K^z{}_{\mathbf{q}}\rangle\tau]$. Neglecting the retardation in the smooth amplitude $\tilde{K}^-{}_{\mathbf{q}}(\tau) \simeq \tilde{K}^-{}_{\mathbf{q}}(t)$ and considering that $|q_{2,2}(q)| \ll |q_{2,1}(q)|$, in the Born-Markoff approximation the operators $a_{2,k}(t)$ and $a^\dagger_{2,k}(t)$ can be represented through the vacuum and interaction parts

$$
\begin{aligned}
& a_{2,k}(t) \simeq a_{2,k}(t_0)\exp\left(-\frac{i}{\hbar}\varepsilon_2(k)(t-t_0)\right) \\
& -\frac{1}{\sqrt{V}}\sum_{\mathbf{q}_1}\{\frac{q_{21}(\mathbf{q}_1)a_{1,k-\mathbf{q}_1}K^-_{\mathbf{q}_1}}{\varepsilon_2(\mathbf{k})-\varepsilon_1(\mathbf{k}-\mathbf{q}_1)-\hbar\omega_{\mathbf{q}_1}-2<b^\dagger_q b_q>\chi_{\mathbf{q}}} \\
& +\frac{q^*_{12}(\mathbf{q}_1)K^\dagger_{\mathbf{q}_1}a_{1,k+\mathbf{q}_1}}{\varepsilon_2(\mathbf{k})-\varepsilon_1(\mathbf{k}+\mathbf{q}_1)+\hbar\omega_{\mathbf{q}_1}+2<b^\dagger_q b_q>\chi_{\mathbf{q}}}\}; \\
& a^\dagger_{2,k}(t) = [a_{2,k}(t)]^\dagger. \qquad (9)
\end{aligned}
$$

It is considered that the interaction between the subsystems takes place for time moments $\varepsilon_2 t/\hbar \gg 1$. Taking in to account that the system temperature, T is so less that $\varepsilon_2(k)-\varepsilon_1(k) >> k_bT$ one can consider that the second band remain unpopulated in the processes of two-quanta interactions of electrons from the ground band

$$
\lim_{t_0\to-\infty}\hat{a}_{2k}(t_0)|0_2\rangle = 0, \quad \lim_{t_0\to-\infty}\langle 0_2|\hat{a}^\dagger_{2k}(t_0) = 0,
$$

where $|0_2\rangle$-is the vacuum state for second band Taking this in to account, after the substitution of the expressions (1) $a_{2,k}(t)$ and $\hat{a}^\dagger_{2,k}(t)$ in equation (7), one can eliminate the $a_{2k}(t)$ and $\hat{a}^\dagger_{2k}(t)$ operators of the second virtual band.

$$
\begin{aligned}
\langle\frac{d\hat{A}(t)}{dt}\rangle \quad &= \quad \frac{i}{\hbar}\langle[\sum_k \varepsilon_1(\mathbf{k})\hat{a}^\dagger_{1k}\hat{a}_{1k} + \sum_{\mathbf{q}}\hbar\omega_q\hat{K}^\dagger_{\mathbf{q}}\hat{K}^-_{\mathbf{q}}, \hat{A}(t)]\rangle \\
&+ \quad \frac{i}{\hbar\sqrt{V}}\sum_{k,\mathbf{q}}\{q_{1,1}(\mathbf{q})\langle[\hat{a}^\dagger_{1,k+\mathbf{q}}(t)\hat{a}_{1,k}(t)\hat{K}^-_{\mathbf{q}}(t), \hat{A}(t)]\rangle
\end{aligned}
$$

$$
\begin{aligned}
- & \; H.c.t.(\hat{A}^{\dagger}(t) \to \hat{A}(t))\} \\
- & \; \frac{i}{\hbar V} \sum_{k} \sum_{\mathbf{q},\mathbf{q}_1} \{ g_{21}(\mathbf{q}) g_{12}(\mathbf{q}_1) \\
\times & \; \frac{\langle \hat{a}^{\dagger}_{1,k+\mathbf{q}-\mathbf{q}_1}(t) \hat{K}^{\dagger}_{\mathbf{q}_1}(t) [a_{1,k}(t) \hat{K}^{-}_{\mathbf{q}}(t), \hat{A}(t)] \rangle}{\varepsilon_2(\mathbf{k}+\mathbf{q}) - \varepsilon_1(\mathbf{k}+\mathbf{q}-\mathbf{q}_1) - \hbar\omega_{\mathbf{q}_1}} \\
+ & \; g_{21}(\mathbf{q}) g_{12}(\mathbf{q}_1) \frac{\langle \hat{K}^{-}_{\mathbf{q}_1}(t) \hat{a}^{\dagger}_{1,k+\mathbf{q}+\mathbf{q}_1}(t) [\hat{a}_{1,k}(t) \hat{K}^{-}_{\mathbf{q}}(t), \hat{A}(t)] \rangle}{\varepsilon_2(\mathbf{k}+\mathbf{q}) - \varepsilon_1(\mathbf{k}+\mathbf{q}+\mathbf{q}_1) + \hbar\omega_{\mathbf{q}_1}} \\
+ & \; g_{21}(\mathbf{q}) g^{*}_{21}(\mathbf{q}_1) \frac{\langle [\hat{a}^{\dagger}_{1,k+\mathbf{q}}(t) K^{-}_{\mathbf{q}}(t), \hat{A}(t)] a_{1,k-\mathbf{q}_1}(t) K^{-}_{\mathbf{q}_1}(t) \rangle}{\varepsilon_2(\mathbf{k}) - \varepsilon_1(\mathbf{k}-\mathbf{q}_1) - \hbar\omega_{\mathbf{q}_1}} \\
+ & \; g_{21}(\mathbf{q}) g^{*}_{21}(\mathbf{q}_1) \frac{\langle [\hat{a}^{\dagger}_{1,k+\mathbf{q}}(t) K^{-}_{\mathbf{q}}(t), \hat{A}(t)] K-^{\dagger}_{\mathbf{q}_1}(t) \hat{a}_{1,k+\mathbf{q}_1}(t) \rangle}{\varepsilon_2(\mathbf{k}) - \varepsilon_1(\mathbf{k}+\mathbf{q}_1) + \hbar\omega_{\mathbf{q}_1}} \\
+ & \; H.c.t.(\hat{A}^{\dagger}(t) \to \hat{A}(t))\} \qquad (10)
\end{aligned}
$$

The two-phonon exchange between the electrons through the virtual bands is similar to many photon effects in quantum optics and in our opinion has a more realistic diagram interpretation. In analogy with atomic system we can reduced the expression (10) to the Heisenberg equation with Hamiltonian which takes in to account two-phonon interaction thought the virtual band:

$$\hat{H} = \hat{H}_0 + \hat{H}_I; \qquad (11)$$

where

$$\hat{H}_0 = \sum_{k} (\varepsilon_1(k) - \mu) \hat{a}^{\dagger}_{1k} \hat{a}_{1k} + \sum_{\mathbf{q}} \hbar\omega_q \hat{K}^{\dagger}_{\mathbf{q}} \hat{K}^{-}_{\mathbf{q}}$$

is the free part of Hamiltonian for the electron subsystem in the lower band "1" and phonon subsystem respectively;

$$\hat{H}_I = \hat{H}_{I1} + \hat{H}_{I2}, \quad \hat{H}_{I2} = \hat{H}^{b}_{I2} + \hat{H}^{2s}_{I2} \qquad (12)$$

is the interaction Hamiltonian between electron and phonon subsystems which takes in to consideration one phonon transitions

$$\hat{H}_{I1} = \frac{1}{\sqrt{V}} \sum_{k,\mathbf{q}} \{ Q_1(\mathbf{q}) \hat{a}^{\dagger}_{1k+\mathbf{q}} \hat{a}_{1k} K^{-}_{\mathbf{q}} + H.c. \}. \qquad (13)$$

Two phonon transitions and scattering effects between two quasi-energetic states of the lower band are described by interaction Hamiltonian parts respectively

$$
\begin{aligned}
\hat{H}^{b}_{I2} &= -\frac{1}{V} \sum_{k} \sum_{\mathbf{q},\mathbf{q}_1} [Q^{b}_2(k, q, q_1) a^{\dagger}_{1k+\mathbf{q}+\mathbf{q}_1} \hat{a}_{1k} \hat{K}^{-}_{\mathbf{q}_1} \hat{K}^{-}_{\mathbf{q}} \\
& \quad + H.c.]; \qquad (14)
\end{aligned}
$$

$$
\begin{aligned}
H^{s}_{I2} &= -\frac{1}{V} \sum_{k} \sum_{\mathbf{q},\mathbf{q}_1} [Q^{s}_2(k, q, q_1) a^{\dagger}_{1k+\mathbf{q}-\mathbf{q}_1} a_{1k} \hat{K}^{\dagger}_{\mathbf{q}_1} K^{-}_{\mathbf{q}} \\
& \quad + H.c.]; \qquad (15)
\end{aligned}
$$

Journal of Physics: Conference Series **338** (2012) 012006 doi:10.1088/1742-6596/338/1/012006

Here $a_{1k}^{\dagger}$ and a_{1k} are creation and annihilation operators of electrons in the lower band of two-band proposed model; $b_{\mathbf{q}}^{\dagger}$ ($b_{\mathbf{q}}$) are creation (annihilation) operator of phonon subsystem. The interaction constant in single and two-phonon interaction with lattice vibration are $G_1(\mathbf{q}) = g_{11}(\mathbf{q})$;

$$Q_2^b(k,q_1,q_2) = \frac{g_{21}(\mathbf{q})g_{12}(\mathbf{q}_1)}{\varepsilon_2(\mathbf{k}+\mathbf{q}) - \varepsilon_1(\mathbf{k}+\mathbf{q}+\mathbf{q}_1) + \hbar\tilde{\omega}_{\mathbf{q}_1}} + \frac{g_{21}(\mathbf{q})g_{21}(\mathbf{q}_1)}{\varepsilon_2(\mathbf{k}+\mathbf{q}_1) - \varepsilon_1(\mathbf{k}) - \hbar\tilde{\omega}_{\mathbf{q}_1}}, \tag{16}$$

$$Q_2^s(k,q_1,q_2) = \frac{g_{21}(\mathbf{q})g_{12}(\mathbf{q}_1)}{\varepsilon_2(\mathbf{k}+\mathbf{q}) - \varepsilon_1(\mathbf{k}+\mathbf{q}-\mathbf{q}_1) - \hbar\tilde{\omega}_{\mathbf{q}_1}} + \frac{g_{21}(\mathbf{q})g_{21}(\mathbf{q}_1)}{\varepsilon_2(\mathbf{k}-\mathbf{q}_1) - \varepsilon_1(\mathbf{k}) + \hbar\tilde{\omega}_{\mathbf{q}_1}}. \tag{17}$$

where $\hbar\tilde{\omega}_{\mathbf{q}} = \hbar\omega_{\mathbf{q}} + 2 < b_q^{\dagger}b_q > \chi_{\mathbf{q}}$. We emphasize that free Hamiltonian part contains only the electron states in the low band and phonon subsystems respectively. The operators $a_{1k}^{\dagger}$ and a_{1k} are creation and annihilation operators of electrons in the lower band of two-band proposed model. The energetic distance between the lower and upper bands satisfies the inequality $\varepsilon_{21} = \varepsilon_2 - \varepsilon_1 > k_bT$ The interaction of electrons in the low with phonon subsystem is described by two types of interaction Hamiltonian $H_I^{(1)}$ and $H_I^{(2)}$ In order to neglect the one quanta interaction in comparison with two-quanta we must consider that interaction part $H_I^{(2)}$ is more large than $H_I^{(1)}$, or

$$\sum_{q_1} \frac{1}{\sqrt{V}} \left| \frac{g_{21}(\mathbf{q})g_{21}^*(\mathbf{q}_1)}{\varepsilon_2(\mathbf{k}+\mathbf{q}) - \varepsilon_1(\mathbf{k}+\mathbf{q}\pm\mathbf{q}_1) \pm \hbar\tilde{\omega}_{\mathbf{q}_1}} \right| \gg |g_{1,1}(\mathbf{q})|. \tag{18}$$

In the next section the two phonon correlation Hamiltonian between the electrons which describe the new type of formation of Cooper pairs it is obtained, taking in consideration the inequality (18) the method of elimination of biboson operators of phonon subsystem is proposed.

3. Projection Operator

According to the Hamiltonian (1) we proposed the equations for the density matrix of the total "electron+phonon" system in the interaction picture

$$i\hbar\frac{\partial\rho(t)}{\partial t} = [\hat{H}_I(t), \rho(t)],$$

where

$$\hat{H}_I(t) = \exp\{\frac{i}{\hbar}\hat{H}_o t\}\hat{H}_I \exp\{-\frac{i}{\hbar}\hat{H}_o t\}.$$

Let $\hat{P}$ be the projection operator for the complete density matrix $\rho(t)$ on the vector basis of a free EMF subsystems, where $\rho_s(t) = \hat{P}\rho(t)$, and $\rho_b(t) = \hat{\bar{P}}\rho(t)$ are the slow and rapidly oscillating parts of density matrix respectively, $\hat{\bar{P}} = 1 - \hat{P}$. It can be shown that $\hat{P}^2 = \hat{P}\hat{\bar{P}} = 0$. Recognizing that for $t = 0$ an electronic subsystem does not interact with lattice vibration, we define the projection operator

$$\hat{P} = \rho_{ph} \otimes Tr_{ph}\{\cdots\}, \tag{19}$$

The equations for the matrix $\rho_s(t)$ and $\rho_b(t)$ are

$$\frac{\partial \rho_s(t)}{\partial t} = -i\lambda \hat{P}\hat{L}_i(t)\{\rho_s(t) + \rho_b(t)\}, \tag{20}$$

$$\frac{\partial \rho_b(t)}{\partial t} = -i\lambda \hat{P}\hat{L}_i(t)\{\rho_s(t) + \rho_b(t)\}, \tag{21}$$

where $\lambda \hat{L}_i(t) = [\hat{H}_i(t), ...]/\hbar$.

Taking into account Hamiltonian parts (12),(14) and (15), we integrate equation (21) with respect to $\rho_b(t)$ and substitute the result in equation (20)

$$\frac{\partial \rho_s(t)}{\partial t} = -\lambda^2 \hat{P} \int_0^t d\tau \hat{L}_i(t)\hat{U}(t, t-\tau)\hat{L}_i(t-\tau)\rho_s(t-\tau), \tag{22}$$

where

$$\hat{U}(t, t-\tau) = T \exp\left\{ -i\lambda \hat{\bar{P}} \int_{t-\tau}^{t} d\tau_1 \hat{L}_i(\tau) \right\}.$$

we obtain the following explicit expressions of collective interaction part of Hamiltonian

$$H_I^{eff} = H_{I1}^{eff} + H_{I2}^{eff} + H_{I12}^{eff}, \tag{23}$$

in which the first term corresponds to BCS interaction Hamiltonian

$$H_{I1}^{eff} = -\frac{1}{V}\sum_{k,k_1}\sum_{\mathbf{q},} \frac{\hbar|G_1(q)|^2\tilde{\omega}_{\mathbf{q}}(1+\chi_{\mathbf{q}}N_{\mathbf{q}})}{[\hbar\tilde{\omega}_{\mathbf{q}}]^2 - [\varepsilon_1(\mathbf{k}_1+\mathbf{q}) - \varepsilon_1(\mathbf{k}_1)]^2} a^{\dagger}_{\mathbf{k}+\mathbf{q}} a^{\dagger}_{\mathbf{k}_1} a_{\mathbf{k}_1+\mathbf{q}} a_{\mathbf{k}}, \tag{24}$$

and second term corresponds to two-phonon interaction Hamiltonian

$$\begin{aligned} H_{I2}^{eff} &= -\sum_{k,k_1}\sum_{\mathbf{q},\mathbf{q}_1}\Big[M^b_{\mathbf{q},\mathbf{q}'}(k,k_1)(1+N_q+N_{q_1})a^{\dagger}_{k+q+q_1}(t)a_{\mathbf{k}}(t)a^{\dagger}_{\mathbf{k}_1}(t)a_{k_1+q+q_1}(t) \\ &+ M^s_{\mathbf{q},\mathbf{q}'}(k,k_1)(N_q - N_{q_1})a^{\dagger}_{k_1+q_1-q}(t)a_{k_1}(t)a^{\dagger}_{k+q-q_1}(t)a_k(t)\Big] \end{aligned} \tag{25}$$

Here $M^b_{\mathbf{q},\mathbf{q}'}(k,k_1)(1+N_q+N_{q_1})$ is exchange energy between two electrons from the states k, and k' in the two-phonon absorption and emission processes

$$\begin{aligned} M^b_{\mathbf{q},\mathbf{q}'}(k,k_1) &= \frac{1}{2V^2}\Bigg[\frac{G_2^b(k,\mathbf{q},\mathbf{q}_1)G_2^{*b}(k_1,\mathbf{q}_1,\mathbf{q})}{\hbar(\omega_{\mathbf{q}_1}+\omega_{\mathbf{q}}) - \varepsilon_1(\mathbf{k}_1+\mathbf{q}_1+\mathbf{q}) + \varepsilon_1(\mathbf{k}_1)} \\ &+ \frac{G_2^b(k,\mathbf{q},\mathbf{q}_1)G_2^{*b}(k_1,\mathbf{q}_1,\mathbf{q})}{\hbar(\omega_{\mathbf{q}_1}+\omega_{\mathbf{q}}) - \varepsilon_1(\mathbf{k}+\mathbf{q}_1+\mathbf{q}) + \varepsilon_1(\mathbf{k})}\Bigg]; \end{aligned} \tag{26}$$

$M^s_{\mathbf{q},\mathbf{q}'}(k,k_1)(N_q - N_{q_1})$ is the exchange energy between two electrons from similar states in two-phonon scattering processes

$$M^s_{\mathbf{q},\mathbf{q}'}(k,k_1) = \frac{1}{V^2}\frac{G_2^s(k,\mathbf{q},\mathbf{q}_1)G_2^{*s}(k_1,\mathbf{q}_1,\mathbf{q})}{\hbar(\omega_{\mathbf{q}_1}-\omega_{\mathbf{q}}) - \varepsilon_1(\mathbf{k}_1+\mathbf{q}_1-\mathbf{q}) + \varepsilon_1(\mathbf{k}_1)}.$$

It is observed here that for elastic scattering process $\varepsilon_1(\mathbf{k}_1 + \mathbf{q}_1 - \mathbf{q})\varepsilon_1(k_1)$ and $q \sim 2k_F$ the contribution of scattering terms in superconduction is positive

$$\frac{N_q - N_{q1}}{\omega_{\mathbf{q}_1} - \omega_{\mathbf{q}}} > 0, \quad N_{q_1} < N_q \; when \; \omega_{\mathbf{q}_1} > \omega_{\mathbf{q}}$$

Finding the commutation expression for imaginary part of third term, we obtain the interferences part of Hamiltonian

$$H^{eff}_{I12} = \frac{2n}{V}\sum_{k,k_1}\sum_{\mathbf{q},}\frac{|G_1(\mathbf{q})G_1(-\mathbf{q})G_2^b(\mathbf{q},-\mathbf{q})(1+\chi_{\mathbf{q}}N_{\mathbf{q}})|\cos[2\varphi_q - \varphi_0]}{(\hbar\tilde{\omega}_q)^2 - [\varepsilon_1(k_1+q) - \varepsilon_1(k_1)]^2} a^\dagger_{k+q}a_{k_1}{}^\dagger a_{k_1+q}a_k. \quad (27)$$

In this section the effective interaction Hamiltonian parts (24), (25) and (27) of the electrons from the first band through the single and two-phonon interactions have been obtained. The cooperative interaction between the electrons through the single and bi-quantum field is reduced not only to the simple processes of simultaneous absorption or emission of the virtual lattice quasi-particles described by operators (4),(5), but it includes the two-quantum exchanges between the electrons, too. As a consequence of the effective interaction Hamiltonian (25) the transitions between two states of the first band can take place with a one quasi-phonon absorption with q wave vector and the emission of another phonon with q_1 wave vector. Following the traditional method proposed in the theory of superconductivity [1]-[9], we can introduce the order parameter $\Delta(T) = \frac{G(T)}{2V}\sum_{k_1}\left\langle a^\dagger_{k_1}a^\dagger_{-k_1}\right\rangle$. Using the method of approximative Hamiltonian proposed in paper[2] in the following form $H = \sum_k \varepsilon_1'(\mathbf{k})a^\dagger_k a_k - \sum_k(\Delta(T)a_{-k}a_k + h.c.)$. In this case we have the possibility to take into account the nonlinearity connected with lattice vibration.

4. Conclusion

In this paper we discussed the effects connected with nonharmonic lattice vibration and nonlinear transitions. The nonlinear effect connect with vibration is taken for the first time in our approach. The nonlinear effects connected with multi-phonon transitions was studied earlier

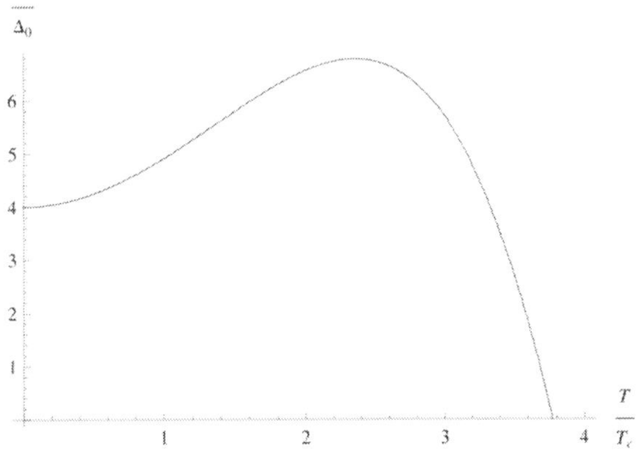

Figure 1. The influence of nonharmonic parameter $\chi_k > 0$ as function of temperature in the relative units $\Delta(T)/\Delta(T_0)$, T/T_C

in the papers [1] and [2]. This paper have taken into account both nonlinearity in the effective Hamiltonian (23). We were interested in the contribution of nonlinearity connected with new approach in the description of lattice vibration. We observed from the Hamiltonian (24) that positive nonlinear positive nonharmonic parameter $\chi_k > 0$ corresponds to the increasing of interaction constant between the cooper electron pairs and brings to the increasing of critical temperature of superconductivity(see Fig.1). The nonlinear lattice vibration corresponds to $SU(1,1)$ algebra and with the excitation of oscillation mode k, the frequency of nonlinear oscillator increase too $\tilde{\omega}_k = \omega_k + 2\chi_k N_k$.

In opposite case when $\chi_k < 0$ with increasing of the temperature the interaction constant in the Hamiltonian (24) decrease. This effect is accompanied with decreasing of the frequency of the nonlinear oscillators $\tilde{\omega}_k = \omega_k - 2|\chi_k N_k|$. The proposed model in the second part give us the possibility to taken into account the cooperative interaction between the cooper pairs through the non-linear lattice vibration in single and two-phonon exchange processes between the electrons. The similar problem it is proposed for quasi-spin systems used in the description of super-radiance and ferromagnetism [2]. This article shows that the nonlinearity in electronic and vibration subsystems can drastically modify the second order phase transition.

References

[1] Enaki N A, Eremeev V 2002 *New J.ournal of Phys.* **4** 80 ; Enaki N A, Prepelitza O B 1992 *Izvestia Academii Nauk Rep. Moldova* **3** 52
[2] ENAKI N A 2010 *JOLT* **160** 157; 2010 *J. of Modern Optics* **57** 1397
[3] DiRienzo A, Rogovin D 1978 *Coherence in Spectroscopy and Modern Physics*(Plenum Press, New York)
[4] Bardeen J, Cooper J, Schrieffer J 1957 *Phys. Rev.* **108** 1175
[5] Silverman R A, Abrikosov A A 1964 *Phys. Today* **17** 78
[6] Ji Kai, Nasu K 2008 *Phys. Lett. A* **372** 524
[7] Bak Z 2004 *Czech J. Phys. 54* D461 Part 2
[8] Levich V G, Vdovin Yu A 1962 *The Course of Theoretical Physics* (Nauka,Moscow)
[9] Bardeen J, Shockly W 1950 *Phys. Rev.* **80** 72
[10] Enaki N A 1988 *Zh. Eksp. Teor. Fiz.* **94** 135; 1988 *Sov. Phys. JETP* **67** 2033
[11] Enaki N A, Eremeev V 2006 *Phys. Lett. A* **357** 104; Enaki N A, Macovei M A 1999 *Rus. Phys. JETP* **88** 633

Generation of photon pairs in hyper-Raman Effects and its connection with transitions symmetries

Nicolae Enaki and Marina Turcan

Abstract. In this paper we propose the model interaction Hamiltonian which takes in-to consideration the four photon process in interaction of dressed field with ^{39}K atoms. Using the method of elimination of virtual states, we obtain this the effective interaction Hamiltonian which describe the simultaneously generation of photon pairs. The master equation for laser field is obtain, taking in to account the good cavity limits in the process of two photon generation. The steady state solution of this equation for threshold case, which takes in to account the quantum fluctuation and photon statistics, is proposed. The cross-correlation function between Stokes and anti-Stokes photon pairs is studied as function of the evolution of the system.

Institute of Applied Physics of Academy of Sciences of Moldova, Academiei str. 5, Chisinau, MD-2028, Republic of Moldova

E-mail: enache@asm.md

1. Introduction

The two-photon coherent generation of light is in the center of attention of many experimental and theoretical studies in the last time. The property of entanglement between the emitted photons in the processes of light generation has a great impact factor today in the applications in quantum information. As a rule this propriety are connected with generation processes, which take place in excited matter. The two-photon amplification of light is possible in multi-level atoms and open the new perspective of application the coherence and entanglement between the photon in the quantum communication.

The experimental realization of two-photons laser was proposed in the papers of J. Gauthier and coauthors [1]. The investigation of two-photon lasing in above experiments consists of spin-polarized and laser-driven $39K$ atoms placed in a high-finesse transverse-mode-degenerate optical resonator and produces a beam with a power of ˜0.2 $Watt$ at a wavelength of 770 nm. In these experimental papers it was observed the complex dynamically instabilities of the state of polarization of the two-photon laser, which are made possible by the atomic Zeeman degeneracy. In the papers [1] the authors have considered laser amplification in a thermal vapor of potassium atoms ^{39}K, when the laser frequency was tuned in the vicinity of the $4S_{1/2} \to 4P_{1/2}$ transition. The two-photon amplification arises in the system in the process of four-quantum Hyper-Raman scattering effect (see Fig. 1). In this four photons process, stimulated by dressed field with the frequency ω_d, the atom makes a transition from the ground state $|g\rangle$ to the new quasi-stable state $|v\rangle$ by absorbing two photons from the dressing field with frequency ω_d and generating two new photons to the probe lasing with frequency ω_a. Taking in to account the energy conservation law, we can found the frequency of new generated light $\omega_p = \omega_d - \Delta_{ge}/2$, where $\hbar\Delta_{ge}$ is the energetically distance between the ground $|g\rangle$ and excited $|e\rangle$ states respectively. The states $|g\rangle$

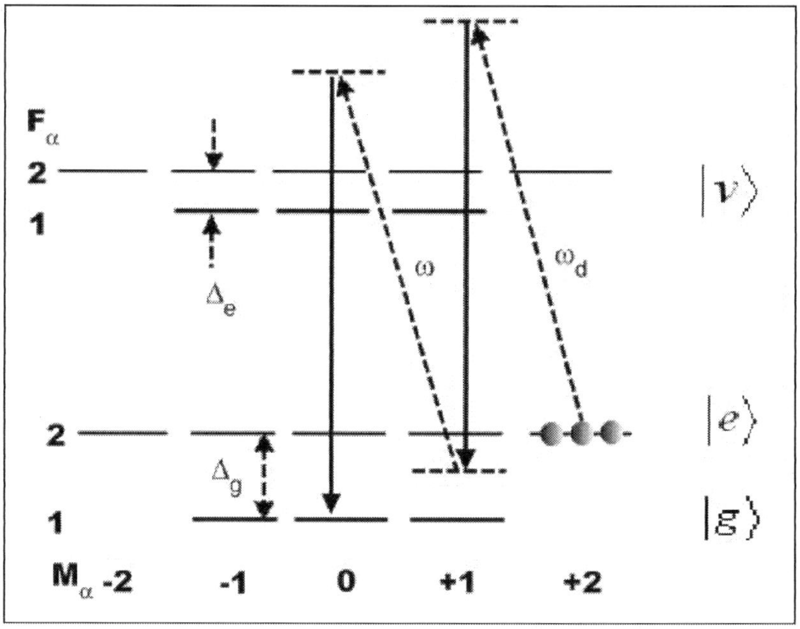

Figure 1. The atomic pump of Hyper-raman scattering process with absorption of two-photon and emission of two Stokes photons

and $|e\rangle$ corresponded to the $4S_{1/2}(F=1)$ and $4S_{1/2}$ $(F=2)$ hyperfine states of ^{39}K atom, respectively. The de-tuning is about $\Delta_{gi}/2\pi = 462MHz$. To obtain two-photon amplification, based on this stimulated emission process, need the condition for which the number of atoms in the ground state N_g is larger then the number of atoms in excited N_e. The authors of the papers [2], [3] and [6] pointed out that $n-$photon Raman scattering processes occur in this system, in which the probe beam frequencies satisfies the condition $\omega = \omega_d - \Delta_{eg}/n$ for $n = 1, 2, 3....$ In the last time using the above condition the authors of the papers [4] and [5] have observed multi-photon Raman scattering in cooled Rb atoms trapped in the potential wells of a $3D$, three-dimensional optical lattice. Following this experimental realization, we propose a model which takes into account the two-photon correlations between the generation and pump field in Hyper-Raman conversion. The model shows that the quantum correlation between Stokes and pump photons contains new peculiarities unstudied today. Following this experimental realization, we propose a model which takes into account the cooperativity between two-photon absorbtion and generation processes of photon pairs from the pump and radiation field in hyper-Raman lasing. The new master equation, which describes the correlation between the pump and generation photon pairs in the cavity is obtain and numerically solved. The possibility of two-photon generation of entangled light in multi-photon hyper-Raman process described in Fig. 1 and connection with the quantum proprieties of pump field is established. We have demonstrated that two-photon amplification arises in four-quantum Hyper-Raman scattering effect represented in Fig. 1 contains the amplitude and phase correlations with pump. In this four photons process stimulated by dressed field with the frequency ω_d the atom makes a transition from the ground state $|g\rangle$ to the new quasi-stable state $|e\rangle$ by absorbing two photons from the dressing field with frequency ω_d and generating two new photons to the probe lasing field with frequency ω_a via virtual intermediate states.

We propose the new interaction picture of hyper-Raman process in which two quanta from pumping field is absorbed an other two entangled photons are generated. Following the modern experimental description of two photon generation of light observed in papers [1], we obtained the effective interaction Hamiltonian for this process. The analytical expressions for two-photon correlation from atom driven in a hyper-Raman configuration are discussed. It is proposed novel two-photon entangled sources, which take into account the coherence and collective phenomena between the these states. The quantum propriety of realistic sources of powerful coherent bi-boson radiation (coherent entanglement of Stokes and anti-Stokes photons) is analyzed. As the lifetime of the atom in the cavity is considered smaller than the cooperative scattering time between the Stokes and anti-Stokes modes, we have eliminated the atomic variables and focused on the transformation of the Stokes mode in anti-Stokes mode and vice versa, as a function of the atomic inversion and initially prepared field states. Taking in to account the process, that takes place with the absorption (emission) of new quanta, we have introduced the bi-boson operators describing such a quantum transformation between the scattering photons into the two cavity modes. The master equation, which describes this process, gives us the possibility to discover the behavior of quantum fluctuations in non stationary process of absorption and radiation of Stokes and anti-Stokes photons respectively. The stationary solution of this master equation reveal the strong quantum correlation between Stokes and anti-Stokes photons. This correlation function gives us the possibility to manipulate with quantum statistics of photons generated in the scattering field as a function of atomic inversion and cavity parameters, like de-tuning, losses, and quality factor of the cavity.

In the next section we derive the model Hamiltonian which describes the two-photon lasing effect, stimulated by dressed field. In the third section using the method of adiabatic elimination, it is derived the master equation for lasing field.

2. Model Hamiltonian and Master Equation

The effective Hamiltonian of potassium atom in interaction with cavity field and external laser field can be obtain from interaction diagrams of three level atom with bimodal electromagnetic field represented in Fig. 1. Let us consider the Hamiltonian of the single atom described by above diagram in interaction with external field. In the two cavity modes of this field we can introduce the creation and annihilation operators, $\hat{d}^{\dagger}$ and $\hat{d}$ of pump field, and generation field of hyper-Raman scattering described by other annihilation and creation operators $\hat{b}^{+}$and $\hat{b}$

$$\hat{H} = \hat{H}_0 + \hat{H}_I, \tag{1}$$

where $\hat{H}_0$ and $\hat{H}_I$ are the free and interaction parts of the Hamiltonian, which can be represented in the following form

$$H_0 = \hbar\omega_v \left|v\right\rangle \left\langle v\right| + \hbar\omega_e \left|e\right\rangle \left\langle e\right| + \hbar\omega_g \left|g\right\rangle \left\langle g\right| + \hbar\omega_a \hat{b}^{+}\hat{b} + \hbar\omega_d \hat{d}^{+}\hat{d}$$

and

$$\begin{aligned} H_I &= -\{[(\mu_{ev}, \mathbf{g}_a)\hat{b}^{+}(t)e^{-ik_a x} + (\mu_{ev}, \mathbf{g}_d)\hat{d}^{\dagger}(t)e^{-ik_d z}] \left|e\right\rangle \left\langle v\right| \\ &+ [(\mu_{gv}, \mathbf{g}_d)(\hat{d}^{\dagger} e^{-ik_d z} + (\mu_{gv}, \mathbf{g}_a)b^{+} e^{-ik_a x}] \left|g\right\rangle \left\langle v\right|\} \\ &\quad -\{[(\mu_{ve}, \mathbf{g}_a)be^{ik_a x} + (\mu_{ve}, \mathbf{g}_d)de^{ik_d z}] \left|v\right\rangle \left\langle e\right| \\ &+ [(\mu_{vg}, \mathbf{g}_d)(de^{ik_d z} + (\mu_{vg}, \mathbf{g}_a)be^{ik_a x}] \left|v\right\rangle \left\langle g\right|\}. \end{aligned} \tag{2}$$

Here $\hbar\omega_v$, $\hbar\omega_e$ and $\hbar\omega_g$ are the energies of the virtual $|v\rangle$, excited $|e\rangle$ and ground $|g\rangle$ states respectively, the expressions $|e\rangle\langle v|$ and $|g\rangle\langle v|$ represent the transition operators between upper virtual state $|v\rangle$ and lower $|e\rangle$, $|g\rangle$ states accompanied with absorbtion and emission of the photons, $\hat{d}^{\dagger}$ and $b^{\dagger}$, of external and scattering fields respectively; k_d and k_a are the photon wave vectors in the dressed and scattered modes respectively. The photon wave vectors of external and scattered modes are directed along the z and x axis in according with μ_{ev} and μ_{gv} are the dipole transition matrix elements between virtual $|v\rangle$ and lower $|e\rangle$, $|g\rangle$ states.

The interaction part of Hamiltonian has rapid oscillatory form. Indeed the terms like $\hat{b}^{+}(t)\,|e(t)\rangle\,\langle v(t)|$ in the Hamiltonian (9) have the time dependent oscillatory functions like exp $[i(\omega_a-\omega_v+\omega_e)t]$, where $\omega_a-\omega_v+\omega_e\neq 0$. In order to obtain the model Hamiltonian, which describes the Hyper-Raman processes in which two-photon from dressed field are absorbed and other two new photons are generated, let us eliminate the operators of virtual atomic state $|v\rangle$ from the Hamiltonian (1). This procedure of elimination must continue till the situation in which the terms of interaction Hamiltonian includes the resonance diagrams like this $(\hat{d}(t))^2(\hat{b}^{+}(t))^2\,|e(t)\rangle\,\langle g(t)|$ represented in the Fig.1. According to this, we solve the solution of Schrodinger equation $\hat{H}\,|v(t)\rangle = -i\hbar\frac{\partial}{\partial t}\,|v(t)\rangle$ for representation of virtual state $|v\rangle$ through $|e\rangle$ and $|g\rangle$ states

$$\begin{aligned}\frac{\partial}{\partial t}\,|v(t)\rangle &= i\omega_v\,|\nu(t)\rangle - \frac{i}{\hbar}[(\mu_{ev},\mathbf{g}_a)\hat{b}^{\dagger}(t)e^{-ik_ax}+(\mu_{ev},\mathbf{g}_d)\hat{d}^{\dagger}(t)e^{-ik_dz}]\,|e(t)\rangle\\ -\frac{i}{\hbar} & [(\mu_{ev},\mathbf{g}_a)\hat{b}^{\dagger}(t)e^{-ik_ax}+(\mu_{ev},\mathbf{g}_d)\hat{d}^{\dagger}(t)e^{-ik_dz}]\,|g(t)\rangle\,. \qquad (3)\end{aligned}$$

According to perturbation theory in first Born-Marcoff approximation, we can represent the solution of equation (3) the following form

$$|v(t)\rangle = |v(0)\rangle\,e^{i\omega_vt} + A^{+}_{v,e}(t,x,z)\,|e(t)\rangle + A^{+}_{v,g}(t,x,z)\,|g(t)\rangle\,. \qquad (4)$$

Here the coefficients $\hat{A}^{+}_{v,e}(t,x,z)$ and $\hat{A}^{+}_{v,g}(t,x,z)$ can be expressed in the Born-Marcoff approximation [9] through the operators of pump and scattering fields

$$\hat{A}^{+}_{v,\alpha}(t,x,z) = \frac{(\mu_{v\alpha},\mathbf{g}_d)\hat{d}^{\dagger}(t)e^{-izk_d}}{\hbar(\omega_{v\alpha}-\omega_d)} + \frac{(\mu_{v\alpha},\mathbf{g}_a)\hat{b}^{+}(t)e^{-ixk_a}}{\hbar(\omega_{v\alpha}-\omega_a)},\quad \alpha=g,e.$$

Introducing the vector $|v(t)\rangle$ in the interaction part of Hamiltonian (9), we obtain the new form of interaction Hamiltonian

$$\begin{aligned}H_i^{eff} &= -[(\mu_{ev},\mathbf{g}_a)\hat{b}^{+}e^{-ik_ax}+(\mu_{ev},\mathbf{g}_d)\hat{d}^{\dagger}e^{-ik_dz}]\\ &\times\ [\hat{A}_{v,e}(t,x,z)\,|e(t)\rangle\,\langle e(t)| + \hat{A}_{v,g}(t,x,z)\,|e(t)\rangle\,\langle g(t)|]\\ &+\ [(\mu_{gv},\mathbf{g}_r)(\hat{d}^{\dagger}e^{-ik_dz}+(\mu_{gv},\mathbf{g}_a)\hat{b}^{+}e^{-ik_ax}]\\ &\times\ [\hat{A}_{v,e}(t,x,z)\,|g(t)\rangle\,\langle e(t)| + \hat{A}_{v,g}(t,x,z)\,|g(t)\rangle\,\langle g(t)|]\\ &+\ H.c. \qquad (5)\end{aligned}$$

As the terms of this Hamiltonian remain rapid oscillatory functions, we must again eliminate the atomic virtual states from the Hamiltonian. In this case we again represent the solutions for excited $|e\rangle$ and ground states $|g\rangle$ through virtual state $|v\rangle$, using the Hamiltonian (2)

$$\begin{aligned}|e(t)\rangle &= \exp[i\omega_et]\,|e(0)\rangle - \frac{i}{\hbar}\int_0^t d\tau\exp[i\omega_e\tau][(\mu_{ev},\mathbf{g}_a)\hat{b}^{+}(t-\tau)e^{-ik_ax}\\ &+\ (\mu_{ev},\mathbf{g}_d)\hat{d}^{\dagger}(t-\tau)e^{-ik_dz}]\,|v(t-\tau)\rangle\,;\\ |g(t)\rangle &= \exp[i\omega_gt]\,|g(0)\rangle - \frac{i}{\hbar}\int_0^t d\tau\exp[i\omega_g\tau][(\mu_{vg},\mathbf{g}_d)(\hat{d}(t-\tau)e^{ik_dz}\\ &+\ (\mu_{vig},\mathbf{g}_a)\hat{b}(t-\tau)e^{ik_ax}]\,|v(t-\tau)\rangle\,,\end{aligned}$$

Now we can easy found the effective Hamiltonian which describes the Hyper-Raman resonances in fours order of perturbation theory relative the small parameter $(\mu_{v\alpha}, \mathbf{g})/\mathbf{h}$. For this we firstly represent the field and atomic operators through smooth and rapid oscillatory parts: $\hat{d}^{+}(t) = \tilde{d}^{+}\exp(i\omega_{d}t)$, $\hat{b}^{+}(t) = \tilde{b}^{+}\exp(i\omega_{s}t)$ and $|\alpha><\beta|(t) = |\tilde{\alpha}><\tilde{\beta}|\exp[i\omega_{\alpha\beta}t]$, where α, β are g and e states. Integrating again the solutions for vectors $|e(t)\rangle$ and $|g(t)\rangle$ in Born-Marcoff approximation

$$\begin{aligned} |e(t)\rangle &= \exp[i\omega_{e}t]\,|e(0)\rangle + \left[\frac{(\mu_{ve}, \mathbf{g}_{d})\hat{d}(t)e^{ik_{d}z}}{\hbar(\omega_{ve} - \omega_{d})} + \frac{(\mu_{ve}, \mathbf{g}_{a})\hat{b}(t)e^{ik_{a}x}}{\hbar(\omega_{ve} - \omega_{s})}\right]|v(t)\rangle\,, \\ |g(t)\rangle &= \exp[i\omega_{g}t]\,|g(0)\rangle + \left[\frac{(\mu_{vg}, \mathbf{g}_{d})\hat{d}(t)e^{ik_{d}z}}{\hbar(\omega_{vg} - \omega_{d})} + \frac{(\mu_{vg}, \mathbf{g}_{a})\hat{b}(t)e^{ik_{a}x}}{\hbar(\omega_{vg} - \omega_{s})}\right]|v(t)\rangle\,, \end{aligned} \tag{6}$$

we obtain the fours photon transitions between the excited $|e\rangle$ and ground $|g\rangle$ states after introducing the solutions (4) in the right hand sides of expressions (6). After the substitution of the formal solutions (6) in the right hand side of the Hamiltonian (5) in the fourth order of small interaction parameter $(\mu_{v\alpha}, \mathbf{g})/\mathbf{h}$, we obtain the following model Hamiltonian with smooth transitions diagrams between the real ground $|g\rangle$ and excited $|e\rangle$ states

$$\begin{aligned} \hat{H}_{eff} &= \hbar\omega_{eg}\sum_{l=1}^{N}(|e\rangle\langle e|_{l} - |g\rangle\langle g|_{l})/2 + \hbar\omega_{s}\hat{b}^{\dagger}\hat{b} + \hbar\omega_{d}\hat{d}^{\dagger}\hat{d} \\ &- \sum_{l=1}^{N}\{q\hat{d}^{2}(t)\hat{b}^{\dagger 2}(t))\exp[2ik_{a}x_{l} - 2ik_{d}z_{l}]\,|e(t)\rangle\langle g(t)|_{l} \\ &+ q^{*}\hat{d}^{2}(t)\hat{b}^{\dagger 2}(t)\exp[-2ik_{a}x_{l} + 2ik_{d}z_{l}]\,|g(t)\rangle\langle e(t)|_{l}\}. \end{aligned} \tag{7}$$

Here the interaction constant q is presented through the small parameter in the following form

$$\begin{aligned} q &= \left(\frac{1}{\omega_{ve} - \omega_{a}} + \frac{1}{\omega_{ve} - \omega_{d}}\right)\left[\frac{(\mu_{ve}, \mathbf{g}_{d})\,(\mu_{vg}, \mathbf{g}_{a})^{2}\,(\mu_{vg}, \mathbf{g}_{d})}{\hbar^{3}\,(\omega_{vg} - \omega_{d})\,(\omega_{vg} - \omega_{a})}\right. \\ &+ \left.\frac{(\mu_{ve}, \mathbf{g})^{2}\,(\mu_{ve}, \mathbf{g}_{r})\,(\mu_{vg}, \mathbf{g}_{r})}{\hbar^{3}\,(\omega_{ve} - \omega_{a})\,(\omega_{vg} - \omega_{d})}\right]. \end{aligned} \tag{8}$$

In the interaction Hamiltonian (7) we have considered an atomic ensemble instead of one atom in the system. In four photon interaction with bimodal cavity field with frequencies ω_{a} and ω_{d}, let us introduce the new bimodal field operators $\hat{J}^{-} = d^{\dagger}b$, $J^{+} = \hat{d}\hat{b}^{\dagger}$ and $\hat{J}_{z} = (\hat{d}^{\dagger}\hat{d} - \hat{b}^{\dagger}\hat{b})/2$ and atomic operators $\hat{R}_{j}^{+} = |e\rangle\langle g|_{j}$, $\hat{R}_{j}^{-} = |g\rangle\langle e|_{j}$ and $\hat{R}_{zj} = (|e\rangle\langle e|_{j} - |g\rangle\langle g|_{j})/2$, which belong to $su(2)$ algebra. In this representation the model Hamiltonian (7) for the ensemble of atoms takes the following form

$$\hat{H}_{I} = \sum_{j=1}^{N}\hbar\omega\hat{R}_{zj} - \hbar\omega\hat{J}_{z} + \sum_{j=1}^{N}\{q(J^{-})^{2}e^{2ik_{a}x_{j} - 2ik_{d}z_{j}}R_{j}^{+} + H.c\}. \tag{9}$$

Here the new operators satisfy the commutation relations belonging to $su(2)$ symmetry

$$\begin{aligned} [\hat{J}^{+}, \hat{J}^{-}] &= 2\hat{J}_{z}, \quad [\hat{J}_{z}, \hat{J}^{\pm}] = \pm\hat{J}^{\pm}; \\ [\hat{R}_{l}^{+}, \hat{R}_{j}^{-}] &= 2\hat{R}_{zj}\delta_{j,l}, \quad [\hat{R}_{zl}, \hat{R}_{j}^{\pm}] = \pm R_{j}^{\pm}\delta_{j,l}; \end{aligned}$$

$\omega_{0} = \omega_{e} - \omega_{g}$. In the next section the master equation approach for the cavity field is proposed in which the atomic variables is eliminated.

3. Master Equation for Hyper-Raman Generation Process and it's solution

Using the method of the elimination of the operators of atomic subsystem from master equation [8] and considering that the polarization and inversion damping rates in the cavity are longer than the generation rate of anti-Stokes photons in scattering process, we can adiabatically eliminate the atomic variables from that equation. Following this method and taking in to account that, the atomic flux enters in the resonator in excited state, we can adiabatically eliminate the atomic operators $\hat{R}_l^+$ and $\hat{R}_l^-$ from Heisenberg equation of mean value of arbitrary field operator $\hat{O}(t)$

$$\begin{aligned}\frac{d}{dt}\left\langle \hat{O}(t)\right\rangle &= i\tilde{\omega}\left\langle \hat{J}_z(t), \hat{O}(t)\right\rangle \\ &- \frac{iq}{\hbar}\sum_{l=1}^{N}\left\{\langle \hat{R}_l^-(t)[\hat{J}^{+2}(t), \hat{O}(t)]\rangle + \langle [\hat{J}^{-2}(t), \hat{O}(t)]\hat{R}_l^+(t)\rangle\right\}, \qquad (10)\end{aligned}$$

where $\hat{J}_z(t)$ is the photons inversion operator between the pump and anti-Stokes modes. In expression (10) we considered that the atomic subsystem has the dimension less than the emission wavelength of radiation so that the exponential function $\exp[-i2k_d z_l + 2ik_a x_l] \simeq 1$. In order to obtain an equation for cavity bimodal field, it is necessary to eliminate the free parts of atomic operators $\hat{R}_l^+(t) = \hat{R}_{lf}^+(t) + \hat{R}_{ls}^+(t)$ where

$$\begin{aligned}\hat{R}_{lf}^+(t) &= \hat{R}_{lf}^+(0)\exp[i\omega_0 t - \gamma_\perp t] \\ \hat{R}_{ls}^+(t) &= \frac{2qi}{\hbar}\int_0^t d\tau \hat{J}^{+2}(t-\tau)\hat{R}_z(t-\tau)\exp[i\omega_0\tau - \gamma_\perp\tau]. \qquad (11)\end{aligned}$$

It is easy to understand that, when the atomic system is in excited states, the free part of the operators $\hat{R}_l^+(t)$ and $\hat{R}_l^-(t)$ is simply eliminated. When the atoms are in the excited state it is necessary to permute the operators $\hat{R}_l^+$ and $\hat{R}_l^-$ according to the definition of the anti-normal correlation product in the equation (10), so that the free part of these operators gives zero contributions $\hat{R}_{l,f}^+(t)|e\rangle = \hat{R}_l^+(0)|e\rangle\exp[-i\omega_0 t] = 0$; $\langle e|\hat{R}_l^-(t) = \langle e|\hat{R}_{l,f}^-(0)\exp[i\omega_0 t] = 0$. If $\hat{O}(t)$ is an arbitrary operator of cavity field subsystem, the mean values of the products are $\langle\hat{O}(t)\hat{R}^+(t)\rangle = \langle\hat{O}(t)\hat{R}_s^+(t)\rangle$. Let us represent solution of the Heisenberg operator (11) in the Born-Marcoff approximation, considering that the polarization de-phasing rate $\gamma_\perp$ is large than the Hyper-Raman scattering rate κ_1, $\kappa_1/\gamma_\perp < 1$

$$\hat{R}_j^+(t) \approx \hat{R}_j^+(0)\exp(i\tilde{\omega}t) + \frac{2qi\hat{R}_{zj}(t)[\hat{J}^+(t)]^2}{[i(\tilde{\omega}-\omega_0)+\gamma_\perp]\hbar}, \qquad \hat{R}_j^-(t) = [\hat{R}_j^+(t)]^+.$$

After the first elimination of free parts of these operators, the generalized equation (10) becomes

$$\begin{aligned}\frac{d}{dt}\langle\hat{O}(t)\rangle &= i(\tilde{\omega}-\omega_0)\left\langle[\hat{J}_z(t), \hat{O}(t)]\right\rangle \\ &+ \frac{2}{\hbar}\sum_{l=1}^{N}q^2\left\{\frac{\left\langle\hat{R}_z^l(t)\hat{J}^{-2}(t)[\hat{J}^{+2}(t), \hat{O}(t)]\right\rangle}{(i(\tilde{\omega}-\omega_0)-\gamma_\perp)\hbar} + \frac{\langle[\hat{J}^{-2}(t), \hat{O}(t)]\hat{J}^{+2}(t)\hat{R}_z^l(t)\rangle}{(i(\tilde{\omega}-\omega_0)+\gamma_\perp)\hbar}\right\}. \qquad (12)\end{aligned}$$

As it was mentioned in papers [10],[8] the relative small second-order terms should be taken into account for the laser stabilization by stimulated emission. In this case, we must eliminate the atomic inversion $\hat{R}_{zj}(t)$ from the generalized equation (12). Indeed, introducing the quasi-stationary solution of Heisenberg equation for $R_{zl}(t)$ operator

$$\hat{R}_z^l(t) = \frac{1}{2} - \frac{iq}{\hbar\gamma_{||}}\left\{\hat{R}_l^-(t)\hat{J}^{+2}(t) - \hat{J}^{-2}(t)\hat{R}_l^+(t)\right\} \qquad (13)$$

in expression (12) and again eliminating the atomic operators $\hat{R}_l^+$ and $\hat{R}_l^-$ from the equation (13) we obtain the following generalized equation for the bimodal cavity field operator $\hat{O}(t)$

$$\begin{aligned}\frac{d}{dt}\langle O(t)\rangle &= i\chi\gamma_\perp\left\langle[\hat{J}_z(t),\hat{O}(t)]\right\rangle+\kappa_1\left\{\left\langle[\hat{J}^{-2}(t)\hat{O}(t),\hat{J}^{+2}(t)]\right\rangle+H.c.\right\}\\ &- \kappa_2\frac{1-\chi^2}{(1+\chi^2)^2}\left[\left\langle\hat{J}^{-2}(t)\hat{J}^{+2}(t)\hat{J}^{-2}(t)[\hat{O}(t),\hat{J}^{+2}(t)]\right\rangle+H.c.\right]\\ &+ \frac{1}{1+\chi^2}\left[\left\langle\hat{J}^{-2}(t)\hat{J}^{-2}(t)[\hat{O}(t),\hat{J}^{+2}(t)]\hat{J}^{+2}(t)\right\rangle+H.c.\right]\\ &+ i\frac{2\chi}{(1+\chi^2)^2}\left[\left\langle\hat{J}^{-2}(t)\hat{J}^{+2}(t)\hat{J}^{-2}(t)[\hat{O}(t),\hat{J}^{+2}(t)]\right\rangle-H.c.\right].\end{aligned} \quad (14)$$

in the fourth order of perturbation theory relatively the small parameter $q/\hbar$. In this equation, the new constants are represented through the interaction parameter, q^2, detuning, $\tilde{\omega}-\omega_0$, number of atoms, N, polarization damping rates $\gamma_\perp$. Here $\kappa_1 = Nq^2/[\hbar^2\gamma_\perp(\chi^2+1)]$ is the explicit expression of emission rate in hyper-Raman scattering process, $\kappa_2 = 2q^4N/[\hbar^4\gamma_{II}\gamma_\perp^2]$ describes the diffusion coefficient in the hyper-Raman generation process of anti-Stokes photons; $\chi = (\tilde{\omega}-\omega_0)\gamma_\perp^{-1}$ depends on the de-tuning between the transition frequency ω_0 and $\tilde{\omega} = 2(\omega_d-\omega_s)$. The first term of equation (14) describes the rate of collective transformation of anti-Stokes photon pairs into Stokes photon pairs. The second order terms proportional to the diffusion coefficient in the generation process of anti-Stokes photons κ_2, correspond to nonlinear attenuation processes of the generation, which, together with first term in equation (14), has the stabilization limit of hyper-Raman scattering process.

Passing from field operator $\hat{O}(t)$ in Heisenberg picture to the density matrix $\hat{W}(t)$ in Schrodinger representation, $Sp\{\hat{W}(t),\hat{O}(0)\} = Sp\{\hat{W}(0),\hat{O}(t)\}$, we obtain the following master equation from equation (14)

$$\begin{aligned}\frac{d\hat{W}(t)}{dt} &= -i\chi\gamma_\perp[\hat{J}_z,\hat{W}(t)]+\kappa_1\left\{[\hat{J}^{+2},\hat{W}(t)\hat{J}^{-2}]+H.c.\right\}\\ &- \kappa_2\{\frac{1-\chi^2}{(1+\chi^2)^2}\{[\hat{J}^{+2},\hat{W}(t)\hat{J}^{-2}\hat{J}^{+2}\hat{J}^{-2}]-\hat{W}(t)\hat{J}^{-2}\hat{J}^{+2}\hat{J}^{-2}\hat{J}^{+2}+H.c.\}\\ &+ \frac{1}{1+\chi^2}\{[\hat{J}^{+2},\hat{J}^{2+}\hat{W}(t)\hat{J}^{2-}\hat{J}^{-2}]+H.c.\}\\ &+ i\frac{2\chi}{(1+\chi^2)^2}\{[\hat{J}^{+2},\hat{W}(t)\hat{J}^{-2}\hat{J}^{+2}\hat{J}^{-2}]-H.c.\}.\end{aligned} \quad (15)$$

This equation describes the collective transformation of pairs of Stokes photons in anti-Stokes in Hyper-Raman generation process. Ignoring the losses from the cavity, we observe, that the steady state solution of this equation is applicable for the time interval $\gamma^{-1} < t < k^{-1}$, where $\gamma^{-1} \sim \gamma_\perp^{-1};\ \gamma_\parallel^{-1}$.

Let us firstly consider that, at the initial time moment, we have 0 photons in the anti-Stokes mode and $n_{ph} = 2j$ photons in the pump field, so that the bimodal cavity fields is described by the state $|\Psi(0)\rangle = |0\rangle_a|n_{ph}\rangle_S$. Let us found a quantum solution for the master equation (15). As it is proposed in papers [11], [12], we can expand the density matrix (15) over the coherent states of generalized of the angular moment states. According to the Q- and P-diagonal representation [12], when the right side of master equation has a relatively biquadrate form of operators $\hat{J}^+$ and $\hat{J}^-$, the solution of master equation can be represented through diagonal bra-ket operators angular momentum states

$$W(t) = \sum_{m=-j}^{+j} P_m(t)|m,j\rangle\langle m,j|. \quad (16)$$

Here, the Hilbert's vectors $|m,j\rangle$ belongs to the set of angular momentum states and obeys to $su(2)$ symmetry; $\hat{J}^+|m,j\rangle = \sqrt{(j-m)(j+m+1)}|m+1,j\rangle$ and $\hat{J}^-|m,j\rangle = \sqrt{(j-m+1)(j+m)}|m-1,j\rangle$. P_m is the occupation probability of the state $|m,j\rangle$. By introducing the representation (16) in equation (15), it is easy to obtain the following system of equations for the probabilities $\{P_m\}$

$$\begin{aligned}\frac{dP_m(t)}{dt} &= 2\kappa_1\{P_{m-2}(t)(j-m+2)(j-m+1)(j+m)(j+m-1)\\ &- P_m(t)(j-m)-1)(j-m)(j+m+2)(j+m+1)\}\\ &- \frac{2\kappa_2(1-\chi^2)}{(1+\chi^2)^2}[P_{m-1}(t)(j-m+2)^2(j-m+1)^2(j+m)^2(j+m-1)^2\\ &- P_m(t)(j-m-1)^2(j-m)^2(j+m+2)^2(j+m+1)^2]\\ &- \frac{2\kappa_2}{1+\chi^2}(j-m+2)(j-m+1)(j+m-1)(j+m)\\ &\times\ [P_{m-4}(t)(j-m+4)(j-m+3)\ (j+m-3)(j+m-2)\\ &- P_{m-2}(t)(j+m+2)(j+m+1)(j-m-1)(j-m)],\quad -j\le m\le -j. \qquad (17)\end{aligned}$$

The set of functions $\{P_m(t)\}$ describes the probability of simultaneous existence of $(2j+1)$ Dicke [13] states in the hyper-Raman scattering process. The initial condition for the system of equations (17) is $P_{-j}=1,\quad P_{-j+1}=P_{-j+2}=...=P_j=0$.

The numerical behaviors of the system of equations (17) is graphically represented in Fig.2. The quantum correlations between number of photons in the pump and anti-Stokes photons play an important role in the hyper-Raman process. As results from the figures presented above, the transformation of 12 Stokes photons into 12 anti-Stokes photons, take places in a very interesting way. At earlier stage of radiation this correlation can be regarded as the probabilistic repartitions between Stokes and anti-Stokes photons: $(0,12)$; $(2,10)$; $(4,8)$; $(6,6)$; $(8,4)$; $(10,2)$; $(12,0)$, which are described by simultaneous absorption and generation of photon pairs $P_6(t)$, $P_4(t)\dots P_0(t)\dots P_{-6}$. In the brackets, the first number (n,m) represents the Stokes photons and the second number-the anti-Stokes photons. The sum of these probabilities is equal to unity: $\sum_{m=-j}^{+j}P_m(t)=1$.

Let's now study the correlations between Stokes and anti-Stokes photons in hyper-Raman process, which can be described by the following function

$$K(t)=\langle\hat{d}^\dagger(t)\hat{d}(t)\hat{b}^\dagger(t)\hat{b}(t)\rangle-\langle\hat{d}^\dagger(t)\hat{d}(t)\rangle\langle\hat{b}^\dagger(t)\hat{b}(t)\rangle$$

According to the definition of the new field operators,the number of photons in Stokes and anti-Stokes modes can be represented through the generators of $su(2)$ algebra $\hat{J}_z$ in the $\hat{d}^\dagger(t)\hat{d}(t)=j+\hat{J}_z(t)$ and $\hat{b}^\dagger(t)\hat{b}(t)=j-\hat{J}_z(t)$. According with this definition the correlation function can be represented through the quantum fluctuation of inversion operator $\hat{J}_z$

$$K(t)=\langle\hat{J}_z^2(t)\rangle-(\langle\hat{J}_z^2(t)\rangle)^2. \qquad (18)$$

We are interested in the behavior of the quantum correlations between bi-photons described by the function (18). In Fig.2B is plotted the time behavior of cross-correlation between the photon numbers absolute of Stokes and anti-Stokes photons Δ^2 and relative square fluctuations δ^2 during the transforation of 12 Stokes photons in anti-Stokes. From this Fig.2A , it follows that in the process of the stabilization of generation process, the square fluctuations remain positive function. The maximal value of quantum fluctuations corresponds to the equal value of photon pairs in Stokes and anti-stokes modes. In this case the transition corresponds to the maximal value of the fluctuations of the normal product of operators $\hat{J}^+$ and $\hat{J}^-$. This fluctuations is accompanied with simultaneous realization of the partition (m,n) with the same magnitude of probabilities P_m represented in Fig.2A.

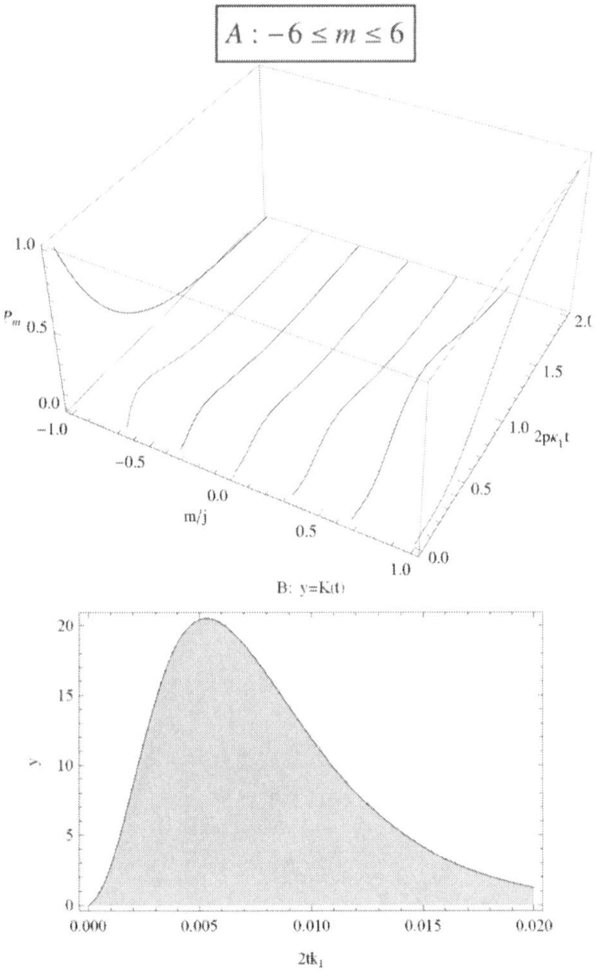

Figure 2. A.The time dependence of probabilities $P_m(t)$ of transforation of pairs of pump photons in anti-stokes photon pairs for following parameters of the system $j = 6$; $\kappa_2/\kappa_1 = 0.00005$; p=100.
B.The time dependence of quantum correlations between the Stokes and anti-Stokes photons in the hyper-Raman scattering. When the number (n,m) of the distribution possibilities between Stokes and anti-Stokes modes archived the increase the fluctuations of cross-correlation function $K(t)$ archives the maximal value.

4. Conclusion

The dynamical transformation of Stokes photons in anti-Stokes and vice versa as the function of the preparation of atomic inversion and field in Hyper-Raman process was the subject of investigation of this article. The proposed $su(2)$ operators for the description of this transformation are similar to Dicke operators in super-radiance [13]. In this case, the role of the number of excited atoms is substituted by the number of Stokes photons, which are scattered into anti-Stokes photons in the process of hyper-Raman interaction. More than this nonlinear interaction is accompanied with annihilation of one pair of photons in Stokes mode and generation of a new pair in anti-Stokes field. The collective bi-operators are introduced which describe the process of generation (annihilation) of Stokes (or anti-Stokes) photons, accompanied by absorbtion of anti-Stokes (Stokes) photon pairs, respectively. The unique features of the field consisting of both Stokes and anti-Stokes photons have been studied by solving the master equation numerically. It has been found the correlation function between Stokes and anti-Stokes modes in the process of cooperative exchanges with pairs of photons in the hyper-Raman interaction between the modes through the atomic subsystem. This function achieved the maximal value when the mean number of photons in Stokes and anti-Stokes modes is approximatively equal.

Acknowledgments
This paper was supported by the projects of the Academy of Sciences of Moldova: No. 10.820.05.05/BF and No. 11.819.05.11F.

References
[1] Gauthier D J, Wu Q, Morin S E, and Mossberg T W 1992 *Phys. Rev. Lett.* **68** 464; Pfister O, Brown WJ, Stenner M D, and Gauthier D J 1999 *Phys. Rev.* A **65**, 031803(R); Pfister O, Brown W J , Stenner M D and Gauthier D J 2001 *Phys. Rev. Lett.* **86** 4512
[2] Concanon H M, Brown W J, Gardiner J R and Gauthier D J 1997 *Phys.Rev.* A **56** 1519
[3] Poelker M and Kumar P 1992, *Opt. Lett.* **17**, 399
[4] Hemmerich A, Zimmerman C and Hansch T W 1994 *Phys.Rev.* A **72**, 625
[5] Cataliotti F S, Scheunemann R, Hanch T V and Weitz M 2002) *Phys. Rev.Lett.* **87** 113601
[6] Agarwal G S 1988 *Opt. Lett.* **13** 482
[7] Drummond P D, Gardiner C W 1980 *J. Phys. A: Math. Gen.* **13** 2353
[8] Eremeev V and Enaki N 2005 *Opt.Commun.* **247** 381
[9] Enaki N A 2010 it J. Modern Optics **57** 2082
[10] Wang Z C and Haken H 1984 *Z. Phys. B-Cond. Matter* **55** 361; **56**, 77 ; 1988 *Z. Phys. B-Cond. Matter* **71** 253
[11] Bonifacio R and Lugiato L A 1975 *Phys. Rev.* A **11** 1507
[12] Drummond P D and Gardiner C W, 1980 *J. Phys.* A bf 13 2353; Gardiner C W, 1996 in the book *Quantum Noise* Springer-Verlag, NewYork
[13] Dicke R H 1954 *Phys. Rev.* **93**, 99

A new method of construction of all sets of mutually unbiased bases for two-qubit systems

Iulia Ghiu

Centre for Advanced Quantum Physics, Department of Physics, University of Bucharest, P. O. Box MG-11, R-077125 Bucharest-Măgurele, Romania

E-mail: iulia.ghiu@g.unibuc.ro

Abstract. Mutually unbiased bases are an important tool in many applications of quantum information theory. We present a new algorithm for finding the mutually unbiased bases for two-qubit systems. We derive a system of four equations in the Galois field GF(4) and show that the solutions of this system are sufficient for obtaining the most general set of mutually unbiased bases. Further, our algorithm is applied to an example and we show that there are three possible solutions of the system of four equations, each solution leading to a different set of mutually unbiased bases.

1. Introduction

The concept of mutually unbiased bases (MUBs) was initially proposed by Julian Schwinger [1]. These states are maximally incompatible, which means that using a basis to obtain optimal outcomes, leads to maximally random results when the other bases are considered in the measurements. If we denote two elements of different orthonormal bases by $\{|\,\psi_j\,\rangle\}$ and $\{|\,\phi_k\,\rangle\}$, then the two bases are called mutually unbiased if

$$|\langle\,\psi_j\,|\,\phi_k\,\rangle|^2 = \frac{1}{d}.$$

where d is the dimension of the Hilbert space [2]. The MUBs constitute now a basic ingredient in many applications of quantum information processing: quantum tomography, quantum key distribution required in cryptography [3], discrete Wigner function [4], quantum teleportation [5], or quantum error correction codes [6].

If the measurements used for the state reconstruction are built with the help of the mutually unbiased bases, then this scheme is called MUB tomography [7]. The simplest example of MUBs can be given for qubits. If the system is prepared in the state $|\uparrow\rangle$ (or $|\downarrow\rangle$), then a measurement of the spin along the $x-$ or $y-$axis generates the outcomes *up* or *down* with the same probability, which is equal to 1/2. The eigenvectors of the Pauli operators $\hat{\sigma}_x$, $\hat{\sigma}_y$, and $\hat{\sigma}_z$ form a set of MUBs for qubits.

It is known that the maximal number of mutually unbiased bases can be at most $d+1$ [8]. This maximal value can be reached in the case when the dimension of the space is a prime ($d=p$) or a power of prime ($d=p^n$) [9].

It was proved in the case when d is a power of a prime number that there is an approach for finding the MUBs, namely one has to generate families of unitary operators, whose eigenvectors

Published under licence by IOP Publishing Ltd

Journal of Physics: Conference Series **338** (2012) 012008 doi:10.1088/1742-6596/338/1/012008

solve the problem of MUBs. In other words, one needs to determine $d+1$ classes of $d-1$ commuting operators. These special operators are called mutually unbiased operators [10].

The paper is organized as follows: In Section 2 we present the standard set of MUBs for qubits, where three bases are obtained from tensor product of $\hat{\sigma}_j$ and $\hat{I}$. Further, some basic notations and definitions used in the Galois fields are recalled in Section 3. The main result of the paper is given in Section 4, where we find a system of four equations in the Galois field GF(4). The solution of these equations leads to the construction of the most general set of MUBs for two-qubit syatems, not only the standard one. Finally, we make some concluding remarks in Section 5.

2. Preliminaries

In this paper we analyze the two-qubit systems. An example of 5 sets of 3 commuting operators for 2 qubits is [11]:

1.	$\hat{\sigma}_z\hat{\sigma}_z$	$\hat{I}\hat{\sigma}_z$	$\hat{\sigma}_z\hat{I}$
2.	$\hat{\sigma}_x\hat{\sigma}_x$	$\hat{I}\hat{\sigma}_x$	$\hat{\sigma}_x\hat{I}$
3.	$\hat{\sigma}_y\hat{\sigma}_y$	$\hat{I}\hat{\sigma}_y$	$\hat{\sigma}_y\hat{I}$
4.	$\hat{\sigma}_x\hat{\sigma}_y$	$\hat{\sigma}_z\hat{\sigma}_x$	$\hat{\sigma}_y\hat{\sigma}_z$
5.	$\hat{\sigma}_y\hat{\sigma}_x$	$\hat{\sigma}_z\hat{\sigma}_y$	$\hat{\sigma}_x\hat{\sigma}_z$

Table 1. The standard set of MU operators for two qubits.

We did not write the tensor multiplication sign in the above Table. In the case of two-qubit systems, there is only one allowed structured of the bases, namely (3,2), which means that three bases are separable, while two are entangled. The three separable bases given by Table 1 are obtained as the common eigenvectors of tensor product of the operators $\hat{\sigma}_z$ and $\hat{I}$ (all the three possibilities: see the first row of Table 1), tensor product of $\hat{\sigma}_x$ and $\hat{I}$ (second row), and tensor product of $\hat{\sigma}_y$ and $\hat{I}$ (third row), respectively. The table of operators with this property are called the standard set of MU operators [7]. The last two rows of Table 1 generate entangled bases which are called *belle* and *beau* [12].

3. Galois fields: notations and definitions

If $d=p^n$ is the power of a prime number, then the $d+1$ MUBs can be constructed with the help of Galois fields GF(d). Wootters [12] and Gibbons *et all.* [13] proposed a method to associate the MUBs to the so-called discrete phase space. The discrete phase space of a d-level system is a $d \times d$ lattice, whose coordinates are elements of the finite Galois field GF(d) [12]. Further a state is associated to a line in the discrete phase space. The set of parallel lines is called a striation. There are $d+1$ striations. It turns out that the MUBs are determined by the bases associated with each striation.

In the following we will recall some basic definitions used in Galois fields. The map trace of a field element $x \in \mathrm{GF}(p^n)$ is as follows:

$$\mathrm{tr}\, x = x + x^2 + ... + x^{p^{n-1}}.$$

A basis $\{\alpha_i\}$ is called self-dual if $\mathrm{tr}(\alpha_i\alpha_j) = \delta_{ij}$.

Let us consider two elements x, $y \in\mathrm{GF}(p^n)$ which can be written with the help of two bases $E = \{e_1, ..., e_n\}$ and $F = \{f_1, ..., f_n\}$ as follows:

$$x \;=\; \sum_{i=1}^{n} x_{e_i}\, e_i;$$

$$y = \sum_{j=1}^{n} y_{f_j} f_j. \tag{1}$$

We denote by $\hat{\sigma}_x, \hat{\sigma}_z$ the generalized Pauli operators [14]:

$$\begin{aligned} \hat{\sigma}_x | n \rangle &= | n+1 \rangle; \\ \hat{\sigma}_z | n \rangle &= \omega^n | n \rangle, \end{aligned}$$

where $| n \rangle$ is the computational basis and $\omega = \exp(2\pi i/d)$ is the dth root of the unity.

To the translation in the discrete phase space by the element (x, y) we associate an operator $T_{(x,y)}$ called the translation operator [13]:

$$T_{(x,y)} := \hat{\sigma}_x^{x_{e_1}} \hat{\sigma}_z^{y_{f_1}} \otimes ... \otimes \hat{\sigma}_x^{x_{e_n}} \hat{\sigma}_z^{y_{f_n}}. \tag{2}$$

In the case when $d = 2^2$, there is only one irreducible polynomial: $P(x) = x^2 + x + 1$. We denote by μ the primitive element of GF(4) and we get

$$\mu^2 + \mu + 1 = 0.$$

The elements of GF(4) are:

$$\{0, 1, \mu, \mu^2\}.$$

If we use the selfdual basis $\{\mu, \mu^2\}$ in GF(4), then the following correspondence given by Eq. (2) between the set of points in the discrete phase space and the set of MU operators can be established:

$$\begin{aligned} (\mu, 0) &\longleftrightarrow \hat{\sigma}_x \hat{I}, \\ (\mu^2, 0) &\longleftrightarrow \hat{I} \hat{\sigma}_x, \\ (0, \mu) &\longleftrightarrow \hat{\sigma}_z \hat{I}, \\ (0, \mu^2) &\longleftrightarrow \hat{I} \hat{\sigma}_z. \end{aligned} \tag{3}$$

4. The new algorithm of construction of MUBs

4.1. The commutation relations

Two operators commute if their associated points in the phase-space (a_1, b_1) and (a_2, b_2) satisfy [14]

$$\mathrm{tr}(a_1 b_2) = \mathrm{tr}(a_2 b_1). \tag{4}$$

The MUBs can be constructing with the help of $d+1$ classes of $d-1$ commuting operators, i.e. MU operators. In the case of two-qubit syatems, we need to generate a table with 5 rows, each row representing the set of 3 commuting operators. We denote the operator which belongs to the row r and column c, where $r = \overline{1,5}$ and $c = \overline{1,3}$ by $O_{r,c}$ as one can see in the left Table of Eq. (5). The point in the phase space $(a_c^{(r)}, b_c^{(r)})$ is associated to the operator $O_{r,c}$:

$$O_{r,c} \longleftrightarrow (a_c^{(r)}, b_c^{(r)}).$$

The correspondence between the set of MU operators and the set of points in the discrete phase space is shown below:

1.	$O_{1,1}$	$O_{1,2}$	$O_{1,3}$
2.	$O_{2,1}$	$O_{2,2}$	$O_{2,3}$
3.	$O_{3,1}$	$O_{3,2}$	$O_{3,3}$
4.	$O_{4,1}$	$O_{4,2}$	$O_{4,3}$
5.	$O_{5,1}$	$O_{5,2}$	$O_{5,3}$

$\longleftrightarrow$

1.	$\left(a_1^{(1)}, b_1^{(1)}\right)$	$\left(a_2^{(1)}, b_2^{(1)}\right)$	$\left(a_3^{(1)}, b_3^{(1)}\right)$
2.	$\left(a_1^{(2)}, b_1^{(2)}\right)$	$\left(a_2^{(2)}, b_2^{(2)}\right)$	$\left(a_3^{(2)}, b_3^{(2)}\right)$
3.	$\left(a_1^{(3)}, b_1^{(3)}\right)$	$\left(a_2^{(3)}, b_2^{(3)}\right)$	$\left(a_3^{(3)}, b_3^{(3)}\right)$
4.	$\left(a_1^{(4)}, b_1^{(4)}\right)$	$\left(a_2^{(4)}, b_2^{(4)}\right)$	$\left(a_3^{(4)}, b_3^{(4)}\right)$
5.	$\left(a_1^{(5)}, b_1^{(5)}\right)$	$\left(a_2^{(5)}, b_2^{(5)}\right)$	$\left(a_3^{(5)}, b_3^{(5)}\right)$

(5)

There are four operators which define the whole Table 1, namely the first two operators of the rows 1 and 2: $O_{1,1}$; $O_{1,2}$; $O_{2,1}$; $O_{2,2}$. The other 11 operators are determine as follows [11]:

$O_{r,c} = O_{r,c-2}O_{r,c-1}$, for $r = 1, 2$; $c = 3$ and

$O_{r,c} = O_{2,c}O_{1,c+r-3}$ for $r > 2$.

The indices r and c are written modulo four.

In our new construction, we consider that the right Table of Eq. (5) of points in the discrete phase space which correspond to the set of MU operators is unique defined by the following four points:

$$\begin{aligned}&(a_1^{(1)}, b_1^{(1)});\ (a_2^{(1)}, b_2^{(1)});\\ &(a_1^{(2)}, b_1^{(2)});\ (a_2^{(2)}, b_2^{(2)}).\end{aligned} \tag{6}$$

Therefore, we need to determine the eight parameters $a_j^{(i)}$ and $b_j^{(i)}$ given by Eq. (6) such that all the 15 points which are obtained correspond to MU operators. In other words, we have to use the commutation relation in the discrete phase space given by Eq. (4) for the set of 3 points of all the 5 rows. For each row we write 3 commutation relations, which lead to a total number of 15 equations for the whole table of points. These 15 equations are written in Appendix A.

4.2. The general algorithm

We proved that only 4 equations of the total number of 15 are independent. These are Eqs. (A.1), (A.4), (A.7), and (A.10) in Appendix A:

$$\begin{aligned}&\mathrm{tr}\left[a_1^{(1)}b_2^{(1)}\right] = \mathrm{tr}\left[a_2^{(1)}b_1^{(1)}\right]\\ &\mathrm{tr}\left[a_1^{(2)}b_2^{(2)}\right] = \mathrm{tr}\left[a_2^{(2)}b_1^{(2)}\right]\\ &\mathrm{tr}[a_1^{(1)}b_2^{(2)} + a_1^{(2)}b_2^{(1)}] = \mathrm{tr}[a_2^{(1)}b_1^{(2)} + a_2^{(2)}b_1^{(1)}]\\ &\mathrm{tr}[a_1^{(2)}b_1^{(1)} + a_1^{(2)}b_2^{(1)} + a_2^{(1)}b_2^{(2)}] = \mathrm{tr}[a_2^{(2)}b_2^{(1)} + a_1^{(1)}b_1^{(2)} + a_2^{(1)}b_1^{(2)}].\end{aligned} \tag{7}$$

The general algorithm of the construction:

- The parameters $b_1^{(1)}, b_2^{(1)}, b_1^{(2)}, a_1^{(1)}, a_1^{(2)}$ are arbitrary chosen.
- The three variables $a_2^{(1)}$, $a_2^{(2)}$, and $b_2^{(2)}$ are obtained from the system of equations (7).
- After obtaining the solution, one has to find the table of points in the discrete phase space analogue to the right table of Eq. (5).
- With the help of the correspondence given by Eq. (3), one generate the table which contain 5 classes of 3 commuting operators.
- The eigenvectors of each set of 3 commuting operators represent the MUBs.

4.3. Example

According to the algorithm presented in Sec. 4.2, we have to fix five parameters:

$$\begin{aligned} (a_1^{(1)}, b_1^{(1)}) &= (\mu^2, \mu) \\ (a_1^{(2)}, b_1^{(2)}) &= (\mu, \mu) \\ b_2^{(1)} &= \mu^2. \end{aligned}$$

The system of Eqs. (7) becomes:

$$\begin{aligned} &\mathrm{tr}\left[\mu\, a_2^{(1)}\right] = 1 \\ &\mathrm{tr}\left[\mu\, b_2^{(2)}\right] = \mathrm{tr}\left[\mu\, a_2^{(2)}\right] \\ &\mathrm{tr}[\mu^2 b_2^{(2)} + 1] = \mathrm{tr}[\mu\, a_2^{(1)} + \mu\, a_2^{(2)}] \\ &\mathrm{tr}[\mu^2 + 1 + a_2^{(1)} b_2^{(2)}] = \mathrm{tr}[\mu^2 a_2^{(2)} + 1 + \mu\, a_2^{(1)}]. \end{aligned} \tag{8}$$

There are three possible solutions of the system (8):

Solution A): $a_2^{(1)} = 1$, $a_2^{(2)} = \mu^2$, and $b_2^{(2)} = \mu^2$. This solution leads to the Table 1, i.e. the standard set of MU operators.

Solution B): $a_2^{(1)} = 1$, $a_2^{(2)} = 1$, and $b_2^{(2)} = \mu$. Therefore we obtain the following table of points in the discrete phase space and accordingly the set of MU operators:

No.	Set of points	MU operators
1.	$(1,\mu^2); (\mu,1); (\mu^2,\mu)$	$\hat\sigma_x\hat\sigma_y;\ \hat\sigma_y\hat\sigma_z;\ \hat\sigma_z\hat\sigma_x$
2.	$(1,\mu); (\mu,\mu); (\mu^2,0)$	$\hat\sigma_y\hat\sigma_x;\ \hat\sigma_y\hat I;\ \hat I\hat\sigma_x$
3.	$(0,1); (1,1); (1,0)$	$\hat\sigma_z\hat\sigma_z;\ \hat\sigma_y\hat\sigma_y;\ \hat\sigma_x\hat\sigma_x$
4.	$(0,\mu); (\mu^2,1); (\mu^2,\mu^2)$	$\hat\sigma_z\hat I;\ \hat\sigma_z\hat\sigma_y;\ \hat I\hat\sigma_y$
5.	$(0,\mu^2); (\mu,\mu^2); (\mu,0)$	$\hat I\hat\sigma_z;\ \hat\sigma_x\hat\sigma_z;\ \hat\sigma_x\hat I$

(9)

Solution C): $a_2^{(1)} = \mu$, $a_2^{(2)} = 1$, and $b_2^{(2)} = \mu$. The table of points in the discrete phase space and the set of MU operators are shown below:

No.	Set of points	MU operators
1.	$(1,1); (\mu,\mu^2); (\mu^2,\mu)$	$\hat\sigma_y\hat\sigma_y;\ \hat\sigma_x\hat\sigma_z;\ \hat\sigma_z\hat\sigma_x$
2.	$(1,\mu); (\mu,\mu); (\mu^2,0)$	$\hat\sigma_y\hat\sigma_x;\ \hat\sigma_y\hat I;\ \hat I\hat\sigma_x$
3.	$(0,1); (\mu,1); (\mu^2,1)$	$\hat\sigma_z\hat\sigma_z;\ \hat\sigma_y\hat\sigma_z;\ \hat\sigma_z\hat\sigma_y$
4.	$(0,1); (0,\mu); (0,\mu^2)$	$\hat\sigma_z\hat\sigma_z;\ \hat\sigma_z\hat I;\ \hat I\hat\sigma_z$
5.	$(1,\mu^2); (\mu,0); (\mu^2,\mu^2)$	$\hat\sigma_x\hat\sigma_y;\ \hat\sigma_x\hat I;\ \hat I\hat\sigma_y$

(10)

5. Conclusions

In this paper we have presented a new algorithm of construction of all sets of MUBs for two-qubit systems. The main result is the system of four equations (7) in the Galois field GF(4). With the help of the solution of this system, one can obtain different sets of MUBs, not only the well known standard set.

Acknowledgments

I wish to thank Gunnar Björk, Andrei B. Klimov, and Luis L. Sánchez-Soto for useful discussions. This work was supported by CNCSIS - UEFISCSU, postdoctoral research project PD code 151, no. 150/30.07.2010 for the University of Bucharest.

Appendix A. The 15 commutation relations

$$\mathrm{tr}\left[a_1^{(1)}b_2^{(1)}\right] = \mathrm{tr}\left[a_2^{(1)}b_1^{(1)}\right] \tag{A.1}$$

$$\mathrm{tr}\left[a_1^{(1)}\left(b_1^{(1)}+b_2^{(1)}\right)\right] = \mathrm{tr}\left[b_1^{(1)}\left(a_1^{(1)}+a_2^{(1)}\right)\right] \tag{A.2}$$

$$\mathrm{tr}\left[a_2^{(1)}\left(b_1^{(1)}+b_2^{(1)}\right)\right] = \mathrm{tr}\left[b_2^{(1)}\left(a_1^{(1)}+a_2^{(1)}\right)\right] \tag{A.3}$$

$$\mathrm{tr}\left[a_1^{(2)}b_2^{(2)}\right] = \mathrm{tr}\left[a_2^{(2)}b_1^{(2)}\right] \tag{A.4}$$

$$\mathrm{tr}\left[a_1^{(2)}\left(b_1^{(2)}+b_2^{(2)}\right)\right] = \mathrm{tr}\left[b_1^{(2)}\left(a_1^{(2)}+a_2^{(2)}\right)\right] \tag{A.5}$$

$$\mathrm{tr}\left[a_2^{(2)}\left(b_1^{(2)}+b_2^{(2)}\right)\right] = \mathrm{tr}\left[b_2^{(2)}\left(a_1^{(2)}+a_2^{(2)}\right)\right] \tag{A.6}$$

$$\mathrm{tr}\left[\left(a_1^{(1)}+a_1^{(2)}\right)\left(b_2^{(1)}+b_2^{(2)}\right)\right] = \mathrm{tr}\left[\left(a_2^{(1)}+a_2^{(2)}\right)\left(b_1^{(1)}+b_1^{(2)}\right)\right] \tag{A.7}$$

$$\mathrm{tr}\left[\left(a_1^{(1)}+a_1^{(2)}\right)\left(b_3^{(1)}+b_3^{(2)}\right)\right] = \mathrm{tr}\left[\left(a_3^{(1)}+a_3^{(2)}\right)\left(b_1^{(1)}+b_1^{(2)}\right)\right] \tag{A.8}$$

$$\mathrm{tr}\left[\left(a_2^{(1)}+a_2^{(2)}\right)\left(b_3^{(1)}+b_3^{(2)}\right)\right] = \mathrm{tr}\left[\left(a_3^{(1)}+a_3^{(2)}\right)\left(b_2^{(1)}+b_2^{(2)}\right)\right] \tag{A.9}$$

$$\mathrm{tr}\left[\left(a_1^{(2)}+a_2^{(1)}\right)\left(b_2^{(2)}+b_3^{(1)}\right)\right] = \mathrm{tr}\left[\left(a_2^{(2)}+a_3^{(1)}\right)\left(b_1^{(2)}+b_2^{(1)}\right)\right] \tag{A.10}$$

$$\mathrm{tr}\left[\left(a_1^{(2)}+a_2^{(1)}\right)\left(b_3^{(2)}+b_1^{(1)}\right)\right] = \mathrm{tr}\left[\left(a_3^{(2)}+a_1^{(1)}\right)\left(b_1^{(2)}+b_2^{(1)}\right)\right] \tag{A.11}$$

$$\mathrm{tr}\left[\left(a_2^{(2)}+a_3^{(1)}\right)\left(b_3^{(2)}+b_1^{(1)}\right)\right] = \mathrm{tr}\left[\left(a_3^{(2)}+a_1^{(1)}\right)\left(b_2^{(2)}+b_3^{(1)}\right)\right] \tag{A.12}$$

$$\mathrm{tr}\left[\left(a_1^{(2)}+a_3^{(1)}\right)\left(b_2^{(2)}+b_1^{(1)}\right)\right] = \mathrm{tr}\left[\left(a_2^{(2)}+a_1^{(1)}\right)\left(b_1^{(2)}+b_3^{(1)}\right)\right] \tag{A.13}$$

$$\mathrm{tr}\left[\left(a_1^{(2)}+a_3^{(1)}\right)\left(b_3^{(2)}+b_2^{(1)}\right)\right] = \mathrm{tr}\left[\left(a_3^{(2)}+a_2^{(1)}\right)\left(b_1^{(2)}+b_3^{(1)}\right)\right] \tag{A.14}$$

$$\mathrm{tr}\left[\left(a_2^{(2)}+a_1^{(1)}\right)\left(b_3^{(2)}+b_2^{(1)}\right)\right] = \mathrm{tr}\left[\left(a_3^{(2)}+a_2^{(1)}\right)\left(b_2^{(1)}+b_1^{(1)}\right)\right]. \tag{A.15}$$

We proved that these 15 equations are not independent, which means that if we start we the Eqs. (A.1), (A.4), (A.7), and (A.10) we obtain all the other 11 equations. The set of four equations is the minimal one, that generates the other equations.

References

[1] Schwinger J 1960 *Proc. Natl. Acad. Sci. USA* **46** 570.
[2] Wootters W K and Fields B D (1989) *Ann. Phys. (N.Y.)* **191** 363.
[3] Bechmann-Pasquinucci H and Peres A (2000) *Phys. Rev. Lett.* **85** 3313.
[4] Wootters W K (1987) *Ann. Phys. (N.Y.)* **176** 1.
[5] Koniorczyk M, Buzek V, and Janszky J (2001) *Phys. Rev. A* **64** 034301.
[6] Paz J P, Roncaglia A J, and Saraceno M (2005) *Phys. Rev. A* **72** 012309.
[7] Klimov A B, Munoz C, Fernandez A, and Saavedra C 2008 *Phys. Rev. A* **77** 060303 (R).
[8] Ivanovic I D (1981) *J. Phys. A* **14** 3241.
[9] Calderbank A R, Cameron P J, Kantor W M, and Seidel J J (1997) *Proc. London Math. Soc.* **75** 436.
[10] Bandyopadhyay S, Boykin P O, Roychowdhury V, and Vatan V 2002 *Algorithmica* **34** 512.
[11] Romeo J L, Björk G, Klimov A B, and Sánchez-Soto L L 2005 *Phys. Rev. A* **72** 062310.
[12] Wootters W K 2004 *IBM J. Res. Dev.* **48** 99.
[13] Gibbons K S, Hoffman M J, and Wootters W K 2004 *Phys. Rev. A* **70**, 062101.
[14] Klimov A B, Romero J L, Björk G, Sánchez-Soto L L 2009 *Ann. Phys. (N.Y.)* **324** 53.

Advanced many-body and statistical methods in mesoscopic systems IOP Publishing
Journal of Physics: Conference Series **338** (2012) 012009 doi:10.1088/1742-6596/338/1/012009

Non-local electron transport and Coulomb effects in three-terminal metallic conductors

D.S. Golubev[1] **and A.D. Zaikin**[1,2]

[1] Institut für Nanotechnologie, Karlsruher Institut für Technologie (KIT), 76021 Karlsruhe, Germany

[2]I.E. Tamm Department of Theoretical Physics, P.N. Lebedev Physics Institute, 119991 Moscow, Russia

Abstract. We demonstrate a close relation between Coulomb effects in non-local electron transport and non-local shot noise in three-terminal metallic conductors. Provided the whole structure is normal, cross-correlations in shot noise are negative and Coulomb interaction tends to suppress both local and non-local conductances of the system. The behavior of normal-superconducting-normal structures at subgap energies is entirely different. In the tunneling limit non-local differential conductance of such systems is found to have an S-like shape and it can turn negative at non-zero bias. At high transmissions crossed Andreev reflection yields positive noise cross-correlations and Coulomb anti-blockade of non-local electron transport.

1. Introduction

It is well known that discreteness of electron charge stays behind a number of fundamentally important physical phenomena such as, e.g., shot noise in mesoscopic conductors [1] and Coulomb blockade of charge transfer in tunnel junctions [2]. Exactly for this reason these two seemingly different phenomena turn out to be closely related to each other: Coulomb blockade is stronger stronger in conductors with bigger shot noise [3, 4]. This fundamental relation was subsequently confirmed in experiments [5]. Subsequently a close relation between shot noise and Coulomb blockade was also identified in hybrid normal-superconducting (NS) structures [6], where doubling of elementary charge due to Andreev reflection becomes important at subgap energies.

Recently the same idea was extended [7] to non-local effects in electron transport across three-terminal normal-superconducting-normal (NSN) systems where entanglement between electrons in different normal terminals can be realized. Non-local electron transport in such systems is determined by an interplay between elastic cotunneling (EC) and crossed Andreev reflection (CAR) and was recently investigated both experimentally [8, 9, 10, 11, 12] and theoretically [13, 14, 15]. While non-interacting theory predicts that CAR never dominates over direct electron transfer, both positive and negative non-local signals have been detected in a number of experiments [8, 9, 11, 12]. Theoretically it was argued that CAR could prevail over EC in the presence of Coulomb interactions [16] or an external ac field [17]. Negative non-local conductance was also predicted in interacting single-level quantum dots in-between normal and superconducting terminals [18].

In Ref. [7] we have already demonstrated that interaction effects in non-local transport and non-local shot noise in NSN systems are intimately related. This relation, however, turns out

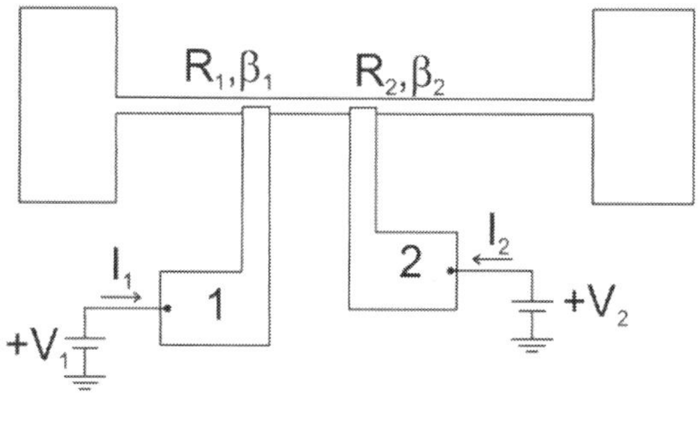

Figure 1. Schematics of the system under consideration.

to be much more complicated than in the local case [3, 4, 6]. The main reason for that is positive cross-correlations in shot noise which may occur in normal-superconducting hybrids [1, 19]. In NSN structures such positive noise cross-correlations were demonstrated theoretically [20, 21, 7] and experimentally [22]. Note that this feature is specific to superconducting systems and is totally absent in normal ones where cross-correlations of fluctuating currents are known to always be negative [1]. Hence, it would be interesting to extend our theory of non-local electron transport in NSN systems in the presence of electron-electron [7] to normal conductors and compare the corresponding results derived for superconducting and normal structures. This is the main goal of the present paper.

Our paper is organized as follows. In Sec. 2 we define our model and outline the key steps of the derivation of the effective action for our three-terminal metallic structure. In Sec. 3 we re-formulate our results in terms of equivalent Langevin equations describing real time dynamics of fluctuating voltages and currents and demonstrate how the expressions for both local and non-local conductances in the presence of Coulomb interaction can be related to the corresponding shot noise correlators. Sec. 4 and 5 are devoted to the analysis of the effect of electron-electron interactions on the conductance matrix respectively for superconducting and normal central electrodes. A brief summary of our main observations is presented in Sec. 6.

2. The model and effective action

Let us consider a thin wire, which can be either normal or superconducting and is connected to two normal metallic leads via two barriers. Quite generally, these barriers are characterized by two sets of channel transmissions $T_k^{(1)}, T_k^{(2)}$. For the sake of simplicity here we assume that the first junction has N_1 conducting channels with the same transmission T_1, while the second one — N_2 channels with transmission T_2. Provided the wire is in the superconducting state, its energy spectrum has the gap Δ. This gap, of course, equals to zero should the wire be in the normal state. In the latter case (i.e. the wire remains normal) it is convenient to define the normal state conductances and the Fano factors of the barriers as follows

$$G_{rr}^{N} = \frac{e^2}{\pi} N_r T_r, \quad \beta_r^N = 1 - T_r, \quad r = 1, 2. \tag{1}$$

If the wire is superconducting, we define Andreev conductances and Andreev Fano factors, which determine the transport properties across NS barriers in the subgap regime $eV_r, T \ll \Delta$:

$$G_{rr}^{NS} = \frac{2e^2}{\pi} N_r \tau_r, \quad \beta_r^{NS} = 1 - \tau_r, \tag{2}$$

where $\tau_r = T_r^2/(2 - T_r)^2$ represent effective Andreev transmissions of NS barriers. Both barriers are supposed to have capacitances C_1, C_2 and the leads 1 and 2 are characterized by large Ohmic conductances $G_1^{\rm sh}$ and $G_2^{\rm sh}$. We also define the corresponding dimensionless conductances of the electromagnetic environment

$$g_r^{NS} = \frac{2\pi}{e^2} G_r^{\rm sh}. \tag{3}$$

The corresponding dimensionless conductances in case of a normal wire, g_r^N, may differ from g_r^{NS} due to the additional contribution from wire resistance.

Weak electromagnetic coupling between two NS barriers (e.g. via modes propagating in the superconductor [16]) will be disregarded. Our main goal here is to evaluate the currents $I_1(V_1, V_2)$ and $I_2(V_1, V_2)$ across barriers 1 and 2. Below we will will consider two most interesting limits: (i) the subgap regime $T, eV_1, eV_2 \ll \Delta$ in the case of a superconducting wire, and (ii) the regime of high energies or a normal wire.

The general Hamiltonian of our system can be expressed in the form [7]

$$H = H_1 + H_2 + H_{\rm wire} + H_{T,1} + H_{T,2}, \tag{4}$$

where

$$H_r = \sum_{\alpha=\uparrow,\downarrow} \int d\boldsymbol{x} \hat\psi^\dagger_{r,\alpha} \left(-\frac{\nabla^2}{2m} - \mu\right) \hat\psi_{r,\alpha}, \quad r = 1, 2,$$

are the Hamiltonians of the normal metals, m is electron mass, μ is the chemical potential,

$$H_{\rm wire} = \int d\boldsymbol{x} \left[\sum_\alpha \hat\chi^\dagger_\alpha \left(-\frac{\nabla^2}{2m} - \mu\right) \hat\chi_\alpha + \Delta \hat\chi^\dagger_\uparrow \hat\chi^\dagger_\downarrow + \Delta^* \hat\chi_\downarrow \hat\chi_\uparrow \right]$$

is the Hamiltonian of a superconducting wire and

$$H_{T,r} = \mathcal{A}_r \sum_{\alpha,\beta=\uparrow,\downarrow} [t_r\, e^{i\varphi_r}\, \hat\psi^\dagger_\beta \hat\chi_{r,\alpha} + t_r^*\, e^{-i\varphi_r}\, \hat\chi^\dagger_{r,\alpha} \hat\psi_\beta] \tag{5}$$

are tunneling Hamiltonians describing transfer of electrons across the contacts with area $\mathcal{A}_r$ and tunneling amplitude t_r. As we already indicated above, we will assume that both barriers are uniform implying that all $N_r = k_F^2 \mathcal{A}_r / 4\pi$ conducting channels in the r-th barrier are characterized by equal transmission values

$$T_r = 4\pi^2 \nu_r \nu_{\rm wire} |t_r|^2 / \left(1 + \pi^2 \nu_r \nu_{\rm wire} |t_r|^2\right)^2, \tag{6}$$

where ν_j (j = 1, 2, wire) is the density of states in the corresponding terminal. Finally, we note that fluctuating phases φ_r introduced in Eq. (5) are linked to the voltage drops across the barriers v_r by means of the standard relation $\dot\varphi_r = ev_r$ and are treated as quantum operators.

We will proceed in a standard manner and eliminate fermionic variables expressing the kernel J of the Keldysh evolution operator via path integral over the phase fields [2]

$$J = \int \prod_{r=1,2} \mathcal{D}\varphi_r^F \mathcal{D}\varphi_r^B \exp(iS_{\rm env}[\varphi_r^F, \varphi_r^R] + iS_T[\varphi_r^F, \varphi_r^R]), \tag{7}$$

where φ_r^F and φ_r^B are fluctuating phases defined respectively on the forward and backward branches of the Keldysh contour, $S_{\rm env}$ is the action of electromagnetic environment and the term iS_T accounts for electron transfer between the terminals. In the case of linear Ohmic environment considered here one has [2]

$$iS_{\rm env} = \sum_{r=1,2} \left[i \int dt \frac{(eV_r - \dot\varphi_r)(-C_r \dot\varphi_r^- + G_r^{\rm sh} \varphi_r^-)}{e^2} - \frac{G_r^{\rm sh}}{2e^2} \int dt dt' \varphi_r^-(t) M(t-t') \varphi_r^-(t') \right], \tag{8}$$

where $\varphi_r = (\varphi_r^F + \varphi_r^B)/2$, $\varphi_r^- = \varphi_r^F - \varphi_r^B$ and

$$M(t) = \int \frac{d\omega}{2\pi} e^{i\omega t} \omega \coth \frac{\omega}{2T} = -\frac{\pi T^2}{\sinh^2(\pi T t)}.$$

The term iS_T is derived from the tunnel Hamiltonians (5) and reads

$$iS_T = \operatorname{tr} \ln \mathcal{G}^{-1}, \quad \mathcal{G}^{-1} = \begin{pmatrix} \check{G}_1^{-1} & \check{t}_1 & 0 \\ \check{t}_1^\dagger & \check{G}_{\rm wire}^{-1} & \check{t}_2 \\ 0 & \check{t}_2^\dagger & \check{G}_2^{-1} \end{pmatrix}, \tag{9}$$

where 4×4 matrices $\check{G}_j^{-1}$ represent the inverse Keldysh Green functions of isolated normal leads ($j = 1, 2$) and the wire, and $\check{t}_r$ is diagonal 4×4 matrix in the Nambu - Keldysh space

$$\check{t}_r = \begin{pmatrix} -t_r e^{-i\varphi_r^F} & 0 & 0 & 0 \\ 0 & t_r e^{-i\varphi_r^B} & 0 & 0 \\ 0 & 0 & t_r e^{i\varphi_r^F} & 0 \\ 0 & 0 & 0 & -t_r e^{i\varphi_r^B} \end{pmatrix}. \tag{10}$$

After some exact manipulations we obtain

$$iS_T = \operatorname{tr} \ln [\check{1} - \check{t}_1^\dagger \check{G}_1 \check{t}_1 \check{G}_{\rm wire} - \check{t}_2^\dagger \check{G}_2 \check{t}_2 \check{G}_{\rm wire}]. \tag{11}$$

While the expression (11) for the action remains formally exact it is still too complicated to be directly employed in our calculations. In order to proceed we will make several additional steps which yield significant simplifications. First, we restrict ourselves to the limit of high conductances

$$g_r^{NS},\ g_r^N \gg 1, \tag{12}$$

in which case phase fluctuations are weak and it suffices to expand the action (11) to the second order in φ_r^-, cf., e.g., [3, 6, 23]. Next we assume that the resistance of the wire segment between the junctions, r_{12}, is small as compared to the junction resistances, $r_{12} \ll 1/G_{rr}^N$. In this case one can additionally expand the action in powers of the wire Green function connecting the two junctions. In case of a superconducting wire and at $T, eV_r \ll \Delta$ the corresponding calculation was elaborated in Ref. [7]. In this case the effective action of our system can be cast to the form

$$iS_T = iS_{11} + iS_{22} + iS_{12}, \tag{13}$$

where the contributions iS_{11} and iS_{12} read

$$iS_{11} = -i\frac{G_{11}}{e^2} \int dt \dot{\varphi}_1 \varphi_1^- - \int dt dt' \frac{\varphi_1^-(t)\, \tilde{\mathcal{S}}_{11}^{tt'}\, \varphi_1^-(t')}{2e^2}, \tag{14}$$

$$iS_{12} = i\frac{G_{12}}{e^2} \int dt\, (\dot{\varphi}_1 \varphi_2^- + \dot{\varphi}_2 \varphi_1^-) - \int dt dt' \frac{\varphi_1^-(t)\, \tilde{\mathcal{S}}_{12}^{tt'}\, \varphi_2^-(t')}{e^2} \tag{15}$$

while the term iS_{22} is obtained by interchanging the indices $1 \leftrightarrow 2$ in Eq. (14). The functions $\tilde{\mathcal{S}}_{rl}^{tt'}$ are defined as

$$\begin{aligned} \tilde{\mathcal{S}}_{11}^{tt'} &= G_{11}^{NS} M(t-t')(1 - \beta_1 + \beta_1 \cos[2\varphi_1^{tt'}]) + 2G_{12}^{NS} M(t-t')(\alpha_1 - \eta_1 \cos[2\varphi_1^{tt'}]) \\ &\quad + (G_{12}^{NS}/2) M(t-t')(\kappa_1^+ \cos[\varphi_1^{tt'} + \varphi_2^{tt'}] + \kappa_1^- \cos[\varphi_1^{tt'} - \varphi_2^{tt'}]), \qquad (16) \\ \tilde{\mathcal{S}}_{12}^{tt'} &= -G_{12}^{NS} M(t-t')(1 - \beta_1 + \beta_1 \cos[2\varphi_1^{tt'}]) - G_{12}^{NS} M(t-t')(1 - \beta_2 + \beta_2 \cos[2\varphi_2^{tt'}]) \\ &\quad + (G_{12}^{NS}/2) M(t-t')(\gamma_+ \cos[\varphi_1^{tt'} + \varphi_2^{tt'}] - \gamma_- \cos[\varphi_1^{tt'} - \varphi_2^{tt'}]). \qquad (17) \end{aligned}$$

Here we denoted $\varphi_r^{tt'} = \varphi_r(t) - \varphi_r(t')$. Other parameters entering in Eqs. (16) and (17) read

$$\begin{aligned} &\alpha_r = \tau_r(1-2\tau_r)/\sqrt{\tau_1\tau_2}, && \eta_r = 2\tau_r(1-\tau_r)/\sqrt{\tau_1\tau_2}, \\ &\kappa_r^{\pm} = \pm(4\tau_r - 3) + 1/\sqrt{\tau_1\tau_2} \ \ (r=1,2), && \gamma_{\pm} = \pm 1 + (1 - 2\tau_1 - 2\tau_2 + 4\tau_1\tau_2)/\sqrt{\tau_1\tau_2}, \end{aligned} \tag{18}$$

while zero bias non-local conductance has the form

$$G_{12}^{NS} = \frac{G_{11}^{NS} G_{22}^{NS}}{2e^2\nu_{\rm wire}} D(2i\Delta, \boldsymbol{r}_1, \boldsymbol{r}_2). \tag{19}$$

Here $D(\omega, \boldsymbol{r}, \boldsymbol{r}')$ is the diffuson, which is defined as a solution of the following diffusion equation

$$\left(-i\omega - D\nabla_{\boldsymbol{r}}^2\right) D(\omega, \boldsymbol{r}, \boldsymbol{r}') = \delta(\boldsymbol{r} - \boldsymbol{r}'). \tag{20}$$

In a simple quasi-one-dimensional geometry of Fig. 1 one finds [15] $G_{12}^{NS} = G_{11}^{NS} G_{22}^{NS} R_\xi e^{-d/\xi}/2$, where d is the distance between two NS barriers and R_ξ is the resistance of the piece of the wire with the length equal to the superconducting coherence length $\xi = \sqrt{D/\Delta}$. It is important to emphasize that *all* order terms in t_r are fully accounted for in Eqs. (14)-(18), i.e. the action applies for arbitrary transmission values $T_{1,2}$ (or $\tau_{1,2}$) ranging from zero to one.

The calculation in the normal case goes along the same lines, however the resulting effective action turns out to be somewhat more complicated because of retardation effects related to diffusion of electrons between the barriers. For the sake of simplicity here we avoid presenting an explicit form of the effective action in the normal case. Rather we will proceed directly to the final expressions for non-local currents which follow from this action. The corresponding expressions will be presented in Sec. 5.

3. Langevin equations and interaction correction to the current

It is well known that the quadratic in $\varphi_{1,2}^-$ effective action can be exactly rewritten in terms of the corresponding Langevin equations [23, 24, 25] which describe the current balance in our system. In case of a superconducting wire these equations read [7]

$$\begin{aligned} C_1\dot{v}_1 + (G_1^{\rm sh} + G_{11}^{NS})v_1 - G_{12}^{NS} v_2 = G_1^{\rm sh} V_1 + \xi_1^{\rm sh} + \xi_1, \\ C_2\dot{v}_2 + (G_2^{\rm sh} + G_{22}^{NS})v_2 - G_{12}^{NS} v_1 = G_2^{\rm sh} V_2 + \xi_2^{\rm sh} + \xi_2. \end{aligned} \tag{21}$$

Here v_r are the fluctuating volatge drops across the junctions, $\xi_r^{\rm sh}$ are stochastic variables with pair correlators

$$\langle \xi_r^{\rm sh}(t)\xi_r^{\rm sh}(t')\rangle = G_r^{\rm sh} M(t-t') \tag{22}$$

describing Gaussian current noise in the shunt resistors, while the variables ξ_r with the correlators

$$\langle \xi_r(t)\xi_l(t')\rangle = \tilde{S}_{rl}^{tt'} \tag{23}$$

describe shot noise in NS barriers.

In case of normal wire the Langevin equation looks similar. One needs to replace G_{rr}^{NS} by G_{rr}^{N}. Besides that, the non-local conductance exhibits retardation effects, i.e. one should replace, for example, $G_{12}^{NS} v_2 \to \int_{-\infty}^{t} dt' G_{12}^{N}(t-t')v_2(t')$. Finally, the correlator of the noises also differs from (23). As we already pointed out, the corresponding expressions are rather cumbersome and for this reason are not presented here.

Let us evaluate the current I_1 across the first barrier. Solving Eqs. (21) perturbatively in $1/g_r \ll 1$, in the lowest non-trivial order in this parameter we get

$$I_1 = G_{11}^{NS} V_1 - G_{12}^{NS} V_2 - \langle \xi_1 \rangle. \tag{24}$$

Here the average $\langle\xi_1\rangle$ does not vanish since according to Eqs. (16), (17) the noise ξ_1 depends on the phases $\varphi_{1,2}$, which, in turn depend on $\xi_{1,2}$ by virtue of Eqs. (21). Hence, we obtain

$$\langle\xi_1\rangle = \langle\delta\varphi_1\, \partial\xi_1/\partial\varphi_1\rangle + \langle\delta\varphi_2\, \partial\xi_1/\partial\varphi_2\rangle, \tag{25}$$

where the phase fluctuations $\delta\varphi_r$ are determined from Eqs. (21). We obtain

$$\delta\varphi_r(t) = e\int_{-\infty}^{t} dt' \frac{1-e^{-(t-t')/\tau_{RC}}}{G_r^{\rm sh}}\xi_r(t'). \tag{26}$$

Here we assumed $G_{12} \ll G_{rr} \ll G_r^{\rm sh}$ and introduced the $RC-$time $\tau_{RC} = C_r/G_r^{\rm sh}$. Substituting this expression into Eq. (25) we get

$$\frac{\langle\xi_1\rangle}{e} = \sum_{r=1,2}\int_{-\infty}^{t} dt' \frac{1-e^{-\frac{t-t'}{\tau_{RC}}}}{G_r^{\rm sh}} \left.\frac{\partial\langle\xi_1(t)\xi_r(t')\rangle}{\partial\varphi_r(t)}\right|_{\varphi_r=eV_rt}. \tag{27}$$

Below we will make use of the above expressions and directly evaluate the non-local currents in both interesting limits of superconducting and normal wires.

4. Superconducting wire

We begin with a superconducting wire. In this case employing Eqs. (16), (17) and performing the time integral in Eq. (27) we arrive at the following expression for the current through the first barrier

$$\begin{aligned} I_1 &= G_{11}V_1 - G_{12}V_2 - \frac{2G_{11}\beta_1 - 4G_{12}\eta_1}{g_1^{NS}}F_0(2V_1) + \frac{2G_{12}\beta_2}{g_2^{NS}}F_0(2V_2) \\ &- G_{12}\left(\frac{\kappa_1^+}{g_1^{NS}} + \frac{\gamma_+}{g_2^{NS}}\right)F_0(V_1+V_2) - G_{12}\left(\frac{\kappa_1^-}{g_1^{NS}} + \frac{\gamma_-}{g_2^{NS}}\right)F_0(V_1-V_2), \end{aligned} \tag{28}$$

where

$$F_0(V) = \mathrm{Re}\left[-V\Psi\left(1+i\frac{eV}{2\pi T}\right) + \left(V - \frac{i}{e\tau_{RC}}\right)\Psi\left(1+\frac{1}{2\pi T\tau_{RC}} + i\frac{eV}{2\pi T}\right)\right]. \tag{29}$$

Taking the derivatives of the current (28) with respect to V_1 and V_2 we define respectively local and non-local differential conductances of the first barrier. They read

$$\begin{aligned} \frac{\partial I_1}{\partial V_1} &= G_{11} - \frac{4G_{11}\beta_1 - 8G_{12}\eta_1}{g_1^{NS}}F(2V_1) \\ &- G_{12}\left(\frac{\kappa_1^+}{g_1^{NS}} + \frac{\gamma_+}{g_2^{NS}}\right)F(V_1+V_2) - G_{12}\left(\frac{\kappa_1^-}{g_1^{NS}} + \frac{\gamma_-}{g_2^{NS}}\right)F(V_1-V_2), \end{aligned} \tag{30}$$

and

$$\begin{aligned} \frac{\partial I_1}{\partial V_2} &= -G_{12}\left[1 - \frac{4\beta_2}{g_2^{NS}}F(2V_2)\right] \\ &- G_{12}\left(\frac{\kappa_1^+}{g_1^{NS}} + \frac{\gamma_+}{g_2^{NS}}\right)F(V_1+V_2) + G_{12}\left(\frac{\kappa_1^-}{g_1^{NS}} + \frac{\gamma_-}{g_2^{NS}}\right)F(V_1-V_2), \end{aligned} \tag{31}$$

where

$$F(V) = \mathrm{Re}\left[\Psi\left(1+\frac{1}{2\pi T\tau_{RC}}+i\frac{eV}{2\pi T}\right)+\left(\frac{1}{2\pi T\tau_{RC}}+i\frac{eV}{2\pi T}\right)\Psi'\left(1+\frac{1}{2\pi T\tau_{RC}}+i\frac{eV}{2\pi T}\right)\right.$$
$$\left.-\Psi\left(1+i\frac{eV}{2\pi T}\right)-i\frac{eV}{2\pi T}\Psi'\left(1+i\frac{eV}{2\pi T}\right)\right], \quad (32)$$

and $\Psi(x)$ is the digamma function. The conductances (30), (31) are displayed in Fig. 2.

We observe that both these conductances are affected by Coulomb interaction which yields non-trivial corrections to the corresponding non-interacting expressions. In the interaction correction to the local conductance in Eq. (30) we recover the Coulomb blockade term [6] $\propto \beta_1$ and, in addition, three non-local contributions. The first of them $\propto \eta_1$ enhances the conductance, while the second one $\propto \delta G_+$ provides additional Coulomb suppression of $\partial I_1/\partial V_1$. The last term $\propto \delta G_-$ can be both positive (at $\tau_{1,2} \ll 1$) and negative (at bigger $\tau_{1,2}$) implying the tendency to Coulomb *anti-blockade* in the latter case. The first term $\propto \beta_2$ in Eq. (31) for the non-local conductance has an opposite sign as compared to G_{12} (thus implying Coulomb blockade), while the second one $\propto \delta G_+$ yields Coulomb *anti-blockade*. Finally, the third term $\propto \delta G_-$ tends to suppress or enhance the absolute value of the non-local conductance respectively for $\delta G_- > 0$ and $\delta G_- < 0$.

As we already argued [7] these non-trivial features of different Coulomb corrections are directly related to the corresponding contributions to shot noise. In particular, negatively and positively correlated noise terms are associated respectively with Coulomb blockade and anti-blockade terms in the above expressions for both local and non-local conductances.

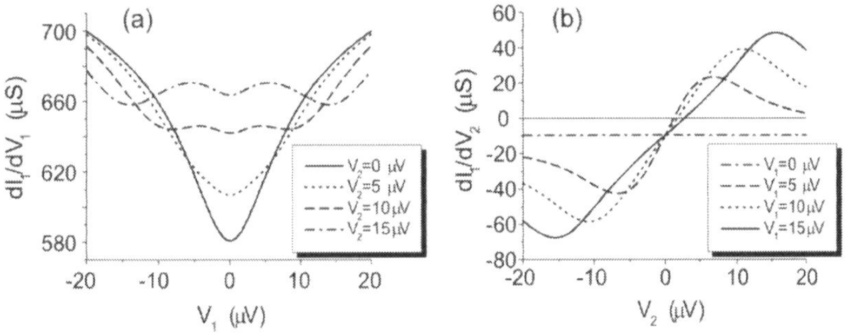

Figure 2. Local (a) and non-local (b) differential conductances of a sample with superconducting wire, which are defined, respectively, in Eqs. (30) and (31). The parameters of the system are: $T = 20$ mK, $G_{11}^{NS} = 1$ mS, $G_{22}^{NS} = 3.2$ mS, $G_{12}^{NS} = 10$ μS, $g_1^{NS} = g_2^{NS} = 516$, $\tau_{RC} = 10^{-12}$ s. The transmissions of the junctions are small, $\mathcal{T}_1 = 0.063$, $\mathcal{T}_2 = 0.11$, the corresponding numbers of channels are, on the contrary, large, $N_1 = N_2 = 6.1 \times 10^6$.

5. Normal wire

Let us now turn to the case of a normal metallic wire. In this case the current I_1 takes the form

$$I_1 = G_{11}^N V_1 - G_{12}^N V_2 - \frac{2G_{11}^N\beta_1^N}{g_1^N}F_0(V_1) + \frac{2G_{12}^N\beta_2^N}{g_2^N}F_0(V_2) - \frac{2G_{12}^N\beta_1^N}{g_1^N}F_0(V_1-V_2), \quad (33)$$

where the normal state non-local conductance is defined as follows

$$G_{12}^N = \frac{D(0,\boldsymbol{r}_1,\boldsymbol{r}_2)}{2e^2\nu_{\rm wire}}G_{11}^N G_{22}^N. \quad (34)$$

Accordingly the differential resistances read

$$\frac{\partial I_1}{\partial V_1} = G_{11}^N - \frac{2G_{11}^N\beta_1^N}{g_1^N}F(V_1) - \frac{2G_{12}^N\beta_1^N}{g_1^N}F(V_1 - V_2), \tag{35}$$

$$\frac{\partial I_1}{\partial V_2} = -G_{12}^N + \frac{2G_{12}^N\beta_2^N}{g_2^N}F(V_2) + \frac{2G_{12}^N\beta_1^N}{g_1^N}F(V_1 - V_2). \tag{36}$$

Here the functions $F_0(V)$ and $F(V)$ are again defined resin Eqs. (29) and (32).
The conductances (35) and (36) are depicted in Fig. 3.

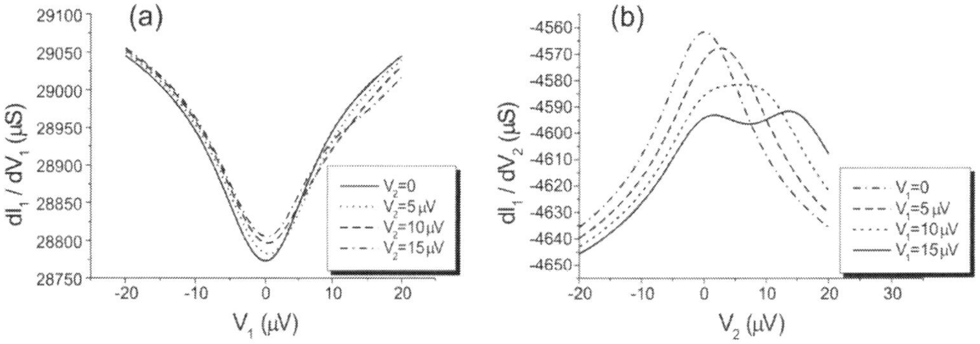

Figure 3. Local (a) and non-local (b) differential conductances of a sample with normal wire, which are defined, respectively, in Eqs. (35) and (36). The system parameters are the same as in Fig. 2, namely $T = 20$ mK, $g_1^N = g_2^N = 516$, $T_1 = 0.063$, $T_2 = 0.11$, $N_1 = N_2 = 6.1 \times 10^6$, $\tau_{RC} = 10^{-12}$ s. Accordingly, the normal conductances take the values $G_{11}^{NS} = 29.8$ mS, $G_{22}^{NS} = 52$ mS, $G_{12}^{NS} = 4.83$ μS.

Comparing these results with those obtained above for NSN systems we observe striking differences between them. The main difference is due to the fact that correlations in non-local shot noise in normal multi-terminal systems are always negative, while positive cross-correlations may occur in structures involving superconductors. Accordingly, in the normal case Coulomb interaction tends to always suppress non-local conductance, and the corresponding interaction correction in Eq. (36) depends on the voltage difference $V_1 - V_2$. In contrast,the expression for non-local conductance in three-terminal NSN system (31) contains Coulomb terms which depend both on $V_1 - V_2$ and $V_1 + V_2$ originating respectively from negative and positive cross-correlations in shot noise. As we already discussed, these terms describe respectively Coulomb blockade and Coulomb anti-blockade of the non-local conductance in the superconducting case.

6. Discussion and summary

In this paper we developed a theory elucidating a non-trivial physical relation between shot noise and Coulomb effects in non-local electron transport in three-terminal metallic structures. We evaluated non-local current-current correlators in such systems at arbitrary interface transmissions and arbitrary frequencies and directly related them to Coulomb effects in non-local electron transport.

One of the important features of NSN systems under consideration is that in the tunneling limit almost no effect of Coulomb interaction on non-local conductance is expected if one of the applied voltages, V_1 or V_2, equals to zero. This effect is directly related to the cancellation between EC and CAR contributions to shot noise in the corresponding limit [20]. For nonzero V_1 and V_2 no such cancellation exists anymore and the non-local conductance $\partial I_1/\partial V_2$ approaches the S-like shape being enhanced at $V_1 \approx V_2$ and partially suppressed at $V_1 \approx -V_2$, see Fig. 2b. Both these features have a clear physical interpretation. Indeed, at $V_1 \approx -V_2$ negative cross-correlations due to EC dominate non-local shot noise leading to Coulomb blockade of non-local conductance while at $V_1 \approx V_2$ positive cross-correlations due to CAR prevail and Coulomb anti-blockade of non-local transport is observed. At higher interface transmissions only Coulomb anti-blockade of non-local conductance remains, which is again related to CAR-induced positive cross-correlations in shot noise.

Comparing these results with those obtained the normal case we observe striking differences. In particular, both local and non-local conductances always tend to be suppressed by Coulomb interaction and anti-blockade effects never occur in normal structures. For fully open normal barriers the Coulomb blockade disappears, while in the case of fully open NS barriers Coulomb anti-blockade sets in. The non-local Coulomb corrections to conductances depend on the voltage difference $V_1 - V_2$, not on their sum $V_1 + V_2$, unlike in the superconducting case. All these features are directly related to the observation that only negative cross-correlations in shot noise occur in normal multi-terminal conductors [1].

It is interesting to point out that S-like shaped non-local signal predicted here was indeed observed in experiments with NSN systems [12, 26]. A good agreement between our theory and the results [26] argues in favor of electron-electron interactions as a physical reason for the observed feature. Some of the features similar to those predicted here have also been observed in experiments [22]. It would be interesting to perform more experiments in systems under consideration both in superconducting and normal states and compare the results with our predictions. Extending experimental investigations to the normal case would hopefully allow for better characterization of the system parameters as well as for clearer demonstration of qualitative differences between normal and superconducting structures outlined above.

Finally, we would like to make one more remark. In some cases non-linearities in both local and non-local differential conductances caused by Coulomb interaction may be combined with the zero bias anomalies resulting from the proximity-enhanced electron interference in diffusive normal leads [27, 28, 29, 15]. In this paper we disregarded this effect for the sake of simplicity. In practice it implies that here we considered the system with weakly disordered or sufficiently thick normal leads and sufficiently resistive barriers. If needed, zero-bias anomaly effects [27, 28, 29, 15] can be included into our analysis in a straightforward manner.

References

[1] Ya.M. Blanter and M. Büttiker 2000 *Phys. Rep.* **336** 1
[2] G. Schön and A.D. Zaikin 1990 *Phys. Rep.* **198** 237
[3] D.S. Golubev and A.D. Zaikin 2001 *Phys. Rev. Lett.* **86** 4887
[4] A. Levy Yeyati, A. Levy Yeyati, A. Martin-Rodero, D. Esteve, and C. Urbina 2001 *Phys. Rev. Lett.* **87** 046802
[5] C. Altimiras, U. Gennser, A. Cavanna, D. Mailly, and F. Pierre 2007 *Phys. Rev. Lett.* **99** 256805
[6] A.V. Galaktionov and A.D. Zaikin 2009 *Phys. Rev. B* **80** 174527
[7] D.S. Golubev and A.D. Zaikin 2010 *Phys. Rev. B* **82** 134508
[8] D. Beckmann, H.B. Weber, and H. v. Löhneysen 2004 *Phys. Rev. Lett.* **93** 197003
[9] S. Russo, M. Kroug, T. M. Klapwijk, and A. F. Morpurgo 2005 *Phys. Rev. Lett.* **95** 027002
[10] P. Cadden-Zimansky and V. Chandrasekhar 2006 *Phys. Rev. Lett.* **97** 237003
[11] A. Kleine , A. Baumgartner , J. Trbovic and C. Schnenberger 2009 *EPL* **87** 27011; A. Kleine, A. Baumgartner, J. Trbovic, D.S. Golubev, A.D. Zaikin, and C. Schönenberger 2010 *Nanotechnology* **21** 274002
[12] J. Brauer, F. Hübler, M. Smetanin, D. Beckmann, and H. v. Löhneysen 2010 *Phys. Rev. B* **81** 024515
[13] G. Falci, D. Feinberg, and F.W.J. Hekking 2001 *EPL* **54** 255
[14] M.S. Kalenkov and A.D. Zaikin 2007 *Phys. Rev. B* **75** 172503; 2007 *ibid.* **76** 224506
[15] D.S. Golubev, M.S. Kalenkov, and A.D. Zaikin 2009 *Phys. Rev. Lett.* **103** 067006

[16] A. Levy Yeyati, F.S. Bergeret, A. Martin-Rodero, and T.M. Klapwijk 2007 *Nat. Phys.* **3** 455
[17] D.S. Golubev and A.D. Zaikin 2009 *EPL* **86** 37009
[18] D. Futterer, M. Governale, M.G. Pala, and J. König 2009 *Phys. Rev. B* **79** 054505
[19] M. P. Anantram and S. Datta 1996 *Phys. Rev. B* **53** 16390
[20] G. Bignon, M. Houzet, F. Pistolesi and F. W. J. Hekking 2004 *EPL* **67**, 110
[21] A. Freyn, M. Flöser, and R. Melin 2010 *Phys. Rev. B* **82** 014510
[22] J. Wei and V. Chandrasekhar 2010 *Nat. Phys.* **6** 494
[23] A. Schmid 1982 *J. Low Temp. Phys.* **49** 609
[24] U. Eckern, G. Schön, and V. Ambegaokar 1984 *Phys. Rev. B* **30** 6419
[25] D.S. Golubev and A.D. Zaikin 1992 *Phys. Rev. B* **46** 10903
[26] D. Beckmann, private communication.
[27] A.F. Volkov, A.V. Zaitsev, and T.M. Klapwijk 1993 *Physica C* **210** 21
[28] F.W.J. Hekking and Yu.V. Nazarov 1993 *Phys. Rev. Lett.* **71** 1625; 1994 *Phys. Rev. B* **49**, 6847
[29] A.D. Zaikin 1994 *Physica B* **203** 255

Dynamical decoupling of superconducting qubits

Jian Li

E-mail: jian.li@aalto.fi

G. S. Paraoanu

Low Temperature Laboratory, Aalto University, PO Box 15100, FI-00076 AALTO, Finland

Abstract. We show that two superconducting qubits interacting *via* a fixed transversal coupling can be decoupled by appropriately-designed microwave field excitations applied to each qubit. This technique is useful for removing the effects of spurious interactions in a quantum processor. We also simulate the case of a qubit coupled to a two-level system (TLS) present in the insulating layer of the Josephson junction of the qubit. Finally, we discuss the qubit-TLS problem in the context of dispersive measurements, where the qubit is coupled to a resonator.

1. Introduction

In the past years, fixed transversal couplings between two superconducting qubits have been extensively studied theoretically [1, 2, 3] and experimentally in systems comprising phase qubits [4], charge qubits [5], flux qubits [6], and in circuit QED systems [7]. Much of the motivation of these studies comes from the need of developing reliable techniques for modulating the coupling between qubits. This is required in order to produce in a controllable way CNOT quantum gates [8] - the basic building blocks of quantum algorithms.

2. Two qubits with fixed coupling

In this paper we consider a system of two transversely coupled qubits under microwave driving. For simplicity we take $\hbar = 1$. The Hamiltonian [6] can be written as

$$\begin{aligned} H &= -\frac{1}{2}\left(\Delta_1\sigma_1^z + \Delta_2\sigma_2^z\right) + J\sigma_1^x\sigma_2^x \\ &\quad + \Omega_1\cos(\omega_d t + \varphi_1)\sigma_1^x + \Omega_2\cos(\omega_d t + \varphi_2)\sigma_2^x, \end{aligned} \tag{1}$$

where Δ_j is the energy splitting (Larmor frequency) of qubit-j, J is the inter-qubit coupling strength, Ω_j and φ_j indicate the amplitude (Rabi frequency) and the relative phase of the driving fields at the frequency ω_d for qubit-j, respectively, and $\sigma_j^{x,y,z}$ are the qubit-j Pauli matrices in the undriven energy eigenbasis.

2.1. *Effective Hamiltonian*

In the simple case when the driving frequency $\omega_d = (\Delta_1 + \Delta_2)/2$, and the Rabi frequencies $\Omega_1 = \Omega_2 \equiv \Omega$, Eq. (1) can be rewritten, following the same procedures as those in Sec. III of

[3], as

$$H_{\rm eff} = \frac{J}{4}\eta\left[\sigma^x_{(1)}\sigma^x_{(2)} + \sigma^y_{(1)}\sigma^y_{(2)} + 2\sigma^z_{(1)}\sigma^z_{(2)}\right], \tag{2}$$

where $\sigma^{x,y,z}_{(j)}$ are the Pauli matrices of qubit-j in the driven energy eigenbasis, see [3]. We obtain the dimensionless qubit-qubit coupling strength η as

$$\eta = \frac{\Omega^2}{\Omega^2 + \Delta^2/4}\cos\phi, \tag{3}$$

where $\Delta = \Delta_1 - \Delta_2$ is the energy difference between the two qubits, and $\phi = \varphi_1 - \varphi_2$ indicates the phase difference between the two driving fields. It can be tuned by either Ω or ϕ independently, as shown in Fig. 1.

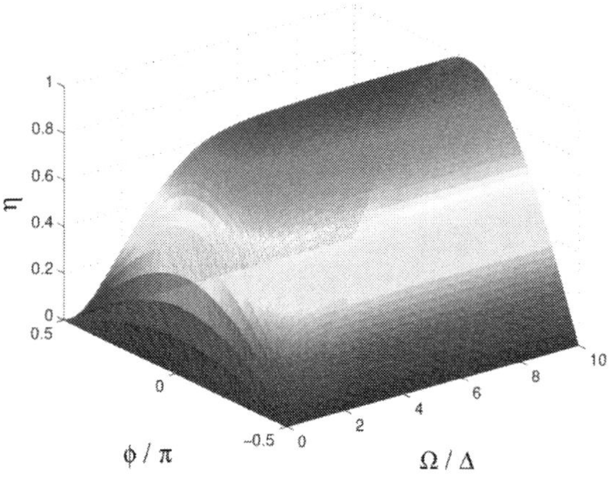

Figure 1. (Color online) The dimensionless coupling strength η as a function of Ω and ϕ.

The time evolution operator generated by $H_{\rm eff}$ is

$$\mathcal{U}_{\rm eff}(t) = \frac{1}{2}\begin{bmatrix} 2\exp(-iJ\eta t/2) & 0 & 0 & 0 \\ 0 & 1+\exp(iJ\eta t) & 1-\exp(iJ\eta t) & 0 \\ 0 & 1-\exp(iJ\eta t) & 1+\exp(iJ\eta t) & 0 \\ 0 & 0 & 0 & 2\exp(-iJ\eta t/2) \end{bmatrix}. \tag{4}$$

2.2. Entanglement

The entangling properties of a system of two qubits can be characterized by calculating an entanglement measure known as *concurrence* [9], which is defined as

$$\mathcal{C}(\psi) = |\langle\psi|\sigma^y_1 \otimes \sigma^y_2|\psi^*\rangle| \tag{5}$$

for a pure two-qubit state $|\psi\rangle$. Here $|\psi^*\rangle$ is the complex conjugate of $|\psi\rangle$. For a general two-qubit state $|\psi\rangle = c_{00}|00\rangle + c_{01}|01\rangle + c_{10}|10\rangle + c_{11}|11\rangle$, the concurrence is

$$\mathcal{C}(\psi) = 2|c_{00}c_{11} - c_{01}c_{10}| \leq 1, \tag{6}$$

where $|mn\rangle \equiv |m\rangle_1 \otimes |n\rangle_2$, and $|0\rangle_j$ ($|1\rangle_j$) represents the ground (excited) state of qubit-j.

If initially the two qubits are in their ground states $|00\rangle$, by using the time evolution operator Eq. (4) and within the approximation that the Rabi frequency is much larger than the energy difference between the two qubits, $\Omega \gg \Delta$, the time-dependent concurrence Eq. (5) can be put in the following approximate form [3]:

$$\mathcal{C}(t) \approx \left|e^{iJ\eta t} - 1\right|/2 = \left|\sin\left[\frac{2J\Omega^2 \cos\phi}{4\Omega^2 + \Delta^2}t\right]\right|, \tag{7}$$

as shown in Fig. 2 (b) below, which is a good approximation to the concurrence numerically calculated by using the original Hamiltonian Eq. (1), as shown in Fig. 2 (a).

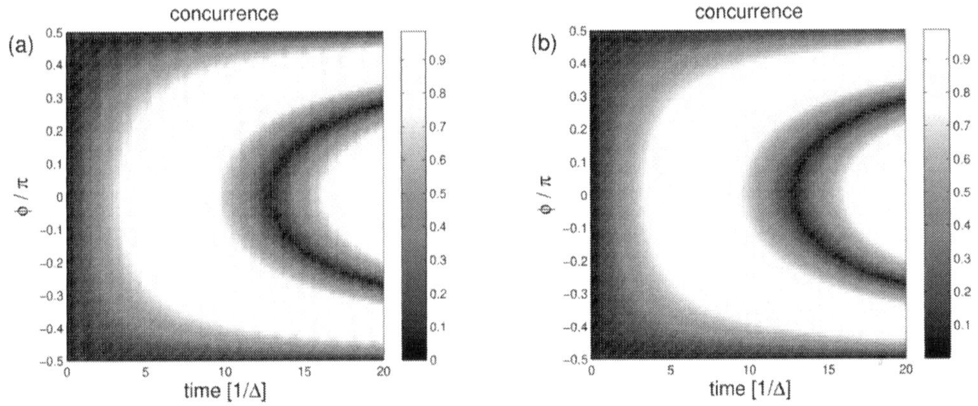

Figure 2. (Color online) The phase-dependent concurrence evaluated (a) by using the original Hamiltonian Eq. (1) and (b) by using the analytical expression Eq. (7). The parameters used in this plot are $\Omega = 5\Delta$ and $J = \Delta/2$.

3. Qubit-TLS systems

A lot of experimental progress has been made recently on phase qubits following the realization that the dielectric insulator forming the Josephson junction contains two-level system (TLS) defects [10, 11]. These defects have been shown to have decoherence times comparable to that of the qubit, thus they can be addressed coherently (*e.g.* by tuning the qubit on- and off-resonance with them). The form of the interaction Hamiltonian between the qubit and the TLS is of the type $\sigma^x\sigma^x$ in the case of phase qubits [11, 12]. The same type of coupling is obtained in the case of charge-based qubits from TLSs located on the island and in the case of flux qubits from pinning centers in the superconductors used for fabricating the qubits.

The interactions between a qubit and a TLS becomes relevant only when $\Delta \equiv |\omega_{\rm qb} - \omega_{\rm TLS}| \lesssim \kappa$, where κ denotes the coupling strength between them, $\omega_{\rm qb}$ and $\omega_{\rm TLS}$ are transition frequencies of the qubit and the TLS, respectively. By assuming that for each single qubit there is only one such TLS near it, the Hamiltonian for this qubit-TLS system is written as

$$H_{\rm qb-TLS} = -\frac{\omega_{\rm qb}}{2}\sigma^z - \frac{\omega_{\rm TLS}}{2}\tau^z + \kappa\sigma^x\tau^x, \tag{8}$$

with the TLS Pauli matrices $\tau^{x,y,z}$. To coherently control the qubit, we apply a transverse microwave field to it. Then the total Hamiltonian reads

$$H(t) = H_{\rm qb-TLS} + \Omega\cos(\omega_d t + \phi)\sigma^x. \tag{9}$$

In Fig. 3(a) we have plotted the fidelity $\mathcal{F}$ [13] of a π-rotation around the X-axis (see also Sec. V of [3]), by simply taking $\omega_d = \omega_{\rm qb} = \omega_{\rm TLS}$, $\phi = 0$, and Ω to be time independent (rectangular pulse). The fidelity of a unitary transformation U applied on the qubit between the initial pure state $|\psi_{\rm in}\rangle$ and the target state $\rho_{\rm out}$ is defined as $\mathcal{F}(U) = \overline{\langle\psi_{\rm in}|U^\dagger\rho_{\rm out}U|\psi_{\rm in}\rangle}$. Since we are not interested in the evolution of the TLS, when calculating the fidelity the output state $\rho_{\rm out}$ was obtained by tracing out the TLS degrees of freedom. Due to the qubit-TLS coupling, $\rho_{\rm out}$ is always a mixed state even if the input states is pure and unentangled. In other words, the entanglement between the qubit and the TLS results in decoherence for the qubit. As we can see from Fig. 3(a) and expected on physical reasons, when the driving amplitude is much larger than the qubit-TLS coupling, the fidelity loss due to the TLS is negligible.

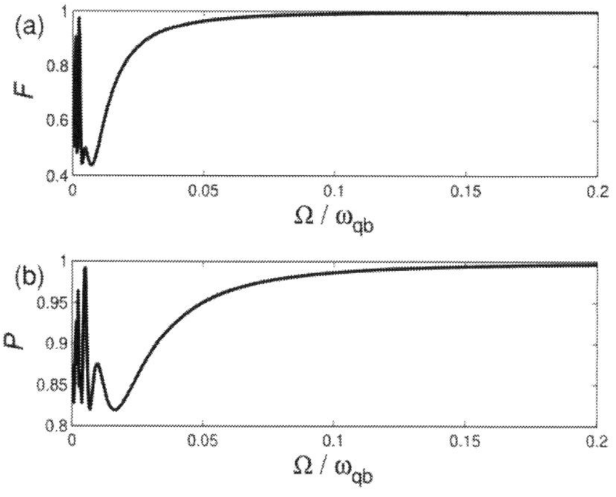

Figure 3. (a) The fidelity $\mathcal{F}$ of the π-pulse as a function of driving amplitude Ω. (b) The gate purity $\mathcal{P}$ as a function of the same variable. The qubit-TLS coupling is taken $\kappa = 5 \times 10^{-3}\omega_{\rm qb}$.

Since the state $\rho_{\rm out}$ is not pure, the gate purity $\mathcal{P} \equiv \overline{{\rm Tr}(\rho_{\rm out}^2)}$ [13] should also be considered. In Fig. 3(b) we show the numerical results of $\mathcal{P}$ for $\kappa/\omega_{\rm qb} = 5 \times 10^{-3}$ [14]. Again, as expected, for relatively large values of the the driving amplitude compared to the coupling κ, we find that the state becomes almost pure.

3.1. Qubit-TLS under dispersive measurement using a resonator

For the qubit dispersively coupled to the resonator, it is possible to decouple the qubit and the TLS by driving the resonator. We take the Jaynes-Cummings form for the system Hamiltonian

$$H = -\frac{\omega_{\rm qb}}{2}\sigma^z - \frac{\omega_{\rm TLS}}{2}\tau^z + \omega_{\rm r}a^\dagger a + g\left(\sigma^+ a + \sigma^- a^\dagger\right) + \kappa(\sigma^+\tau^- + \sigma^-\tau^+), \tag{10}$$

with $\sigma^\pm$ and $\tau^\pm$ the raising/lowering operators for the qubit and the TLS respectively, $\omega_{\rm r}$ is the resonance frequency of the resonator, and g is the coupling strength between the qubit and the cavity mode.

In the dispersive regime the Rabi frequency $\Omega \equiv 2g\sqrt{n}$ ($n = \langle a^\dagger a\rangle \gg 1$ indicates the number of photons) is much smaller than the detuning $\delta = \omega_{\rm r} - \omega_{\rm qb}$. To eliminate the qubit-photon coupling to leading order, we transform the Hamiltonian by using the Schrieffer-Wolff transformation operator $A = g(\sigma^+ a - \sigma^- a^\dagger)/\delta$. Expanded to second order in g/δ, the

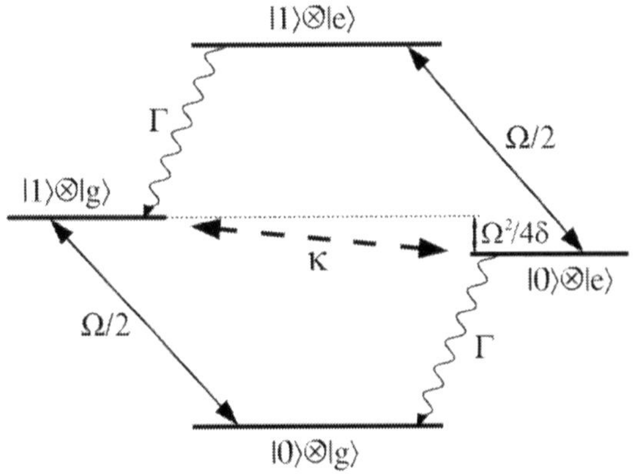

Figure 4. Energy level configuration for the qubit-TLS system. $|0\rangle$ ($|g\rangle$) and $|1\rangle$ ($|e\rangle$) denote the ground and the excited states of the qubit (TLS). The solid arrows indicate qubit transitions with a rate $\sim g\sqrt{n}$ due to the coherent driving. The spontaneous decays of the TLS is indicated by the wiggly arrows, and the decay rate Γ is assumed to be negligible compared with Ω. The dashed double arrow denotes transitions due to the qubit-TLS coupling κ.

Hamiltonian is approximately

$$\begin{aligned} H' &= e^{-A}He^{A} \approx H + [H, A] + \frac{1}{2}[[H, A], A] = \\ &= -\left[\frac{\omega_{\rm qb}}{2} - \frac{g^2}{\delta}(a^\dagger a + 1/2) - \frac{kg}{\delta}(\tau^+ a + \tau^- a^+)\right]\sigma^z - \frac{\omega_{\rm TLS}}{2}\tau^z + \\ &\quad + \omega_{\rm r} a^\dagger a + \kappa(\sigma^+\tau^- + \sigma^-\tau^+). \end{aligned} \tag{11}$$

We now assume for the simplicity of the argument that the resonator is in a photon number state $|n\rangle$; then the term $\tau^+ a + \tau^- a^+$ (which can be interpreted as a qubit-mediated exchange of quanta between the resonator and the TLS) can be neglected and, up to a constant energy shift $n\omega_{\rm r}$, we obtain

$$H' \approx -\left[\frac{\omega_{\rm qb}}{2} - \frac{g^2}{\delta}(n + 1/2)\right]\sigma^z - \frac{\omega_{\rm TLS}}{2}\tau^z + \kappa(\sigma^+\tau^- + \sigma^-\tau^+). \tag{12}$$

From this expression one sees that the qubit transition frequency is ac-Stark shifted by the quantity $\sim \Omega^2/4\delta = g^2 n/\delta$ due to the presence of n photons in the resonator. When $\Omega^2/4\delta \gg \kappa$, the transitions between the states $|0\rangle \otimes |e\rangle$ and $|1\rangle \otimes |g\rangle$, as illustrated in Fig. 4, are suppressed. Therefore, in order to decouple the qubit and TLS, the driving field must satisfy $\kappa \ll \Omega \ll \delta$.

4. Conclusions

We have shown that by an appropriate choice of the amplitudes and phases of the microwave signals applied to a system of two qubits, the coupling between them can be modulated. In the case of a spurious coupling between a qubit and a TLS residing for example in the insulating layer of the junction, this technique can be used for eliminating the decohering effect of the defect.

Acknowledgements

This work was supported from NGSMP and Academy of Finland (projects 135135 and 141559).

Journal of Physics: Conference Series **338** (2012) 012010 doi:10.1088/1742-6596/338/1/012010

References

[1] Rigetti C, Blais A and Devoret M 2005 *Phys. Rev. Lett.* **94** 240502.

[2] Ashhab S, Matsuo S, Hatakenaka N and Nori F 2006 *Phys. Rev. B* **74** 184504; Ashhab S and Nori F 2007 *Phys. Rev. B* **76** 132513.

[3] Li J, Chalapat K and Paraoanu G S 2008 *Phys. Rev. B* **78** 064503.

[4] Berkeley A J , Xu H, Ramos R C , Gubrud M A, Strauch F W, Johnson P R, Anderson J R, Dragt A J, Lobb C J and Wellstood F C 2003 *Science* **300** 1548.

[5] Paskin Yu A, Yamamoto T, Astafiev O, Nakamura Y, Averin D V and Tsai J S 2003 *Nature (London)* **421** 823; Yamamoto T, Pashkin Yu A, Astafiev O, Nakamura Y and Tsai J S, *Nature (London)* 2003 **425** 941.

[6] de Groot P C, Lisenfeld J, Schouten R N, Ashhab S, Lupascu A, Harmans C J P M and Mooij J E 2010 *Nature Physics* **6** 763.

[7] Majer J, Chow J M, Gambetta J M, Koch J, Johnson B R, Schreier J A, Frunzio L, Schuster D I, Houck A A, Wallraff A, Blais A, Devoret M H, Girvin S M and Schoelkopf R J 2007 *Nature (London)* **449** 443.

[8] Paraoanu G S 2006 *Phys. Rev. B* **74** 140504(R).

[9] Wootters W K 1998 *Phys. Rev. Lett.* **80** 2245; Hill S and Wootters W K 1997 *Phys. Rev. Lett.* **78** 5022.

[10] Martinis J M, Nam S, Aumentado J and Urbina C 2002 *Phys. Rev. Lett.* **89**, 117901.

[11] Cooper K B, Steffen M, McDermott R, Simmonds R W, Oh S, Hite D A, Pappas D P and Martinis J M 2004 *Phys. Rev. Lett.* **93** 180401; Martinis J M, Cooper K B, McDermott R, Steffen M, Ansmann M, Osborn K D, Cicak K, Oh S,Pappas D P, Simmonds R W and Yu C C 2005 *Phys. Rev. Lett.* **95** 210503; Zagoskin A M, Ashhab A, Johansson J R, and Nori F 2006 *Phys. Rev. Lett.* **97** 077001.

[12] Cole J H, Müller C, Bushev P, Grabovskij G J, Lisenfeld J, Lukashenko A, Ustinov A V and Shnirman A 2010 *Appl. Phys. Lett.* **97** 252501; Grabovskij G J, Bushev P, Cole J H, Müller C, Lisenfeld J, Lukashenko A and Ustinov A V 2011 *New J. Phys.* **13**, 063015.

[13] Poyatos J F, Cirac J I and Zoller P 1997 *Phys. Rev. Lett.* **78** 390.

[14] Simmonds R W, Lang K M, Hite D A, Nam S, Pappas D P, and Martinis J M 2004 *Phys. Rev. Lett.* **93** 077003.

Journal of Physics: Conference Series **338** (2012) 012011 doi:10.1088/1742-6596/338/1/012011

Embedding method for the scattering phase in strongly correlated quantum dots

Rafael A. Molina
Instituto de Estructura de la Materia, CSIC, Serrano 123, 28006 Madrid, Spain.
E-mail: rafael.molina@csic.es

Peter Schmitteckert
Institute of Nanotechnology, Karlsruhe Institute of Technology, 76344 Eggenstein-Leopoldshafen, Germany
Center of Functional Nanostructures, Karlsruhe Institute of Technology, 76131 Karlsruhe, Germany

Dietmar Weinmann and Rodolfo A. Jalabert
Institut de Physique et Chimie des Matériaux de Strasbourg, UMR 7504, CNRS-UdS, 23 rue du Loess, BP 43, 67034 Strasbourg Cedex 2, France

Philippe Jacquod
Physics Department, University of Arizona
1118 E. Fourth Street, P.O. Box 210081, Tucson, AZ 85721, USA
Département de Physique Théorique Université de Genève 24, Quai Ernest Ansermet 1211 Genève, Switzerland

Abstract. The embedding method for the calculation of the conductance through interacting systems connected to single channel leads is generalized to obtain the full complex transmission amplitude that completely characterizes the effective scattering matrix of the system at the Fermi energy. We calculate the transmission amplitude as a function of the gate potential for simple diamond-shaped lattice models of quantum dots with nearest neighbor interactions. In our simple models we do not generally observe an interaction dependent change in the number of zeroes or phase lapses that depend only on the symmetry properties of the underlying lattice. Strong correlations separate and reduce the widths of the resonant peaks while preserving the qualitative properites of the scattering phase.

1. Introduction

The scattering theory of quantum transport has demonstrated to be a very successful tool for the study and design of novel nanodevices. The information about the transport properties of a quantum mechanical system is encoded in the complex transmission amplitude. The conductance of such a system in a two-terminal configuration can be obtained from the trace of the transmission matrix t multiplied by its Hermitian conjugate $t^\dagger$ according to the Landauer-Büttiker formula $G = (e^2/h)\mathrm{Tr}\left(t\ t^\dagger\right)$ [1, 2]. In a single channel device the conductance reduces to the modulus of the transmission probability T through the channel and is easily measurable in experiments. On the other hand, even if the phases of the scattering matrix of a mesoscopic system show an interesting behaviour [3], their direct measurement cannot be done. One possible way to extract the phase of the transmission coefficient is to embed the system to measure in one of the arms of an Aharonov-Bohm (AB) interferometer. Under the appropiate conditions it was shown to be possible to extract the transmission phase from the phase of the conductance oscillations of the whole structure as a function of the magnetic flux threading the AB ring [4, 5].

The pioneering AB experiments that tried to measure the transmission phase of transport through a quantum dot in the Coulomb Blockade (CB) regime used a two-terminal geometry. In this situation, Onsager relations due to conservation of current and time-reversal symmetry imply that the conductance must be an even function of the magnetic flux and the measured transmission phase can only be 0 or π [2, 6]. Later experiments used a six-terminal configuration with additional leads that modify the reprocity relations for transmission coefficients [5]. The obtained transmission phase presented a smooth increase of π as a function of a gate potential in the dot V_g when a resonance peak was crossed while a sudden jump of $-\pi$ (a phase lapse following the experimentalists convention) was observed between resonances. In a more recent experiment the transition between this regime, termed universal regime, and a mesoscopic regime where phase lapses and resonances occur randomly with respect to each other was studied as a function of the number of electrons in the dot [7].

These observations have given rise to a whole body of literature devoted to study the transmission phase of quantum dots. Many theoretical works [8, 9, 10, 11, 12] were concerned with explaining the experimental results, in particular, the universal regime for the transmission phase which is at odds with the standard models for CB phenomena. Recent progress in that direction was obtained studying the statistical properties of the parity correlations between succesive wave functions in ballistic chaotic quantum dots [13]. Another interesting path followed for the analysis of the experimental results is a many-body extension of the constant-interaction model (CIM) for the description of resonances in quantum dots [14, 15, 16]. In the regime where the distance between resonances is much smaller than the width of the resonant levels Karrasch et al. found that the number of phase lapses could be increased by moderate interactions. Although the results obtained do not seem to be generic, it has a lot of theoretical and practical interest to understand the effects of strong correlations on the behavior of the full complex scattering quantities that characterize the electronic transport properties.

A complete theoretical understanding of transport through systems with strong correlations is generally missing, however. The problem can be formulated using the Keldysh approach [17, 18]. However, the calculations using Keldysh Green functions for interacting systems can usually only be performed within some approximation [19]. There are also many numerical methods in the literature but none are without problems. The more generally applicable ones rely on extensive real time simulations, see [20] and references therein. However, the computing power needed for the application of these methods limits their practical implementations for strongly correlated systems. A more recent approach is based on the exact diagonalization of teh full but isolated, many-body system, and accordingly treats tunneling perturbatively [21]. It obviously is restricted to small systems. Other types of approaches rely on the calculation of an equilibrium quantity that can be related to the conductance. Although these approaches are usually limited to the linear response regime they can be extremely useful to understand the effects of strong correlations on electronic transport [22, 23, 24, 25, 26, 27, 28]. The purpose of this work is to present one of these methods, the embedding method, in a selfcontained manner. This numerical method in combination with Density Matrix Renormalization Group (DMRG) allows the computation of the transmission amplitude of a system with strong correlations attached to one-dimensional leads. As an illustration, we apply the method to simple lattice models of small quantum dots.

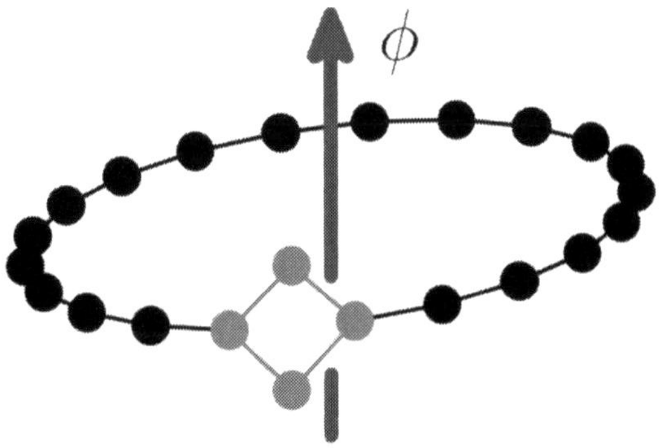

Figure 1. Sketch of the system considered within the embedding approach: The interacting region of length L_s (red) is embeded in a one-dimensional ring formed by a non-interacting lead (black) of length L_L. The ring is threaded by an Aharonov-Bohm flux ϕ. The interacting system we consider consists of four sites in a diamond-shaped configuration with nearest-neighbor interaction.

2. Embedding method for the transmission through a quantum dot

During the last decade, the embedding method has been used quite successfully to calculate the transmission of electrons through simple lattice models of strongly correlated systems [22, 23, 24, 25, 26, 27]. The system under consideration is embedded in a non-interacting ring. The ground state properties of the combined system pierced by a magnetic flux can be related to the scattering properties of the embedded system. In this section, we show the main steps of the derivation of the method. We follow closely the approach that allows to extract the modulus of the transmission amplitude presented in Appendix A of Ref. [29]. Moreover, we extend the method to calculate also the phase of the transmission amplitude. We will limit ourselves to the case of a single channel in the leads which is the simplest one although generalizations of the method to more than one channel have been reported [30]. The derivation is only rigorously valid for non-interacting impurities. However it was shown numerically that an interacting system in the limit of infinite lead length can be described by an effective scattering matrix whose matrix elements are accessible via the embedding method [29].

We can write the quantization condition of the single-particle states of the ring in terms of the transfer matrix of the non-interacting part (the lead) and the transfer matrix of the scatterer (see the sketch in Fig. 1) as

$$\det\left(I - M_{\mathrm{L}} M_{\mathrm{S}}\right) = 0 \,, \tag{1}$$

where M_{S} and M_{L} are the transfer matrices of the system and the lead, respectively. In the presence of time-reversal symmetry, the transfer matrix of a one-dimensional scatterer can be expressed in terms of three independent angles $\hat{\alpha}$, θ and φ:

$$\begin{aligned} M_{\mathrm{S}} &= \begin{pmatrix} 1/t^* & r^*/t^* \\ r/t & 1/t \end{pmatrix} \\ &= \frac{1}{\sin\varphi} \begin{pmatrix} e^{i\hat{\alpha}}/\sin\theta & -i\cot\theta + \cos\varphi \\ i\cot\theta + \cos\varphi & e^{-i\hat{\alpha}}/\sin\theta \end{pmatrix} , \end{aligned} \tag{2}$$

where the two components correspond to right and left moving particles while r and t are the reflection and transmission amplitudes, respectively. The transmission amplitude is given by $t = e^{i\hat{\alpha}} \sin\theta \sin\varphi = |t| \exp(\mathrm{i}\alpha)$. This relation between the transmission phase α and the scattering phase $\hat{\alpha}$ was at the origin of some misunderstandings about the interpretation of the experimentally observed phase lapses until the issue was clarified by Taniguchi and Büttiker [31]. α is the relevant experimental quantity and is not well defined when there is a zero of the transmission, resulting in a phase lapse whenever the transmission vanishes changing sign as a function of some parameter. $\hat{\alpha}$ is the relevant phase for the observance of the Friedel sum rule and only presents trivial 2π jumps.

The transfer matrix of a lead of length $L_{\rm L}$ for a state with wave number $k \geq 0$ reads

$$M_{\rm L} = \exp\left(\mathrm{i}\Phi\right) \begin{pmatrix} \exp(\mathrm{i}kL_{\rm L}) & 0 \\ 0 & \exp(-\mathrm{i}kL_{\rm L}) \end{pmatrix}, \tag{3}$$

taking into account that the flux threading the ring can be included by twisting the boundary condition of the system by a gauge transformation. Note that the scatterer embedded in the ring is not affected by this artificial flux such that we always work at zero magnetic field in the system of interest.

Inserting the transfer matrices (2) and (3), the eigenvalue condition (1) yields

$$\cos(\Phi) = \frac{1}{|t(k)|} \cos\left(kL + \delta\hat{\alpha}(k)\right) . \tag{4}$$

The phase shift $\delta\hat{\alpha} = \hat{\alpha} - kL_{\rm S}$ is the phase of the scattering region relative to a perfect lead of the same length $L_{\rm S}$. The solution of (4) yields the quantized momenta k of the energy eigenstates in the lead.

Since both t and $\delta\hat{\alpha}$ are functions of k, it is in general impossible to obtain an analytic solution of (4). However, one can study the asymptotic limit of large L, which was done by Gogolin and Prokof'ev in their study of the persistent current of a one-dimensional ring with a defect [32]. We use a general dispersion relation $\epsilon(k)$ in the lead that allows us to discuss continuum and tight-binding models at the same time.

The eigenvalue condition (4) can be rewritten as

$$k = k_n^0 + \frac{1}{L} f_{\pm}(k, \Phi) . \tag{5}$$

Here, $k_n^0 = 2\pi n/L$ with $n \geq 0$ denotes the eigenvalues in the case of perfect transmission with $|t| = 1$ and $\delta\hat{\alpha} = 0$. Following the notation of Ref. [32], we introduce the function

$$f_{\pm}(k, \Phi) = \pm \mathrm{Arccos}\left(|t(k)| \cos\Phi\right) - \delta\hat{\alpha}(k) . \tag{6}$$

Arccos denotes the principal branch of the inverse cosine function that takes values in the interval $[0, \pi]$. As k should be positive, $f_{-}(k, \Phi)$ cannot be used for the case $n = 0$. The splitting of the solutions of (5) corresponding to "+" and "-" cannot exceed the spacing $2\pi/L$ between the k_n^0, provided that $\delta\hat{\alpha}(k)$ is smooth on this scale. This is the case in the limit $L \to \infty$ and ensures that the order of the solutions with respect to energy is given by n.

Iterating (5) and expanding $f_{\pm}$ for large systems, we obtain the expansion

$$\begin{aligned} k_n^{\pm} =& k_n^0 + \frac{1}{L} f_{\pm}(k_n^0, \Phi) \\ &+ \frac{1}{L^2} f_{\pm}(k_n^0, \Phi) \left(\frac{\partial f_{\pm}(k, \Phi)}{\partial k}\right)_{k=k_n^0} + O\left(\frac{1}{L^3}\right) \end{aligned} \tag{7}$$

for the solutions of (5) in powers of $1/L$. Such an expansion is problematic in the vicinity of resonances where there is a rapid variation of the function f with k. In that case, the expansion is valid only for sufficiently large L.

We now calculate the ground state energy of the system as a function of the flux to order $1/L^2$. Using (7), the expansion of the one-particle energies in powers of $1/L$ can be shown to be

$$\begin{aligned}\epsilon(k_n^{\pm}) =& \epsilon(k_n^0) + \frac{1}{L}\left(\frac{\partial \epsilon}{\partial k} f_{\pm}(k,\Phi)\right)_{k=k_n^0} \\ &+ \frac{1}{2L^2}\frac{\partial}{\partial k}\left(\frac{\partial \epsilon}{\partial k} f_{\pm}^2(k,\Phi)\right)_{k=k_n^0} + O\left(\frac{1}{L^3}\right).\end{aligned} \tag{8}$$

We consider the simplest case of an odd number of particles in the ring. The leading order results for even number of particles give rise to the same final expression for the calculation of the transmission modulus [29]. For an odd number of particles N in the ring, all occupied states n come in pairs ([n,-] and [n,+]), except for the one corresponding to $n = 0$. The total ground state energy is

$$\begin{aligned}E_0^{\mathrm{odd}}(\Phi) &= \epsilon(k_0^+) + \sum_{n=1}^{n_{\mathrm{F}}}[\epsilon(k_n^+) + \epsilon(k_n^-)] \\ &= \epsilon(0) + \frac{1}{2L^2}\left(\frac{\partial^2 \epsilon}{\partial k^2}[\mathrm{Arccos}(|t|\cos\Phi) - \delta\hat{\alpha}]^2\right)_{k=0} \\ &+ \sum_{n=1}^{n_{\mathrm{F}}}\left\{2\epsilon(k_n^0) - \frac{2}{L}\left(\frac{\partial \epsilon}{\partial k}\right)_{k=k_n^0}\delta\hat{\alpha}(k_n^0)\right. \\ &\left.+\frac{1}{L^2}\frac{\partial}{\partial k}\left(\frac{\partial \epsilon}{\partial k}[\mathrm{Arccos}^2(|t|\cos\Phi) + \delta\hat{\alpha}^2]\right)_{k=k_n^0}\right\} + O\left(\frac{1}{L^3}\right).\end{aligned} \tag{9}$$

The sum runs up to $n_{\mathrm{F}} = (N-1)/2$. We have assumed $(\partial\epsilon/\partial k)_{k=0} = 0$ and kept all terms which can give rise to contributions up to order $1/L$. The first term in the sum is the ground state energy in the absence of scattering. It is proportional to L when N is of order L. The second term represents the energy change due to the scattering potential and is of order 1. From this second term, we can obtain the scattering phase shift by changing the number of particles as it is discussed below. The third term is the leading flux-dependent correction which allows to obtain the modulus of the transmission.

We start by considering the leading order terms for the calculation of the modulus of the transmission (see App. A of Ref. [29] for next order corrections and for the case of an even number of particles). Converting the sums over n into integrals, these flux-dependent contributions can be expressed as

$$\begin{aligned}&\frac{1}{2\pi L}\int_{\pi/L}^{k_F+\pi/L} \mathrm{d}k \frac{\partial}{\partial k}\left(\frac{\partial \epsilon}{\partial k}\mathrm{Arccos}^2(|t|\cos\Phi)\right) \\ &= \frac{\hbar v_{\mathrm{F}}}{2\pi L}\mathrm{Arccos}^2\big(|t(k_F)|\cos(\Phi)\big) + O\left(\frac{1}{L^2}\right).\end{aligned} \tag{10}$$

Here, $k_{\mathrm{F}} = 2\pi n_{\mathrm{F}}/L$ is the Fermi wave number and $v_{\mathrm{F}} = (\partial\epsilon/\hbar\partial k)_{k=k_{\mathrm{F}}}$ is the Fermi velocity. The leading flux-dependent term of the ground state energy is

$$E_0^{\mathrm{odd}(1)}(\Phi) = \frac{\hbar v_{\mathrm{F}}}{2\pi L}\mathrm{Arccos}^2\big(|t(k_F)|\cos(\Phi)\big), \tag{11}$$

where the superscript indicates the order of the result in $1/L$. In order to calculate the transmission modulus we need the leading order of the charge stiffness

$$\begin{aligned} D^{(1)} &= -\frac{L}{2}\left(E_0^{\mathrm{odd}(1)}(0) - E_0^{\mathrm{odd}(1)}(\pi)\right) \\ &= \frac{\hbar v_{\mathrm{F}}}{2}\left[\frac{\pi}{2} - \mathrm{Arccos}(|t(k_F)|)\right] . \end{aligned} \tag{12}$$

This last result is independent of the parity of the number of particles. Then, the transmission modulus can be obtained as

$$|t(E_F)|^2 = \lim_{L\to\infty} \sin^2\left(\frac{\pi D}{2D^0}\right), \tag{13}$$

where D^0 is the charge stiffness of a clean ring without scatterer.

We now keep the flux fixed. Then, the lowest order change of the ground state energy (9) in $1/L$ with repect to the change of the particle number N depends on $\delta\hat{\alpha}$. It is thus possible to express the scattering phase as

$$\delta\hat{\alpha}(k_{\mathrm{F}}) = -\lim_{L\to\infty}\frac{L}{2}\left(\frac{E(N) - E(N-2) - 2\epsilon(k_{n_{\mathrm{F}}}^{(0)})}{\mathrm{d}\epsilon/\mathrm{d}k|_{k=k_{n_{\mathrm{F}}}^{(0)}}}\right) \tag{14}$$

in terms of the many-body energies at particle numbers N and $N-2$ and the properties of the dispersion relation at the Fermi energy. In the special case of a tight binding chain with unit hopping and lattice spacing at half filling, this formula reduces to

$$\delta\hat{\alpha}(k_{\mathrm{F}}) = -\lim_{L\to\infty}\frac{L}{4}\left[E(L/2+1) - E(L/2-1)\right] . \tag{15}$$

Eqs. 14 and 15 show that the scattering phase is related to the change in total ground state energy in large rings when the particle number is changed. It is therefore possible to use the embedding method to extract the scattering phase.

From the embedding method, we can thus obtain $\hat{\alpha}$. However, the quantity of interest from the experimental point of view is α with its phase lapses coming from the different branches of the relation $|t|\exp(i\alpha) = \sin\theta\sin\varphi\exp(i\hat{\alpha})$. To achieve this goal, we need to access the sign of φ. For that purpose we go back to the general expression (2) of the transfer matrix and see that in order to have left-right symmetry we need $r = r'$, and therefore $\sin\theta\cos\varphi = 0$.

We have three possibilities: i) $\sin\theta = 0$, ii) $\varphi = \pi/2$, iii) $\varphi = -\pi/2$. Thus, in a left-right symmetric case we have two branches ($\sin\varphi = 1$ and $\sin\varphi = -1$) and we can pass from one to the other only when $t = 0$.

The embedding method allows to follow the branch-switching since the phase sensitivity is given by [29]

$$\Delta E = \left(E(\pi) - E(0)\right) = \frac{\hbar v_{\mathrm{F}}}{L}\left[\frac{\pi}{2} - \mathrm{Arccos}(\sin\theta\sin\varphi)\right] , \tag{16}$$

and therefore

$$\sin\theta\sin\varphi = \sin\left(\Delta E\frac{L}{\hbar v_{\mathrm{F}}}\right) = \sin\left(\frac{\pi}{2}\frac{\Delta E}{|\Delta E^{(0)}|}\right), \tag{17}$$

where $\Delta E^{(0)}$ is the phase sensitivity of a perfectly transmitting scatterer. Therefore, the sign changes of ΔE determine the branch-switch of φ and thus the jumps in α. One should also notice that there can also be transmission zeroes without phase lapses in cases when there is a zero of ΔE without sign change. This possibility does occur for particular values of the parameters in some of the models we have analyzed.

For a strictly one-dimensional system, the sign of ΔE is fixed by Leggett's theorem [33, 34]. For an odd number of particles we are in the branch of positive φ, t never vanishes, and there cannot be branch-switches. For a quasi-one dimensional scatterer this is no longer true, there can be parameter values where t vanishes and branch-switches appear. The embedding method for the phase of the transmission has been checked for a one-dimensional chain and its predictions compared with the Friedel sum rule. Its validity has been confirmed for interacting scatterers attached to reservoirs via non-interacting leads, where it describes the effective one-body scattering at the Fermi-energy [16, 29].

3. Numerical results for diamond-shaped quantum dots

A one-dimensional scatterer does not exhibit zeroes of the transmission and thus no phase lapses are observed. We therefore study a minimal model of a scatterer in which transmission zeroes occur, and where phase lapses are expected. In this context, the diamond lattice quantum dots were introduced by Levy-Yeyati and Büttiker as a simple example to understand the physics behind the phase lapses of the transmission [35]. In the non-interacting case the form of the transmission as a function of the gate voltage V_g can be obtained analytically [35]. The results for the non-interacting case can serve as a guide for the convergence properties of the embedding method for quantum dot lattice models beyond one dimension.

The Hamiltonian of the system under study including interactions reads

$$\begin{aligned} H &= -t\left(c_1^\dagger c_2 + c_1^\dagger c_3 + c_2^\dagger c_4 + c_3^\dagger c_4 + \text{H.C.}\right) + \epsilon n_2 - \epsilon n_3 \\ &\quad + V_g \sum_{i=1}^{4} n_i + U\left(\tilde{n}_1\tilde{n}_2 + \tilde{n}_1\tilde{n}_3 + \tilde{n}_2\tilde{n}_4 + \tilde{n}_3\tilde{n}_4\right), \end{aligned} \tag{18}$$

where $\tilde{n}_i = n_i - 1/2$. The nearest neighbor interaction is added in that form in order to enforce charge neutrality at half-filling. A sketch of the lattice used in the model is shown in Fig. 1. Site 1 is connected to the left lead and to sites 2 and 3 while site 4 is connected to the right lead and also to sites 2 and 3. The single particle energies of sites 2 and 3 are the same but with opposite signs while the single particle energies at the sites connected to the leads 1 and 4 are taken to be 0. This symmetry ensures that a zero of the transmission appears exactly at $V_g = -2U$.

In the non-interacting case with $\epsilon = 2$ the conductance for this model has two symmetric resonant peaks at $V_g = 2.75$ and $V_g = -2.75$. The zero of the transmission is associated with a phase lapse at $V_g = 0$. There are two more symmetric resonances around the transmission zero at $V_g = -1.01$ and $V_g = 1.01$ but corresponding to very badly coupled wave-functions so they hardly surpass the transmission background from the bigger resonances. We compare the exact results to calculations using the embedding method. For obtaining results from the embedding method an extrapolation to infinite lead lengths is needed. A finite size scaling analysis of the results is made using increasing lead lengths and linear or quadratic extrapolation formulas as explained in detail in Ref. [29]. This procedure is performed for the transmission modulus and for the transmission phase in exactly the same fashion. We show in Fig. 2 the values for g and α obtained with total lengths of the system plus the auxiliary ring of $L = 28, 46, 86$ and compare them with the exact result. The actual value of the length that we need to use in the extrapolation formulas (Eq. 12 and Eq. 15) is $L - 1$ due to the geometrical form of the system (sites 2 and 3 are just to alternative ways for the electron to go through the system and count as a single site for the purpose of the extrapolation formulas). The results for all the three cases are very good around the zero of the transmission. There are small deviations close to the resonances for $L = 28$ and $L = 46$ while the results for $L = 86$ coincide almost exactly with the exact results. The qualitative behavior is correctly reproduced even for the smallest ring size $L = 28$. The deviation from the exact results close to the resonances is characteristic of the

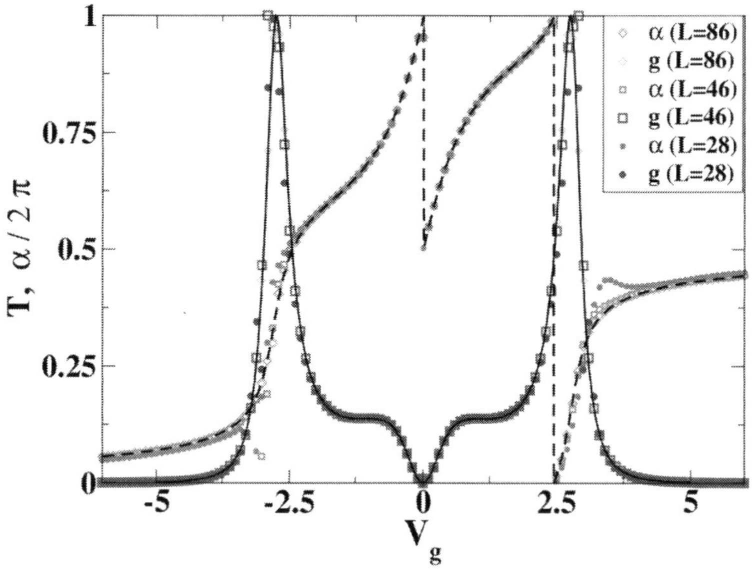

Figure 2. Transmission probability and transmission phase for the non-interacting diamond dot model with $\epsilon = 2$ calculated using the embedding method. The solid lines represent the exact results.

method when full convergence in the length of the auxiliary ring has not been reached. Similar deviations are observed when interactions are switched on.

When $U > 0$, the ground state properties of the ring including the diamond-shaped quantum dot are calculated using the Density Matrix Renormalization Group (DMRG) algorithm [36]. We keep up to 650 states in order to get a good accuracy in the calculation of the charge stiffness of the ring for the larger sizes.

From very general arguments based on the Green's function approach to non-interacting transport it can be shown that a zero of the transmission appears between two consecutive resonant peaks depending on the relative symmetry properties of the wave functions corresponding to the resonances with respect to the positions of the leads [35]. If consecutive resonances have the same symmetry a zero appears between peaks and when they have opposite symmetry the zero is missing. The four resonances of the diamond quantum dot Hamiltonian Eq. 18 have even, odd, odd and even symmetry with respect to sites 1 and 4 so there is only one zero in the middle.

The effect of the interactions is to change the position of the zero of the conductance (due to the charge neutrality requirements) and the position of the resonance peaks while deforming their overall shape. The resonances get more isolated and symmetric. However, the phase lapse of α when we go through the transmission zero is the same as in the non-interacting case. Within our model, The symmetries of the resonances are not modified by the many-body correlations that are induced through the increase of the interaction strength U. In the different panels of Fig. 3 we show the behavior of the complex transmission coefficient as a function of V_g. We compare the exact results for $U = 0$ with embedding method results calculated with an extrapolation up to $L = 76$ for the cases of $U = 2$ and $U = 20$.

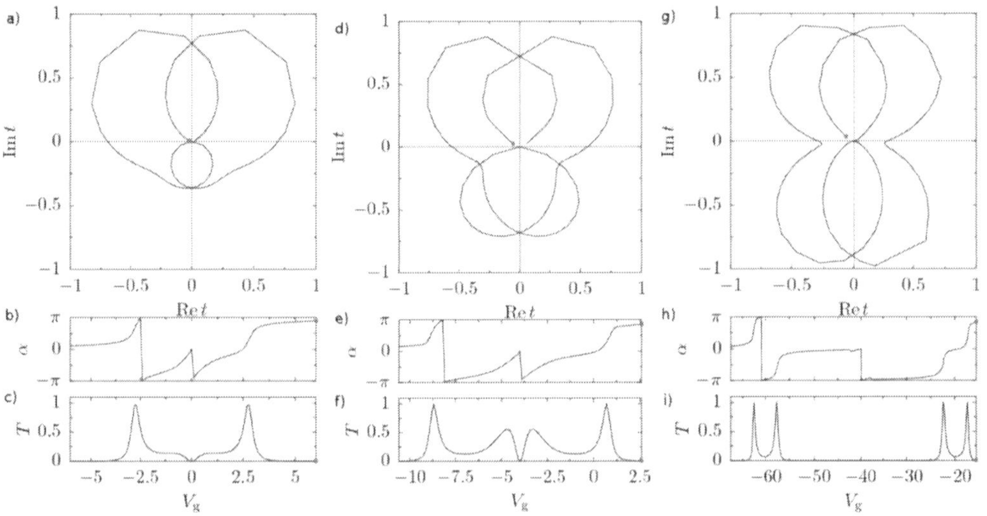

Figure 3. Results for the diamond quantum dot model with $\epsilon = 2$ and different values of the interaction strength U. The left, middle, and right panels show the trajectory of the complex transmission amplitude (a, d, and g), transmission phase as a function of V_g (b, e, and h) and transmission probability T as a function of V_g (c, f, and i) for $U = 0$, $U = 2$, and $U = 20$, respectively. The red dot marks the same V_g point in each of the three graphs with the same value U so it is easier to follow the trajectory of the transmission amplitude in the complex plane.

Due to the symmetry of the problem, the transmission zero is always placed between the two groups of resonances. As we increase the interaction the symmetric resonances close to the zero of transmission that are barely visible in the case $U = 0$ fully develop into lorentzian peaks and a wide gap opens between them. This is due to the transition between extended wave functions for small U to spatially localized wave-functions for large U. However, there is no qualitative change in the behavior of the transmission phase as we increase the interaction U. The trajectories of the complex transmission amplitude as we vary V_g shown in the top panel of the figure show similar features. The mechanisms proposed for the change of the number of phase lapses in interacting models like population switching[14, 15, 37] are not at work in this simple geometrical model of a quantum dot precisely because the geometrical properties of the lattice dominate the symmetry properties of the wave functions even in the many-body case.

4. Conclusions

The embedding method for the calculation of the transmission through a strongly correlated system connected to non-interacting leads can be successfully extended to calculate also the phase of the transmission amplitude. The phase of the effective transmission amplitude of a correlated system connected to non-interacting one-dimensional leads can be extracted from differences of the ground state energy of an auxiliary ring at different electron numbers. We have illustrated the application of this method for the calculation of the full complex transmission coefficient of a diamond-shaped quantum dot with nearest-neighbor interactions. The calculations of the ground state energy differences have been performed using the DMRG algorithm that allows for great accuracy in the models considered. For the small systems that we treated in the work, interactions modify substantially the shape of the resonances although the number and position of the zeros between resonances are not changed and depend only on the symmetry properties of the quantum dot model.Larger structures which represent a harder computational challenge may present richer phenomena concerning the scattering phase. These and further issues will be explored in future work.

Acknowledgments
We acknowledge support from the Spanish MICINN through project FIS2009-07277, the NSF under grant No DMR-0706319, the ANR through grant ANR-08-BLAN-0030-02, and the Swiss NCCR MANEP.

References

[1] Landauer R 1970 *Philos. Mag.* **21** 863
[2] Büttiker M 1986 *Phys. Rev. Lett.* **57** 1761
[3] Jalabert R A and Pichard J-L 1995 *J. Phys. (France)* **5** 287; Mucciolo E R, Jalabert R A and Pichard J-L 1997 *J. Phys. (France)* **7** 1267
[4] Jacoby A, Heiblum M, Mahalu D, Umansky V and Strikman H 1995 *Phys. Rev. Lett.* **74** 4047
[5] Schuster R, Buks E, Heiblum M, Mahalu D, Umansky V and Strikman H 1997 *Nature* **385** 417
[6] Onsager L 1931 *Phys. Rev.* **38** 2265
[7] Avinum-Kalish M, Heiblum M, Zarchin O, Mahalu D and Umansky V 2005 *Nature* **436** 529
[8] Baltin R and Gefen Y 1999 *Phys. Rev. Lett.* **83** 5094
[9] Hackenbroich G 2001 *Phys. Rep.* **343** 463
[10] Aharony A, Entin-Wohlman O, Halperin B I and Imry Y 2002 *Phys. Rev.* B **66** 115311
[11] Golosov D I and Y. Gefen 2006 *Phys. Rev.* B **74** 205316
[12] Bertoni A and Goldoni G 2007 *Phys. Rev.* B **75** 235318
[13] Molina R A, Jalabert R A, Weinmann D and Jacquod P 2011 *Preprint* ArXiv:1111.0443
[14] Karrasch C, Hecht T, Weichselbaum A, Oreg Y, von Delft J and Meden V 2007 *Phys. Rev. Lett.* **98** 186802
[15] Karrasch C, Hecht T, Weichselbaum A, von Delft J, Oreg Y and Meden V 2007 *New J. Phys.* **9**, 123
[16] Molina R A, Schmitteckert P, Jalabert R A, Weinmann D and Jacquod Ph unpublished.
[17] Meir Y and Wingreen N S 1992 *Phys. Rev. Lett.* **68** 2512
[18] Datta S 1992 *Phys. Rev.* B **46** 9493
[19] Schoeller H and König J 2000 *Phys. Rev. Lett.* **84** 3686
[20] Branschädel A, Schneider G and Schmitteckert P 2010 *Ann. Phys. (Berlin)* **522** 657
[21] Bergfield J P and Stafford C A 2009 *Phys. Rev.* B **79** 245125; Bergfield J P, Jacquod Ph and Stafford C A 2010 *Phys. Rev.* B **82** 205405
[22] Favand J and Mila F 1998 *Eur. Phys. J.* B **2** 293
[23] Sushkov O P 2001 *Phys. Rev.* B **64** 155319
[24] Molina R A, Weinmann D, Jalabert R A, Ingold G-L and Pichard J-L 2003 *Phys. Rev.* B **67** 235306
[25] Meden V and Schollwöck U 2003 *Phys. Rev.* B **67** 193303
[26] Rejec T and Ramsak A 2003 *Phys. Rev.* B **68** 035342
[27] Moliner M and Schmitteckert P, *EPL* **96** 10010
[28] Oguri A, Nisikawa Y and Hewson A C 2005 *J. Phys. Soc. Japan* **74** 2554 ; Louis K and Gros C 2003 *Phys. Rev.* B **68** 184424 ; Bohr D and Schmitteckert P and Wölfle P 2006 *EPL* **73** 246 ; Bohr D and Schmitteckert P 2007 *Phys. Rev.* B **75** 241103(R) ; Schmitteckert P and Evers F 2006 *Phys. Rev. Lett.* **100** 086401 ; Karrasch C, Enss T and Meden V 2006 *Phys. Rev.* B **73** 235337 ; Büsser C A, Anda E V, Lima A L, Davidovich M A and Chiappe G 2000 *Phys. Rev.* B **62** 9907
[29] Molina R A, Schmitteckert P, Weinmann D, Jalabert R A, Ingold G-L and Pichard J-L 2004 *Eur. Phys. J.* B **39** 107
[30] Freyn A, Vasseur G, Schmitteckert P, Weinmann D, Ingold G-L, Jalabert R A and Pichard J-L *Eur. Phys. J.* B **75** 253
[31] Taniguchi T and Büttiker M 1999 *Phys. Rev.* B **60** 13814
[32] Gogolin A O and Prokof'ev N V 1994 *Phys. Rev.* B **50** 4921
[33] Leggett A J 1991 *Granular Nanoelectronics* ed D K Ferry (New York: Plenum Press) pp 297-311
[34] Waintal X, Fleury G, Kazymyrenko K, Houzet M, Schmitteckert P and Weinmann D 2008 *Phys. Rev. Lett.* **101** 106804
[35] Levy Yeyati A and Büttiker M 1995 *Phys. Rev.* B **52** R14360; Levy Yeyati A and Büttiker M 2000 *Phys. Rev.* B **62** 7307
[36] White S R 1992 *Phys. Rev. Lett.* **69** 2863; Schmitteckert P 1996 *Interplay between interaction and disorder in one-dimensional Fermi systems* (PhD. Thesis: Universität Augsburg)
[37] Golosov D I and Gefen Y 2007 *New Journal of Physics* **9** 120

Advanced many-body and statistical methods in mesoscopic systems IOP Publishing
Journal of Physics: Conference Series **338** (2012) 012012 doi:10.1088/1742-6596/338/1/012012

Reduction of ballistic spin scattering in a spin-FET using stray electric fields

G. A. Nemnes

University of Bucharest, Faculty of Physics,
"Materials and Devices for Electronics and Optoelectronics" Research Center,
P.O. Box MG-11, 077125 Magurele-Ilfov, Romania

E-mail: nemnes@solid.fizica.unibuc.ro

A. Manolescu

School of Science and Engineering,
Reykjavik University, Menntavegur 1, IS-101 Reykjavik, Iceland

V. Gudmundsson

Science Institute,
University of Iceland, Dunhaga 3, IS-107 Reykjavik, Iceland

Abstract. The quasi-bound states which appear as a consequence of the Rashba spin-orbit (SO) coupling, introduce a strongly irregular behavior of the spin-FET conductance at large Rashba parameter. Moreover, the presence of the bulk inversion asymmetry, i.e. the Dresselhaus SO coupling, may compromise the spin-valve effect even at small values of the Rashba parameter. However, by introducing stray electric fields in addition to the SO couplings, we show that the effect of the SO induced quasi-bound states can be tuned. The oscillations of the spin-resolved conductance become smoother and the control of the spin-FET characteristics becomes possible. For the calculations we employ a multi-channel scattering formalism, based on the R-matrix method extended to spin transport, in the presence of Rashba and Dresselhaus SO couplings.

1. Introduction

The practical realization of the spin field effect transistor (spin-FET) suggested by Datta and Das [1] in the 90's was first met with serious challenges concerning the efficient spin injection using metallic ferromagnetic contacts and the attainment of the ballistic regime. Since then, a number of studies, both theoretical and experimental, reported significant progress.

Some theoretical studies suggested several improvements: enhanced spin control due to interband coupling [2] or large magnetocurrent due to the spin-filter effect of Schottky junctions [3]. Other design elements that are shown to play a key role are channel orientation and boundary effects [4] and the effects of stray magnetic fields [5]. Recently, the operation of the Datta-Das transistor was investigated in the ballistic [6] and nonballistic regimes [7, 8, 9].

An experimental realization of a transit-time spin-FET [10], making use of magnetic fields for spin precession has been achieved. Other spin-FETs with different operation principles than the Datta-Das transistor were reported, indicating a giant negative magnetorezistance effect

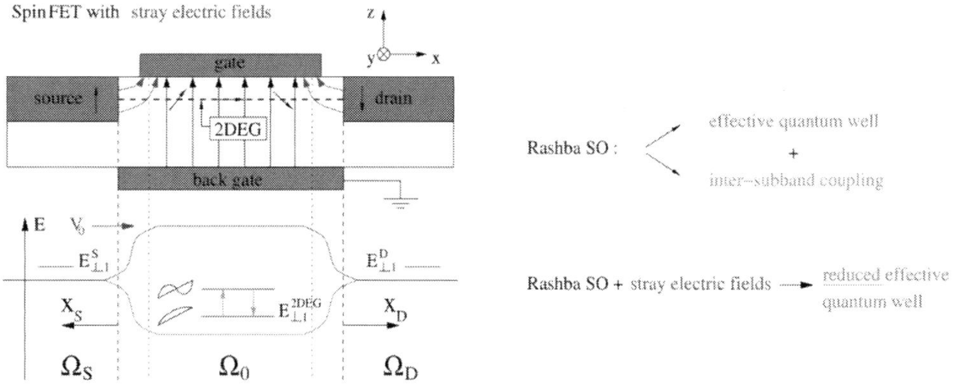

Figure 1. Schematics and functionality of a spin-FET with stray electric fields. The device structure is depicted together with the diagram of the first energy levels inside the leads ($\Omega_{S/D}$) and the scattering region (Ω_0). The combined action of the Rashba SO coupling and stray electric fields is indicated.

[11], tunneling through a magnetic domain wall [12], or the possibility of optical control of a quantum dot transistor [13]. Experimental devices investigated electric field tuning of the spin-orbit coupling [14], and the performance of a spin-valve transistor [15]. Recently, the concrete integration of spin-FETs with magnetic tunnel junctions to produce nonvolatile static random access memory has been analyzed [16] and the electrical control in a spin-FET using Rashba spin-orbit coupling was achieved in the experimental setup of Koo et al. [17].

From the theoretical point of view, some performance limitations were predicted, which are due to the inter-band ballistic scattering and spin interference effects in the large SO coupling regime [18]. This is due to the coupling of propagating modes to the quasi-bound states, which result following the SO interaction, leading to an uncontrollable sequence of minima and maxima in the conductance.

In this study we point out the possibility of using stray electric fields in order to diminish the effects introduced by the quasi-bound states and thus recover a controllable characteristics for the spin-FET, including the spin-valve regime. Calculations including the Dresselhaus SO coupling and an embedded point contact are also discussed, and they support the idea of using stray electric fields for obtaining a reasonable spin control.

2. Model

The transistor model considered is based on the spin-FET configuration proposed by Datta and Das [1]. As depicted in Fig. 1, the device is a usual field effect transistor, with the major difference that the source and drain contacts are feromagnetic materials. The spins of electrons injected by the source contact will have the same orientation with the source contact magnetization, while the drain electrode acts as a spin filter and collects electrons with spins aligned with the drain contact magnetization. The electrons passing from source to drain are subject of spin-orbit interaction, also called the Rashba effect [19]: the electric field produced by the gate, perpendicular on the channel direction, causes a "structural inversion asymmetry", which leads to the precession of the electron-spins. In this way, the orientation of the spins at the drain contact can be controlled by electrical means, simply by applying a gate voltage.

It is known that the Rashba SO coupling produces an effective quantum well as indicated schematically in Fig. 1, and it also provides the inter-subband coupling. The two effects combined are detrimental for the source-drain current, as inter-band scattering between the free traveling states and the quasi-bound states accommodated in the effective quantum well lead to uncontrollable characteristics. One way to overcome this undesired result is to introduce a potential step, to counterbalance the effect of the Rashba SO coupling. This is achieved by means of stray electric fields which naturally occur in a spin-FET structure between the source/drain contacts and the gate. The electric field components along the transport direction can be adjusted by varying the potential difference between source/drain and the gate, while the electric field which generates the Rasbba effect is controlled by the potential difference between gate and back-gate.

We assume the device is based on a InAs heterojunction, which is a typical material for spintronic applications, due to a relatively high Rashba parameter $\alpha \approx 10$ meV nm. The effective mass is $m^* = 0.023m_0$. The two dimensional electron gas (2DEG) has a parabolic confinement in the y-direction (see Fig. 1), given by the level spacing $\hbar\omega_0 = 0.01$ meV, which leads to an effective confinement length $l_0 = \hbar/\sqrt{m^*\hbar\omega_0}$. The length of the Rashba active region is $8l_0$.

The key element in this approach – the stray electric fields – introduces a shift (V_0) in the energy levels inside the active region (2DEG), which is opposite to the effect produced by the Rashba SO coupling. The value V_0 is set by matching the first Wigner-Eisenbud energy ($E_{\perp,1}^{2DEG} = \epsilon_1$, see next section) with the first subband in the semi-infinite source/drain leads ($E_{\perp,1}^S = E_{\perp,1}^D$). This matching condition implicitely means that V_0 must depend on α:

$$V_0(\alpha) = E_{\perp,1}^S - E_{\perp,1}^{2DEG}(\alpha). \tag{1}$$

3. The R-matrix scattering formalism

The spin-resolved conductance is obtained using the R-matrix scattering formalism, which has already proved its efficiency in a number of studies regarding charge transport in mesoscopic systems [20, 21, 22, 23].

The scattering problem is described by the two-dimensional Schrödinger equation defined in the (x, y) plane of the 2DEG, with asymptotic boundary conditions corresponding to incoming electrons from either source or drain:

$$\mathcal{H}\Psi(\vec{r}) = E\Psi(\vec{r}), \tag{2}$$

where $\mathcal{H} = T + W + H_R + H_D + V_0(\alpha)$ is the total Hamiltonian composed by the following terms: kinetic energy $T = -\hbar^2/(2m^*)\triangle$, potential energy $W = W(\vec{r})$ – which includes the parabolic confinement in the y-direction, Rashba SO interaction $H_R = \alpha/\hbar(p_y\sigma_x - p_x\sigma_y)$, Dresselhaus SO interaction $H_D = \beta/\hbar(p_x\sigma_x - p_y\sigma_y)$ and the add-on potential due to the stray electric fields $V_0(\alpha)$.

The device is partitioned as usual in leads (Ω_s), which are uniform in the transport direction, and the scattering region (Ω_0), which includes the channel and the regions containing the stray electric fields. The wavefunctions inside the leads take the general form of traveling waves modulated by the confinement potential:

$$\Psi_s(\vec{r} \in \Omega_s, E) = \sum_i \Psi_\nu^{in} \exp\left(-ik_\nu x_s\right)|\Phi_\nu\rangle + \sum_i \Psi_\nu^{out} \exp\left(ik_\nu x_s\right)|\Phi_\nu\rangle, \tag{3}$$

with $|\Phi_\nu\rangle = \sum_\sigma \Phi_{\nu,\sigma}(\vec{r}_{\perp;s})|\sigma\rangle$, where σ denotes the electron spin. Following notations from Ref. [23], the wavevectors are given by $k_\nu = \sqrt{\frac{2m^*}{\hbar^2}(E - E_\perp^\nu)}$, where $\nu = (s, i)$ is a composite index denoting channel i from lead $s = S, D$ and $E_\perp^\nu$ are the energies of transverse modes Φ_ν inside the leads.

The scattering matrix relates the complex coefficients of incoming and outgoing functions, $\vec{\Psi}^{\rm out} = \mathsf{S}\vec{\Psi}^{\rm in}$, and it is given by the R-matrix

$$\mathsf{S} = -\frac{1 + \frac{i}{m^*}\mathsf{Rk}}{1 - \frac{i}{m^*}\mathsf{Rk}}, \tag{4}$$

where k is the diagonal matrix given by the wavevectors k_ν. The R-matrix is defined as

$$(\mathsf{R})_{\nu\nu'}(E) = \sum_{\sigma,\sigma'} \int_{\Gamma_s} d\Gamma_s \int_{\Gamma_{s'}} d\Gamma_{s'}\ \Phi_{\nu,\sigma}(\vec{r}_{\perp;s})\Phi_{\nu',\sigma'}(\vec{r}_{\perp;s'}) R_{\sigma,\sigma'}(\vec{r}\in\Gamma_s; \vec{r}\,'\in\Gamma_{s'}; E), \tag{5}$$

where $\Gamma_s = \Omega_s \cap \Omega_0$ are the interfaces between the leads and the scattering region. One should note that the R-function has now an explicit dependence on the electron spin:

$$R_{\sigma,\sigma'}(\vec{r};\vec{r}\,';E) = -\frac{\hbar^2}{2}\sum_{l=1}^{\infty}\frac{\chi_{l,\sigma}(\vec{r})\chi^*_{l,\sigma'}(\vec{r}\,')}{E-\epsilon_l}, \quad \vec{r},\vec{r}\,' \in \Omega_0, \tag{6}$$

where χ_l and ϵ_l are the Wigner-Eisenbud functions and energies, respectively. The Wigner-Eisenbud problem is defined as usual:

$$\mathcal{H}\chi_l(\vec{r}) = \epsilon_l \chi_l(\vec{r}), \quad \vec{r} \in \Omega_0, \tag{7}$$

with the boundary conditions on Γ_s:

$$\left[\frac{\partial \chi_l}{\partial x_s}\right]_{\Gamma_s} = 0. \tag{8}$$

Using the S-matrix one can extract directly the total transmission between leads s and s':

$$T_{ss'}(E) = \sum_{i,i'} \Theta(E - E_\perp^\nu)\Theta(E - E_\perp^{\nu'})|\tilde{S}_{\nu\nu'}(E)|^2, \tag{9}$$

where $\tilde{S} = \mathsf{k}^{1/2}\mathsf{S}\mathsf{k}^{-1/2}$. The Θ-functions ensure that the summation is performed on the open channels only.

4. Results and discussion

The spin-FET conductance as a function of the Rashba parameter was analyzed for three spin polarizations of the contacts in the z direction: unpolarized, parallel, and anti-parallel.

In a first run, by varying the Rashba parameter α, the source-drain conductance is determined for the three contact polarizations and the positions of the first Wigner-Eisenbud energies are recorded. Then, using Eq. (1), one obtains the parametric dependence $V_0(\alpha)$, which is added to the Hamiltonian, and after that a second run is performed.

In Fig. 2(a) and 2(b) the results for the source-drain conductance are depicted, for one and five propagating modes, respectively. The number of participating channels N is set by the energy of the incident electrons. For one mode one obtains, as in Ref. [18], a chaotic behavior (upper plot) above a value of the Rashba parameter $\alpha \approx \hbar\omega_0 l_0 = 5.75$ meV nm, for all three contact polarizations. When the additional potential, $V_0(\alpha)$, is included in the Hamiltonian, the irregular oscillations are replaced by a more predictable characteristics. Increasing the number of modes to five (see Fig. 2(b)), one obtains a similar behavior as in Ref. [18]: the conductance oscillations drop in amplitude, although the onset of the irregular regime is still visible and, in the limit of large α, the conductance approaches the value of $N/2$. Including the term $V_0(\alpha)$ the

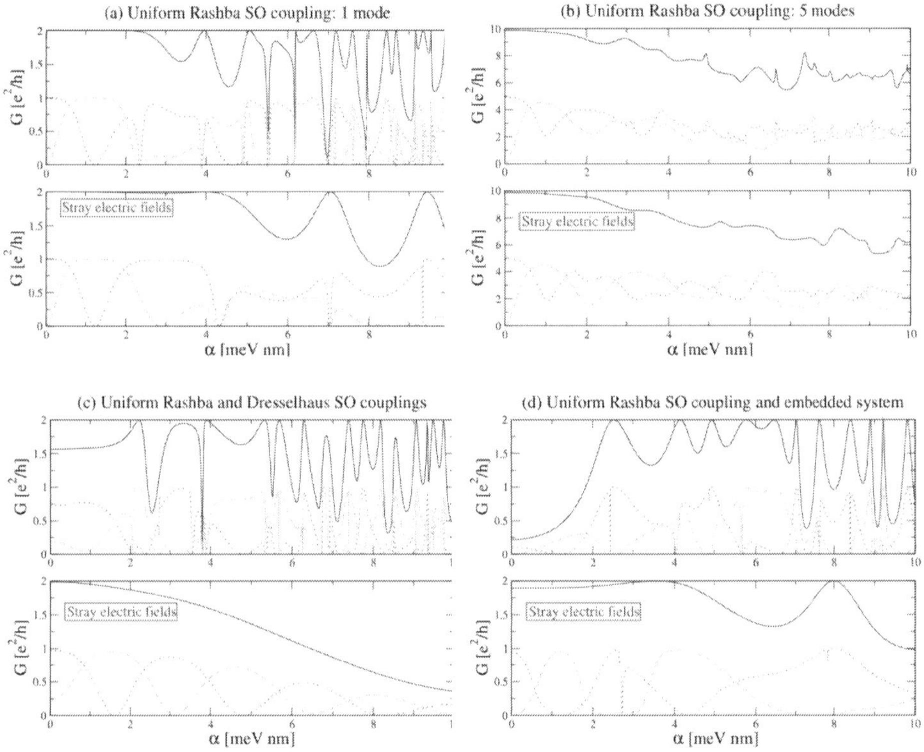

Figure 2. Conductance as function of the Rashba parameter, for the following test cases: (a) one mode, (b) five modes, (c) additional Dresselhaus SO coupling and (d) embedded point contact, for unpolarized (black/solid), parallel polarized (red/dotted), anti-parallel polarized (green/dashed) contacts. In each case, in the lower plots are depicted the improved characteristics, which follow from the inclusion of the stray electric fields, i.e. the term $V_0(\alpha)$.

oscillations are further smoothed and for the systems with polarized contacts an enhancement of the spin-valve effect can be observed.

For some materials, the Dresselhaus spin-orbit coupling becomes equally important. For the one mode case, by introducing a Dresselhaus SO parameter $\beta = 3.5$ meV nm (as for InAs), one finds that the spin-valve effect is seriously affected at even smaller values of the Rashba parameter α (see Fig. 2(c)). The two SO couplings, i.e. Rashba and Dresselhaus, act in a similar way, i. e. they both produce an effective quantum well and, in addition, they generate the coupling between different subbands. By including the stray electric fields, i.e. $V_0(\alpha)$, according to Eq. (1), one obtains a nearly ideal characteristics, indicating a high quality spin valve-effect. The low conductance for both unpolarized and polarized contacts is a consequence of the relatively high potential shift V_0, which now compensates the effective quantum well produced by both Rashba and Dresselhaus SO couplings.

Including embedded systems is commonly done for improving spin injection or designing efficient spin filters. Here, a sharp point contact of length $4l_0$ and aperture $2l_0$, placed in the middle of the Rashba active region was considered. Fig. 2(d) (upper plot) shows the conductance in the presence of the point contact, which acts as an effective potential step, as the wire confinement is locally increased. Consequently, it follows that for all contact polarizations the conductance is low at small α, while overall, the oscillations are better defined as compared to Fig. 2(a). Applying the same recipe of additional stray electric fields in this case, one also obtains an improvement in the spin control.

5. Conclusions

The undesired effects introduced by the quasi-bound states are diminished by adding proper stray electric fields. The spin-FET characteristics is improved in single and multi-channel case, without adjusting the wire confinement. The procedure is robust even in the presence of Dresselhaus spin-orbit coupling or after including an embedded system like a sharp point contact. The spin-valve effect is also restored.

Acknowledgments

This work was supported by the Romanian National Authority for Scientific Research, CNCS-UEFISCDI, project number PN-II-RU-PD-2011-3-0044 and by the Icelandic Research Fund. We are thankful to Roxana and Paul Racec for instructive discussions on the R-matrix theory.

References

[1] Datta S and Das B *Appl. Phys. Lett.* 1990 **56** 665
[2] Egues J C, Burkard G and Loss D 2003 *Appl. Phys. Lett.* **82** 2658
[3] Sugahara S and Tanaka M 2004 *Appl. Phys. Lett.* **84** 2307
[4] Liu M H and Chang C R 2006 *Phys. Rev.* B **73** 205301
[5] Koo H C, Eom J, Chang J and Han S H 2009 *Solid-State Electronics* **53** 1016
[6] Jiang K M, Yang J, Zhang R and Wang H 2008 *J. Appl. Phys.* **104** 053722
[7] Schliemann J, Egues J C, and Loss D 2003 *Phys. Rev. Lett.* **90** 146801
[8] Cartoixa X 2003 *Appl. Phys. Lett.* **83** 1462
[9] Ohno M and Yoh K 2008 *Phys. Rev.* B **77** 045323
[10] Huang B, Monsma D J and Applebaum I 2007 *Appl. Phys. Lett.* **91** 072501
[11] Gurzhi R N et al. 2003 *Appl. Phys. Lett.* **83** 4577
[12] Vignale G and Flatte M E 2002 *Phys. Rev. Lett.* **89** 098302
[13] Lu H F, Guo Y, Zu X T and Zhang H W 2009 *Appl. Phys. Lett.* **94** 162109
[14] Nakamura H and Kimura T 2009 *Phys. Rev.* B **80** 121308
[15] Huang Y W et al. 2005 *J. Appl. Phys.* **97** 10D504
[16] Shuto Y, Yamamoto S and Sugahara S 2009 *J. Appl. Phys.* **105** 07C933
[17] Koo H C, Kwon J H, Eom J, Chang J, Han S H and Johnson M 2009 *Science* **325** 1515
[18] Gelabert M M, Serra L, Sanchez D and Lopez R 2010 *Phys. Rev.* B **81** 165317
[19] Yu A Bychkov et al. 1984 *J. Phys.* C: *Solid State Phys.* **17** 6039
[20] Nemnes G A, Wulf U and Racec P N 2004 *J. Appl. Phys.* **96** 596
[21] Nemnes G A, Wulf U and Racec P N 2005 *J. Appl. Phys.* **98** 084308
[22] Nemnes G A, Ion L and Antohe S 2009 *J. Appl. Phys.* **106** 113714
[23] Nemnes G A, Ion L and Antohe S 2010 *Physica* E **42** 1613

Journal of Physics: Conference Series **338** (2012) 012013 doi:10.1088/1742-6596/338/1/012013

Nonadiabatic generation of spin currents in a quantum ring with Rashba and Dresselhaus spin-orbit interactions

Marian Niţă[1], D. C. Marinescu[2], Bogdan Ostahie[1], Andrei Manolescu[3,*], Vidar Gudmundsson[4]

[1]National Institute of Materials Physics, P.O. Box MG-7, Bucharest-Magurele, Romania

[2]Department of Physics and Astronomy, Clemson University, Clemson, South Carolina 29634, USA

[3]School of Science and Engineering, Reykjavik University, Menntavegur 1, IS-101 Reykjavik, Iceland

[4]Science Institute, University of Iceland, Dunhaga 3, IS-107 Reykjavik, Iceland

[*] E-mail: manoles@ru.is

Abstract. When subjected to a linearly polarized terahertz pulse, a mesoscopic ring endowed with spin-orbit interaction (SOI) of the Rashba-Dresselhaus type exhibits non-uniform azimuthal charge and spin distributions. Both types of SOI couplings are considered linear in the electron momentum. Our results are obtained within a formalism based on the equation of motion satisfied by the density operator which is solved numerically for different values of the angle ϕ, the angle determining the polarization direction of the laser pulse. Solutions thus obtained are later employed in determining the time-dependent charge and spin currents, whose values are calculated in the stationary limit. Both these currents exhibit an oscillatory behavior complicated in the case of the spin current by a beating pattern. We explain this occurrence on account of the two spin-orbit interactions which force the electron spin to oscillate between the two spin quantization axes corresponding to Rashba and Dresselhaus interactions. The oscillation frequencies are explained using the single particle spectrum.

PACS numbers: 73.23.Ra, 71.70.Ej, 72.25.Dc, 73.21.Hb

1. Introduction

The controlled induction of spin and charge currents in mesoscopic rings represents a longstanding interest of experimentalists and theorists alike. By taking advantage of the right-left asymmetry of the electron states, various methods have been proposed over the years to this end [1, 2, 3]. In particular, by applying an ultrashort, terahertz frequency laser pulse with spatial asymmetry, the charge and spin excitations are realized in a non-adiabatic regime, since their excitation occurs on a time scale that is much shorter than the electron relaxation lifetime [3, 4, 5]. More recently, these ideas have been reconsidered in the case of rings made out of zinc-blend semiconductor structures that exhibit spin-orbit interaction (SOI). The linear coupling between the electron momentum and spin appears as a result of broken inversion symmetry that results from spatial confinement (Rashba - R) [6] or from bulk properties (Dresselhaus - D) [7]. Under these general circumstances, it was shown that the spin and charge flow simultaneously [8, 9, 10] and special configurations, such as hybrid structures [11] or a number constraint (odd number of particles) [12] are needed to isolate the spin contribution. Recently, we demonstrated that by applying a two-component terahertz frequency pulse, for certain values of the phase difference angle ϕ between the two components, a pure spin current is non-adiabatically induced in the ring. Physically, this situation is a consequence of the interplay between SOI which rotates the electron spin around the ring and the asymmetry of the external perturbation embodied by ϕ.

The present paper discusses a follow-up to this analysis by studying the effect of a terahertz laser pulse on a ring endowed with both Rashba and Dresselhaus spin-orbit couplings. The coexistence of these couplings introduces precession effects of the electron spin along different directions, a phenomenon bound to affect the spin and charge distributions around the ring. Our formalism is based on obtaining the stationary solutions of the equation of motion satisfied by the particle-density operator which are later employed in estimating the spin and charge currents as as statistical averages. The three independent parameters of the problem are the phase of the laser excitation, ϕ, and the coupling strengths of the Rashba and Dresselhaus interactions. Our main result is that, for comparable R and D intensities, the spin current, calculated along the $\hat{z}$ direction, showcases an oscillatory pattern that results from the nutation of the electron spin between the two natural directions associated with the two spin-orbit interactions.

2. The Equilibrium Hamiltonian

We consider a one-dimensional (1D) quantum ring of radius r_0 in the presence of both Rashba and Dresselhaus spin-orbit interactions. The ring is discretized in N sites with a lattice constant $a = 2\pi r_0/N$. The polar coordinates are chosen such that the site with index n has an azimuthal angle $\theta_n = 2n\pi/N$. We denote by $c^+_{n,\sigma}$ and $c_{n,\sigma}$ the creation and annihilation operators for an electron with spin σ on the site $n = 1, ..., N$.

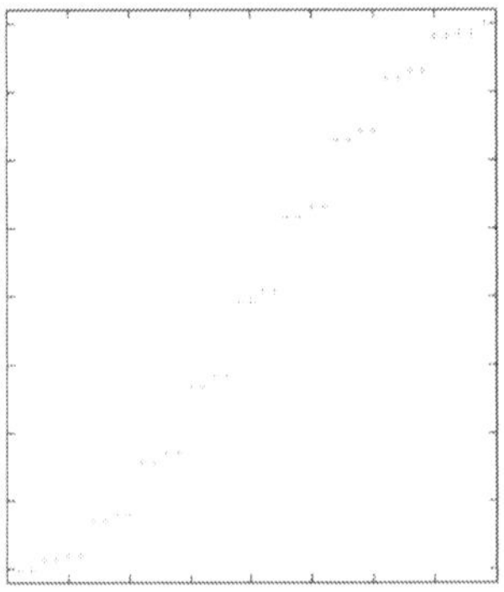

Figure 1. The energy spectrum of a ring with SOI. The Rashba energy $V_\alpha = 0.1$ and the Dresselhaus energy $V_\beta = 0.05$. On the horizontal axis is the rank of the energy value within the spectrum of the Hamiltonian. All states are spin degenerate. The energy unit is $V = \hbar^2/2m^*a^2$. At the initial time only the first 6 energy levels are occupied.

In the following considerations we adopt the discrete representation of the Hamiltonian $H_{\rm ring} = H_0 + V_R + V_D$, that describes the non-interacting electron dynamics around the ring in the presence of the Rashba (R) and Dresselhaus (D) SOI couplings. Its components are:

$$H_0 = 2V \sum_{n,\sigma} c^+_{n,\sigma} c_{n,\sigma} - V \sum_{n,\sigma} c^+_{n,\sigma} c_{n+1,\sigma} - V \sum_{n,\sigma} c^+_{n,\sigma} c_{n-1,\sigma}, \tag{1}$$

$$V_R = -iV_\alpha \sum_{n,\sigma,\sigma'} \sigma_r(\theta_{n,n+1})^{\sigma,\sigma'} c^+_{n\sigma} c_{n+1\sigma'} + \text{h.c.}, \tag{2}$$

$$V_D = -iV_\beta \sum_{n,\sigma,\sigma'} \sigma^*_\theta(\theta_{n,n+1})^{\sigma,\sigma'} c^+_{n\sigma} c_{n+1\sigma'} + \text{h.c.}. \tag{3}$$

Three different energy scales are introduced: the hopping energy $V = \hbar^2/2m^*a^2$, Rashba energy $V_\alpha = \alpha/2a$ and Dresselhaus energy $V_\beta = \beta/2a$, with α and β being SOI parameters and m^* the effective mass of the electron in the host semiconductor material. It is customary to introduce the radial and azimuthal spin matrices σ_r and σ_θ

$$\sigma_r(\theta) = \sigma_x \cos\theta + \sigma_y \sin\theta\,, \quad \sigma_\theta(\theta) = -\sigma_x \sin\theta + \sigma_y \cos\theta\,. \tag{4}$$

In the presence of only one type of SOI, either Rashba or Dresselhaus, the eigenstates of the Hamiltonian can be obtained analytically [9, 10, 8]. In both cases, the spin quantization axis is tilted relatively to the z axis by angles Ω_α and Ω_β respectively, given by

$$\tan\Omega_{\alpha,\beta} = \frac{V_{\alpha,\beta}}{V\sin(\Delta\theta/2)}\,, \tag{5}$$

where $\Delta\theta = 2\pi/N$. Consequently, if either $V_\beta = 0$ or $V_\alpha = 0$, there exist a corresponding natural direction of the spin, given by $\mathbf{e}_\alpha = \cos\Omega_\alpha \mathbf{e}_z - \sin\Omega_\alpha \mathbf{e}_r$ or $\mathbf{e}_\beta = \cos\Omega_\beta \mathbf{e}_z + \sin\Omega_\beta \mathbf{e}_{-\theta}$, respectively, along which the projection of the electron spin is conserved and equal to $\pm\hbar/2$. The directions $\mathbf{e}_\alpha$ and $\mathbf{e}_\beta$ are called the principal spin axes, or Rashba and Dresselhaus spin quantization axes, respectively. The difference between the Rashba and the Dresselhaus SOIs is reflected by the orientation of the spin precession relatively to the orbit: in the Rashba case, the precession and the orbital angular speed have the same sign, whereas in the Dresselhaus case they are opposite [13, 10].

When both interactions are present the diagonalization of the Hamiltonian can be performed only by numerical methods. The energy spectrum for $N = 20$ sites and with $V_\alpha = 0.1V$ and $V_\beta = 0.05V$ is shown in Fig. 1. Each state is double (spin) degenerated. For a ring radius $r_0 = 50$ nm our SOI strengths become comparable to those in experimental materials [13]. One obtains $\alpha \approx 21$, $\beta \approx 10$ meVnm for InAs (i. e. using $m^* = 0.023m_e$), or $\alpha \approx 34$, $\beta \approx 17$ meVnm for InSb (i. e. using $m^* = 0.014m_e$). Our time unit is $t = \hbar/V$, corresponding to ≈ 0.1 ps for InAs and ≈ 0.06 ps for InSb. We will consider $n_e = 6$ (noninteracting) electrons in the ring. The occupied states have energies $E_{1,2} = -0.01184, E_{3,4} = 0.07311, E_{5,6} = 0.1003$. The next three energy values are $E_{7,8} = 0.3484, E_{9,10} = 0.3967$, and $\mathrm{E}_{11,12} = 0.3967$ (units of V). The even and odd labels correspond to positive and negative spin chiralities, respectively.

The combination of the two types of SOI breaks the circular symmetry of the electron distribution and creates a charge deformation as shown in Fig. 2. This effect has been discussed in the recent literature both for one-dimensional [10] and two-dimensional [14] quantum rings. In the one-dimensional ring model the charge density has two minima along the direction $y = x$ and two maxima along $y = -x$. The effect can be explained by an intrinsic periodic potential that develops around the ring [10],

$$U(\theta) = 2V_\alpha V_\beta \sin(2\theta)\,. \tag{6}$$

3. Time dependent perturbation

At time $t = 0$ the quantum ring is excited by a short terahertz laser pulse linearly polarized along a direction making the angle ϕ with the y axis. The pulse has a lifetime of $1/\Gamma$ and an angular frequency ω and can be represented by the time dependent Hamiltonian

$$H_{\text{pulse}}(t) = Ae^{-\Gamma t}\sin(\omega t)\cos(\theta + \phi)\ . \tag{7}$$

The time evolution of the system is given by the Liouville equation,

$$i\hbar\dot{\rho}(t) = [H + H_{pulse}(t),\ \rho(t)]\ , \tag{8}$$

where $\rho(t)$ is the time dependent statistical operator. The time evolution of the electron states, assumed to be initially in the minimum energy configuration, is described by the numerical solutions of the Liouville equation obtained with the Crank-Nicolson finite

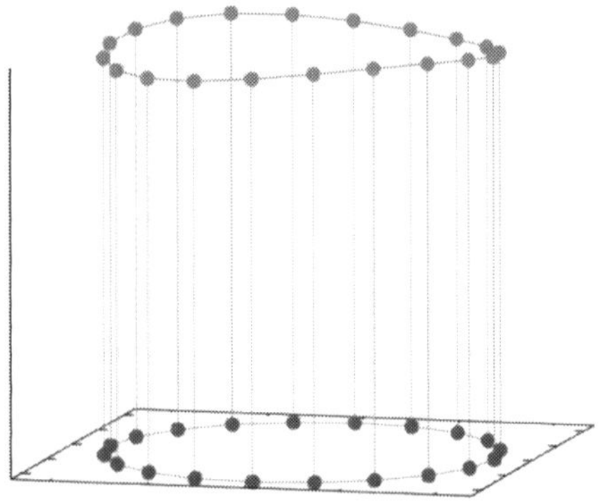

Figure 2. The electron density around the ring. Due to the presence of both SOIs the distribution of electrons is nonuniform. Minima occur along the direction $y = x$, or at angles $\theta = \pi/4$ and $\theta = 5\pi/4$, and maxima along $y = -x$ or at angles $\theta = 3\pi/4$ and $\theta = 7\pi/4$. The blue dots indicate the $N = 20$ sites on the ring.

difference method [3] with small time steps $\delta t \ll \Gamma^{-1}$. The expectation value of any observable O is then calculated as $\langle O \rangle = \mathrm{Tr}\,[\rho(t)O]$.

4. Numerical results

In the following example the pulse parameters are $A = 2.3V$, $\omega = 0.3V/\hbar$, and $\Gamma = \omega$, generating a pulse lifetime of $3.3\hbar/V$. The time evolution of the charge current $I_c(t) = ev(t)$ for $\phi = 0$ is shown in panel (a) of Fig. 3. After the pulse vanishes, the charge current exhibits a large period oscillatory behavior. At high frequencies, superimposed secondary oscillations are noticeable. These findings can be explained on account of single electron excitations. The large period is correlated with transitions between the second and the third states on the energy scale, $\{3,4\}$ and $\{5,6\}$ respectively, that generate $2\pi/(E_6 - E_4) \approx 231\hbar/V$. Transitions between the states $\{3,4\}$ and $\{1,2\}$, occur with period $74\hbar/V$, i. e. one third of the large period, have less weight and cannot be distinguished from the large oscillations. Excitations over larger energy intervals, like $2\pi/(E_8 - E_6) \approx 23\hbar/V$, or $2\pi/(E_{10} - E_2) \approx 15\hbar/V$, etc. are responsible for the finer oscillations, with periods varying from $50\hbar/V$ down to $15\hbar/V$ or less.

Next, in Fig. 3(b) we illustrate the time evolution of the spin current, defined by $I_{S_z} = \frac{\hbar}{2}(\sigma_z v_\theta + v_\theta \sigma_z)$, i. e. polarized along the $\hat{z}$ direction, for $\phi = 0$. This instance indicates more clearly that the time evolution of the system is not periodic even long after the external pulse has vanished. A striking beating pattern is observed, on a background of fast oscillations, faster and much more pronounced than for the charge current. To explain the fast oscillations we refer to Fig. 4, which shows the time dependence of the

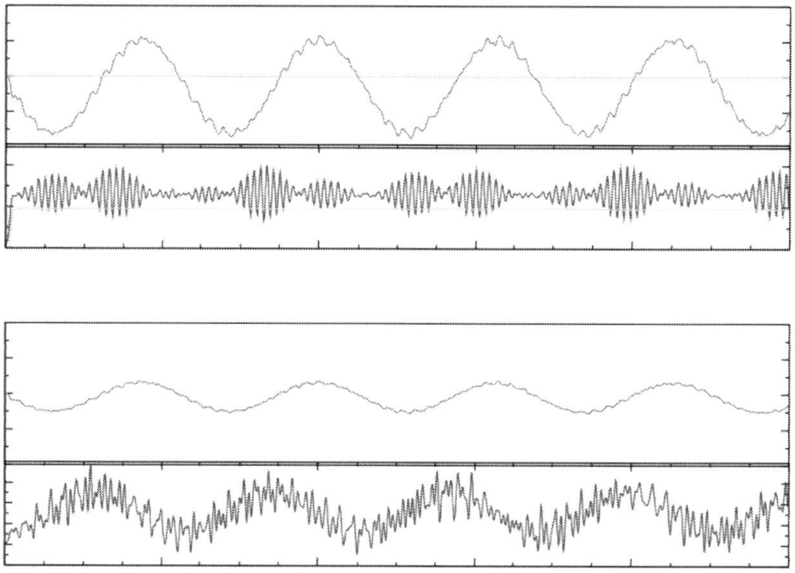

Figure 3. Charge and spin currents in the presence of both Rashba and Dresselhaus SOI. $V_\alpha = 0.1V$ and $V_\beta = 0.05V$. The upper panels show the results for the relative angle $\phi = 0$, and the lower panels for $\phi = \pi/5$. The oscillations are explained by the Bohr frequencies of the energy spectrum, see text. The green lines in the upper panels show the results obtained in [5] with only one type of SOI interaction. The time is in units of $\hbar/V$

spin projection on the z axis within a time interval of 200 units. A series of double peaks is noticeable created by the two SOI couplings that generate spin torque in opposite directions. The result is a superposition of Rashba and Dresselhaus precessions, with a certain relative phase shift. Consequently, the fast oscillations seen in the charge current are doubled for S_z and for the the spin current. The average time between consecutive maxima in Fig. 4 is about 7.4 $\hbar/V$ which is about half of the time corresponding to the fast oscillations of the charge current within the same time interval.

We find, therefore, that the spin current is much more sensitive to the single electron transitions between states with distant energies, implicating high frequencies, than the charge current. The beating pattern is a result of that. Since the SOI parameters are different, $V_\alpha \neq V_\beta$, the principal spin axes $\mathbf{e}_\alpha$ and $\mathbf{e}_\beta$ have different angles with the z direction, $\Omega_\alpha \neq \Omega_\beta$, and the electron spin executes rapid oscillations between them. In physical terms the orbital motion is accompanied by spin nutation.

At equal SOI parameters, $V_\alpha = V_\beta$, the principal directions do not coincide, but make the same angle with the z axis, $\Omega_\alpha = \Omega_\beta$. The spin precessions cancel each other and the corresponding spin current vanishes. However the charge current remains

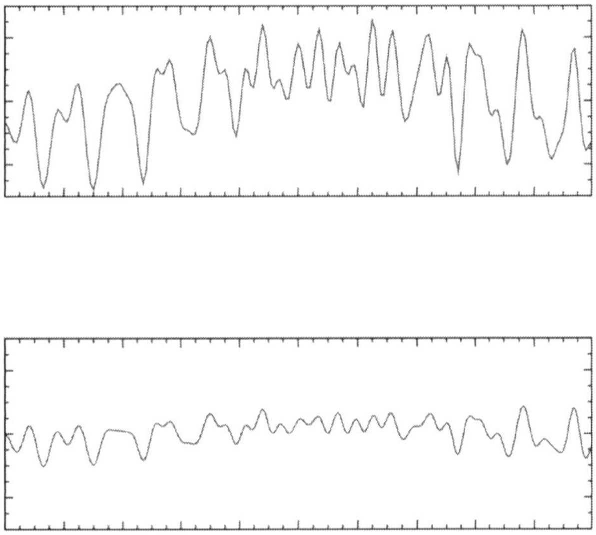

Figure 4. The spin projection S_z within a time subinterval. The fluctuations look like a series of split peaks. The time is in units of $\hbar/V$

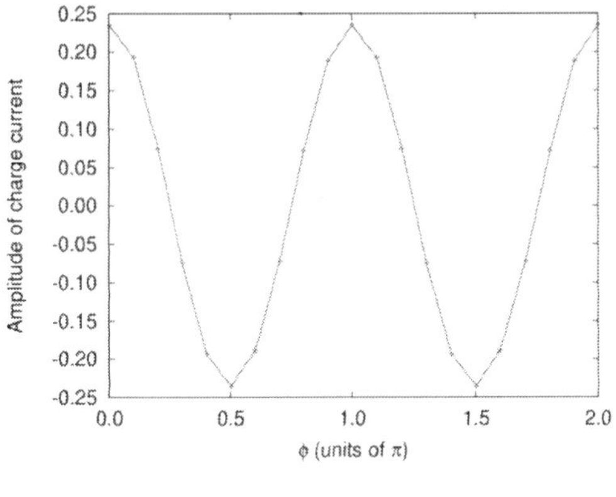

Figure 5. Amplitude of the charge current vs relative angle

qualitatively as in Fig. 3. If only one type of SOI is present, the spin current is not zero, but constant, whereas the charge current vanishes [5], as shown by the green lines in Fig. 3 (a) and (b).

In the lower panels of Figs. 3, (c) and (d), we show the results for the charge and spin currents calculated at $\phi = \pi/5$. When compared with the previous case, the amplitude of the slow oscillations of the charge current decreases, while that of the spin current increases. Fast oscillations are still pronounced in the spin current. Moreover, the spin current shows similar oscillations as the charge, with a phase shift of $\pi/2$, which dominate now the nutation effects. This effect can be explained on account of the anisotropy of the charge distribution shown in Fig. 2. The symmetry axis of the ring corresponds to $\theta = \pi/4$, and therefore if the pulse is oriented along this direction no charge current is induced. This is seen further in Fig. 5 where we show the amplitude of the charge current vs. the angle ϕ, which vanishes for $\phi = \pi/4$. In this case a spin current of maximum amplitude is induced due to the opposite chiralities of the states. The slow oscillations overlap with the fast fluctuations described earlier. At $\phi = 0$, the charge current has a maximum amplitude, whereas that of the spin current is minimal and reduced to the fast fluctuations related to spin nutation. The angle $\phi = \pi/5$ is used in Fig. 3(c) and (d) as the closest value to $\pi/4$ which can be obtained on our discretized ring with 20 sites.

5. Conclusion

The simultaneous effect of Rashba and Dresselhaus spin-orbit interactions upon spin and charge currents are estimated numerically in a ring geometry. A linearly polarized terahertz laser pulse is used for a non-adiabatic generation of the currents. In the presence of the dual spin-orbit coupling, the charge and spin currents exhibit a quasi-periodic behavior whose amplitude depends on the relative angle between the pulse and the charge deformation created by the SOI effects. The spin current reflects a nutation of the electron spin between the Rashba and Dresselhaus spin quantization axes, superimposed as a fast beating pattern on the main oscillation.

Acknowledgments

This work was supported by the Icelandic Research Fund and the Romanian PNCDI2 Research Programme and Core Programme (Contract No. 45N/2009).

References

[1] M. Moskalets and M. Büttiker 2003 *Phys. Rev. B* **68** 075303
[2] M. Moskalets and M. Büttiker 2003 *Phys. Rev. B* **68** 161311
[3] V. Gudmundsson, C. S. Tang, and A. Manolescu 2003 *Phys. Rev. B* **67** 161301(R)
[4] S. S. Gylfadottir, M. Niţă, V. Gudmundsson, and A. Manolescu 2005 *Physica E* **27** 278
[5] M. Niţă, D. C. Marinescu, A. Manolescu, and V. Gudmundsson 2011 *Phys. Rev. B* **83** 155427
[6] Y. Bychkov and E. I. Rashba 1984 *JETP Lett.* **39** 78
[7] G. Dresselhaus 1955 *Phys. Rev.* **100**, 580
[8] J. Splettstoesser, M. Governale, and U. Zülicke 2003 *Phys. Rev. B* **68** 165341
[9] S. Souma and B. K. Nikolić 2004 *Phys. Rev. B* **70** 195346
[10] J. S. Sheng and K. Chang 2006 *Phys. Rev. B* **74**, 235315
[11] Q.-f. Sun, X. C. Xie, and J. Wang 2007 *Phys. Rev. Lett.* **98**, 196801
[12] G.-Y. Huang and L.Shi-Dong 2009 *EPL* **86**, 67009

[13] T. Ihn 2010 *Semiconductor nanostructures. Quantum states and electronic transport* (Oxford University Press)
[14] M. P. Nowak and B. Szafran 2009 *Phys. Rev. B* **80** 195319

Ab-initio investigation of point-like defects in AlN nanowires

Adela Nicolaev[1], T.L. Mitran[1], G.A. Nemnes[1], L. Ion[1,2] and S. Antohe[1]

[1] University of Bucharest, Faculty of Physics, str. Atomistilor nr. 405, Magurele-Ilfov, 077125, Romania

E-mail: lucian@solid.fizica.unibuc.ro

Abstract. We have studied the relaxed atomic configurations and formation energies of intrinsic and some extrinsic point-like defects in atomic-sized würtzite AlN wires, by means of density-functional calculations, using local basis sets. The results indicate that the most likely native defects in small AlN nanowires are interstitial N atoms, while in the case of doping with Si, the impurity is most likely to substitute an Al atom.

1. Introduction

It is well known that structures at the nanometer scale exhibit peculiar effects that make them interesting candidates for future quantum information technology [1], UV [2] and field emitters [3] or chemical [4] [5] and temperature sensors. The fabrication of quantum gates and qubit manipulation is an essential part for such technology. Recently, single molecule, atomic chain and break-junction devices have attracted a great deal of interest, from both experimental and theoretical point of view. As such, mechanically controlled break-junction techniques are used for producing mono-atomic chains. Unlike their bulk counterparts, the influence of structural point-like defects on the physical properties of nanostructured systems is much less known, both from the experimental and theoretical point of view. The changes in the conduction properties produced by some intrinsic or extrinsic defects which are likely to appear in structures such as nanowires with a cross section of one or more lattice constants were addressed in [6, 7]. Since the results in [7] show that the transport properties are drastically influenced by the presence of impurities, a more detailed study of defects in AlN nanowires is considered in the present work. Up to this point, most theoretical studies have either investigated defects in bulk AlN [8], surface defects in AlN nanowires [9] or point defects in single walled AlN nanotubes [10], but very few have been focused on the study of point defects in AlN nanowires.

In this study we investigate atomic configurations and formation energies of small-diameter AlN nanowires containing native defects (cation and anion vacancies and interstitials) and Si impurities using *ab-initio* density-functional calculations.

[2] corresponding author

Journal of Physics: Conference Series **338** (2012) 012014 doi:10.1088/1742-6596/338/1/012014

2. Calculation Method

Atomic-sized würtzite AlN nanowires with longitudinal axis along the [001] direction were investigated. All calculations were based on the density-functional theory (DFT) method, in the local-density approximation (LDA) as implemented in the SIESTA method [11], with the exchange-correlation term parametrized as proposed by Ceperley and Alder [12]. The wavefunctions were expanded using a localized double-ζ polarized basis set; such localized basis orbitals allow to achieve a linear scaling of the computational time with the number of atoms involved. The Monkhorst-Pack [13] scheme with ($1 \times 1 \times 5$) mesh points in k-space was used to perform integrations over the Brillouin zone.

A super-cell containing six pairs of Al-N atomic layers (equivalent to three unit cells along [001] crystalline direction) was used, with sufficient vacuum space in the two transverse directions (60.2 Åby 60.2 Å) to avoid interaction between wires in adjacent cells. While considering as initial input the experimental values for the würtzite bulk lattice constants, $a_{AlN} = 3.112$ Å, $c_{AlN} = 4.982$ Å, a structural relaxation was performed until the inter-atomic forces were less than 0.01 eV/Å. For all considered neutral point-like defects, the formation energy was calculated as in [19]:

$$E_f(X, q) = E_{def}^{(tot)} - E_{perf}^{(tot)} - n_{Al}\mu_{Al} - n_N\mu_N - n_X\mu_X, \quad (1)$$

where $E_{def}^{(tot)}$ is the total energy of the relaxed, defected nanowire, $E_{perf}^{(tot)}$ is the total energy of the nanowire without defects, μ_{Al}, μ_N and μ_X are the chemical potentials of Al,N and impurity X, and n_{Al}, n_N and n_X are the number of exchanged particles between the super-cell and the reservoir; $n > 0$ if atoms are added to the nanowire and $n < 0$ if atoms are removed from it. Another important parameter used when computing the formation energy is the chemical potential reference. The results presented here were obtained for both technologically limited conditions: the case of a nitrogen rich or that of an aluminum rich growth environment. As indicated in [19], the upper bounds for the chemical potential are given by: $\mu_{Al} \leq \mu_{Al}^{bulk}$ and $\mu_N \leq \mu_N^{bulk}$. The reason for choosing as reference the chemical potential of solid nitrogen is because these calculations were carried out for 0 K, where AlN, Al and N are found in their solid phase. The lower bounds of the chemical potential are given by :

$$\mu_{Al} + \mu_N = E_{AlN}^{tot} \quad (2)$$

where E_{AlN}^{tot} is the total energy for a Al-N pair in the defect-free nanowire and μ_{Al} or μ_N are found by fixing either one at a time. The bulk chemical potentials for N is $\mu_N^{bulk} = -269.80eV$ and for Al, $\mu_{Al}^{bulk} = -56.97eV$. In the case of Si defects, the chosen reference chemical potential is that of bulk silicon ($\mu_{Si}^{bulk} = -160.26eV$) because the main chemical component that can form in the presence of Si is silicon nitride which has a molar enthalpy of formation of 829 kJ, according to [20], being more than double that of aluminum nitride (320 kJ [21]).

3. Results and discussions

As previously observed for boron nitride in [14], the perfect $R\,1$ relaxed nanowire is of graphite-like type (Fig. 1 - images obtained with xcrysden [15]); the Al-N bonds lengths are indicated in Table 1. The $R\,1$ nanowire can also exist in a collapsed, graphite-like phase (Fig. 1) with bond lengths also indicated in Table 1. An important piece of information is that the collapsed $R\,1$ nanowire has a slightly more stable structure than the cylindrical one, having an energy lower by 0.02 eV per Al-N pair. In order to determine if the increased stability of the collapsed phase is valid for larger radii, a single walled nanotube with 40 atoms/super-cell was also investigated. The collapsed nanotube, which is basically a two layer AlN ribbon, has an energy 0.15 eV lower

	Bond between atoms	Bond length (Å)
R 1 nanowire		
in the transversal plane		1.97
along the [001] direction		1.88
collapsed R 1 nanowire		
in the transversal plane		2.03
along the [001] direction		1.88
R 2 nanowire		
in the transversal plane		
	1-2	1.82
	2-3	1.87
	3-4	1.99
along the [001] direction		
	1	1.79
	2	1.79
	3	1.94
	4	1.95
R 3 nanowire		
in the transversal plane		
	a-1	1.80
	1-2	1.87
	2-3	2.00
	3-b	1.92
along the [001] direction		
	1	1.78
	2	1.92
	3	1.95

Table 1. Bond lengths in the defect-free nanowires

per Al-N pair than the cylindrical one, indicating a result which is in agreement with the ones presented in [16], [17] and [18] for carbon nanotubes. The bond lengths of the defect-free R 2 nanowire (Fig. 2) and the defect-free R 3 nanowire (Fig. 1) are also indicated in Table 1.

Because the nanowire has a broken translational symmetry in the transversal plane, Al and N vacancies have more nonequivalent locations than in the bulk material, as can be seen in Fig. 1 for the R 1 nanowire (in both cylindrical and collapsed phase) and the R 3 nanowire, and in Fig. 2 for the R 2 nanowire. When indicated by "center" it means that in the relaxed configuration the impurity resides in the interior of the nanowire and if no indication is given, the impurity is found in the plane or on the exterior surface of the nanowire, forming bonds with atoms of the nanowire wall.

It is worth mentioning that the collapsed R 1 nanowire with an interstitial Al atom reaches a single relaxed configuration, independent on the initial positions, with the Al atom forming a 1.97 Ålong bond with an N atom, and a 134.6 ° angle with both walls of the nanowire. On the other hand, the N interstitial can be found in three relaxed final configurations, forming

Journal of Physics: Conference Series **338** (2012) 012014 doi:10.1088/1742-6596/338/1/012014

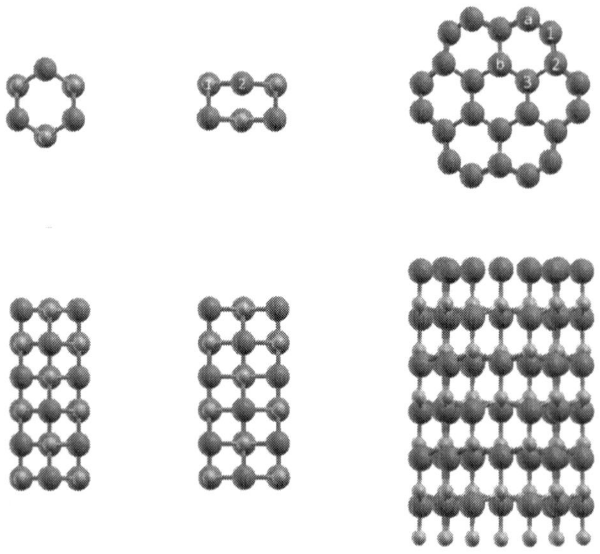

Figure 1. Relaxed R 1, R 1-collapsed and R 3 nanowires (where indexes a and b are only used to indicate bond lengths)

distorted bonds on either one of the different nanowire walls, or inside in its interior.

An interesting result is also obtained for an Al vacancy in position 1 (Fig. 2) of the *R* 2 nanowire because only for this type of defect the system relaxes into a graphite-like structure (Fig. 2), similar to the one found for *R* 1. The cell suffers a contraction by 25.9% along the nanowire axis from 5.09 Åto 3.77 Å. Since the defected graphite-like phase is more energetically stable than the defect-free nanowire, the formation energy for the Al vacancy in position 1 is negative for both N (-11.51 eV) and Al (-10.31 eV) rich growth conditions. This means that nanowires of this type are very unlikely to exist, especially in the wurtzite phase, since Al vacancies have a natural tendency to appear.

The formation energies of the defects described above were computed in the case of a neutral charge, in both N and Al rich environments. From the defect formation energy it can be seen that in the case of the *R* 1 nanowire (Table 2 and Fig. 3) interstitial N is the most probable intrinsic defect in the N rich limit, while interstitial Al is the most probable in the case of an Al rich growth environment. In the case of Si doping, the most probable defect is a substitution of Al.

In the case of the collapsed *R* 1 nanowire (Table 3 and Fig. 4), the most probable intrinsic defect is interstitial N in the N rich condition, and interstitial Al for the Al rich environment, while Si substitutes Al the easiest.

For the *R* 2 nanowire, the Al vacancy that leads to the graphite-like phase has a decisive role and indicates an unstable structure. As such, the subsequent calculations for the formation energies of other defects were not included.

The largest diameter nanowire, *R* 3 (Table 4 and Fig. 5), is most likely to contain interstitial Al in both N rich and Al rich conditions. Also, in this case, a Si impurity is most likely to appear as a substitutional defect of Al in both N and Al rich conditions.

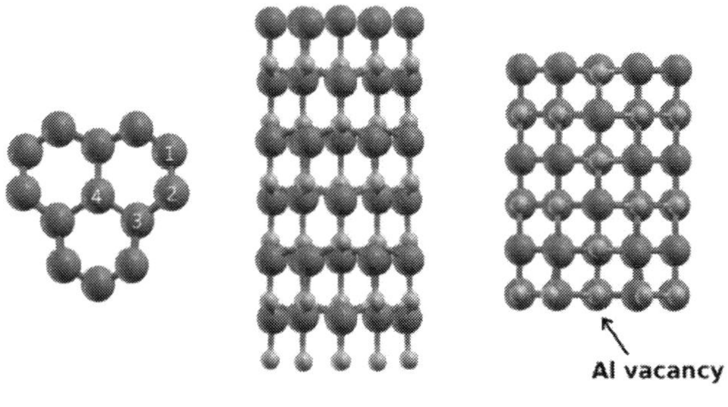

Figure 2. Relaxed R 2 nanowire

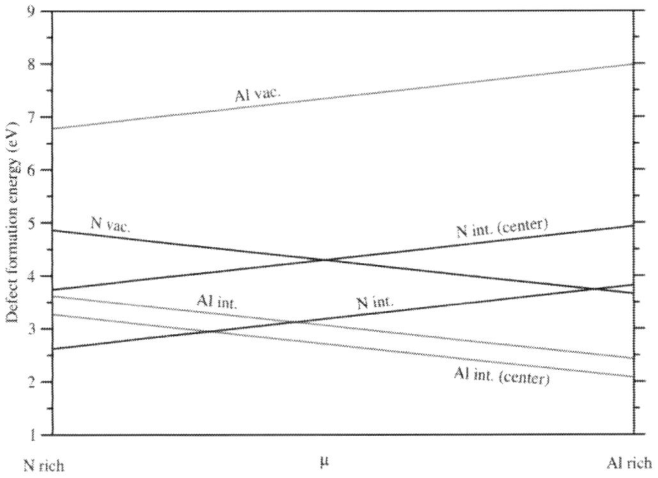

Figure 3. Intrinsic defect formation energies for the R 1 nanowire

Defect type	E_f (eV) - N rich	E_f (eV) - Al rich
Si interstitial	2.61	2.61
Si interstitial (center)	9.26	9.26
Si substitutional (Al)	1.108	2.31
Si substitutional (N)	4.39	3.19

Table 2. Extrinsic defect formation energies for the R 1 nanowire

4. Conclusions

The relaxed configurations and formation energies of selected defect were obtained for four types of AlN nanowires. It was found that single walled nanotubes can also be found in a more energetically favorable collapsed configuration. An indication for the instability of R 2 type

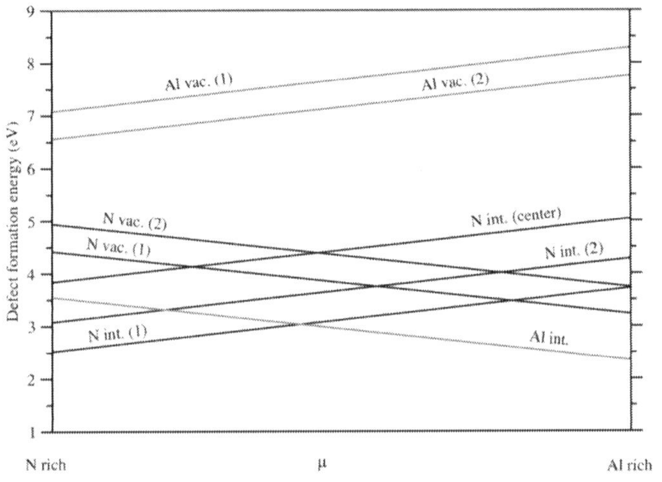

Figure 4. Intrinsic defect formation energies for the R 1 collapsed nanowire

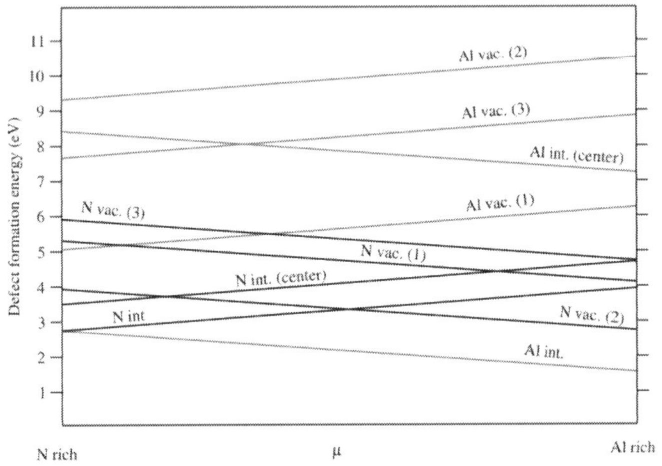

Figure 5. Intrinsic defect formation energies for the R 3 nanowire

Defect type	E_f (eV) - N rich	E_f (eV) - Al rich
Si interstitial	2.69	2.69
Si interstitial (center)	3.64	3.64
Si substitutional Al (1)	1.69	2.89
Si substitutional Al (2)	1.28	2.48
Si substitutional N (1)	4.02	2.82
Si substitutional N (2)	5.31	4.11

Table 3. Extrinsic defect formation energies for the collapsed R 1 nanowire

Defect type	E_f(eV) - N rich	E_f (eV) - Al rich
Si interstitial	1.80	1.80
Si interstitial (center)	9.86	9.86
Si substitutional Al (1)	0.17	1.37
Si substitutional Al (2)	0.71	1.91
Si substitutional Al (3)	1.03	2.23
Si substitutional N (1)	5.98	4.78
Si substitutional N (2)	4.08	2.88
Si substitutional N (3)	6.71	5.51

Table 4. Extrinsic defect formation energies for the R 3 nanowire

nanowires is given by the negative formation energy of Al vacancies. The formation energies indicate the most probable intrinsic defect to appear in R 1 nanowires is that of interstitial N, in N rich environment, and interstitial Al in the Al rich environment, while in the case of R 3 nanowires interstitial Al is most probable. For all the nanowires studied in this paper Si atoms will most easily substitute Al.

5. Acknowledgments

This work was partially supported by the Romanian National Authority for Scientific Research, project ID_1900/2008. A.N. and T.L.M. acknowledge support from ESF through the project POSDRU 107/1.5/S/80765.

References

[1] Babinec T M, Hausmann B J M, Khan M, Zhang Y, Maze J R, Hemmer P R, Lonar M, 2010 Nat. Nanotech. **5** 195
[2] Xu C, Xue L, Yin C, Wang G, 2003 Phys. Stat. Sol. **198** 329
[3] Tang Y B, Cong H T, Zhao Z G, Cheng H M, 2005 Appl. Phys. Lett. **86** 153104
[4] Chattopadhyay S, Shi S C, Lan Z H, Chen C F, Chen K H, Chen L C, 2005 J. Am. Chem. Soc. **127** 2820
[5] Zhou Z, Zhao J, Chen Y, Schleyer P R, Chen Z, 2007 Nanotechnology **18** 424023
[6] Nemnes G A, Visan C, Antohe S, 2011 doi:10.1016/j.physe.2011.07.014
[7] Visan C, Mitran T L, Nicolaev A, Nemnes G A, Ion L, Antohe S 2011 Dig. J. Nanomat. Biostruct. **6** 1173
[8] Mattila T, Nieminen R M, 1996 Phys. Rev. B **54** 23
[9] Ye H, Chen G, Wu Y, Zhu Y, 2009 Phys. Rev. B **80** 033301
[10] Zhukovskii Y F, Popov A I, Balasubramanian C, Bellucci S, 2007 J. Phys.: Conf. Series **93** 012005
[11] Soler J M, Artacho E, Gale J D, Garca A, Junquera J, Ordejn P, Snchez-Portal D 2002 J. Phys.: Cond. Mat. **14** 2745
[12] D.M. Ceperley and B.J Alder, 1980 Phys. Rev. Lett., **45** 566
[13] Monkhorst H J, Pack J D, 1976 Phys. Rev. B **13** 5188
[14] Mitran T L, Nicolaev A, Nemnes G A, Ion L, Antohe S, 2011 Comp. Mat. Sci. **50** 2955
[15] Kokalj A, 2003 Comp. Mater. Sci. **28** 155
[16] Xiao J, Liu B, Huang Y, Zuo J, Hwang K C, Yu M F, 2007 Nanotechnology **18** 395703
[17] Liu H, Cho K, 2004 Appl. Phys. Lett. **85** 5 807
[18] Chopra N G, Benedict L X, Crespi V H, Cohen M L, Louie S G, Zettl A, 1995 Nature **377** 135
[19] Walle C, Neugebauer J, 2004 Appl. Phys. Lett. **95** 3851
[20] Tomaszkiewicz I, Thermal J, 2001 Anal. Cal. **65** 425
[21] Neugebauer C A, Margrave J L, 1957 Z. Anorg. Allg. Chem. **290** 82

Local operation on identical particles systems.

O Rendón[1], and E Medina[2].

[1]Departamento de Física – FACYT, Universidad de Carabobo, Valencia, Edo. Carabobo, Venezuela.

[2]Instituto Venezolano de Investigaciones Científicas, IVIC, Apartado 21827, Caracas 1020 A, Venezuela.

E-mail: orendon@uc.edu.ve

Abstract. We describe identical particles through their extrinsic physical properties and operationally with an operator of selective measure M_m. The operator M_m is formed through non-symmetrized tensor product of one-particle measurement operators, so that it does not commute with the permutation operator P. This operator of selective measure M_m is a local operation (LO) if it acts on physical systems of distinguishable particles, but when M_m acts on the Hilbert sub-space of a system with a constant number of indistinguishable particles this can generate entanglement in the system. We will call entanglement by measurement (EbM), this way of producing entanglement. In this framework, we show entangle production examples for systems of two fermions (or bosons) when the operator M_m has two-particle events that are not mutually exclusive.

1. Introduction

In the basic texts of the quantum mechanics it is very well-known that the identical particles are indistinguishable, so that an observable that distinguishes the intrinsic labels of each particle cannot exist when the wave functions of the individual particles are overlapped, [1]. The indistinguishable approach consists in not to privileging the presence of each particle in one of the quantum states of one-particle, then it is fundamentally ignored which particle is in each one-particle quantum state. In consequence, for indistinguishable particles the wave function is antisymmetrized (fermions) and symmetrized (bosons) and therefore the definition of entangled states as in the case of distinguishable particles does not apply, [2].

Bi-partite entanglement has been studied with different focuses, see [3]: An approach based on the structure of the tensorial product of composite systems ([4], [5] and [6]), one based on the occupation-number representation ([7] and [8]) and another on the mixture of both [9]. In spite of these efforts the indistinguishable particles issue is hard to kill: [3] Horodecki quote: "However, it seems that there is still controversy concerning the meaning of entanglement for indistinguishable particles, [9]"

In the ideal Fermi system, spin entanglement between two distant particles in the ground state, at T= 0 K, has been studied by Vedral ([10] and [11]). He concludes:

- No correlation for spin electrons with different momentum.
- Correlation for spin electrons located in different regions of space, $V \cap V' = \varnothing$. This entanglement depends on the dimensionality and decrease with the distance among disjoint regions.

We develop the concept of "local operation" for two identical particles with possibility of being distinguished for their extrinsic properties following the ideas of: 1) Wiseman and Vaccaro [9] to project a state of N identical particles $|\psi\rangle$ into two distinct well-defined subspaces, and 2) Herbut [12] on the possibility of using both indistinguishable and distinguishable approach to describe a system of two identical particles that are well separated, e.g. $V \cap V' = \varnothing$ or distinct spin orientations along the z axis (assuming these properties don´t change during the time of interest).

Here, we use the distinguishable particle framework and in future studies [13] we show a more comprehensive approach, considering the approach of distinguishable particles as indistinguishable and within a theoretical development based on quantum information concepts [14].

In the following we present the conditions that a "local operation" should have so that it doesn't change the entanglement of a system of two identical particles, then two examples are shown where the entanglement generation for non local operation is evident.

2. "Local Operation" for two identical particles

We define the local operation (LO) for a system of two identical particles, within of the distinct-particle approach. This formalism is valid for fermions and bosons. In the simple version to distinguish two identical particles, we follow to Herbut [12]:

"Actually, we want to generalize the well-know idea that if two particles can be distinguished either because they are known to be localized in well-separated spatial regions, e.g., V and V´, or because they have, e.g., distinct spin orientations along the z axis (and these properties not change during the time of interest). Then we can treat the identical particles as if they were distinct."

Let a distinguishable two-particles Hilbert space $H^{(2)}=H^{(1)}\otimes H^{(1)}$, consisting of one-particle Hilbert space factorable in $H^{(1)}=H'\otimes H''\otimes\ldots$, be defined for example: spatial versus spin component along the z axis ($H^{(1)}=H_r\otimes H_{Sz}$) or momentum versus spin component along the x-axis ($H^{(1)}=H_k\otimes H_{Sx}$).

The operator of selective measure M_m distinguishes particles and has its domain and codomain in $H^{(2)}$, is a local operation (LO) if the following two conditions are satisfied:

a) M_m is a tensor product of one-particle measure selective operator

$$M_m = M_{m1} \otimes M_{m2}, \tag{2.1}$$

generally with arbitrary domain and codomain in one-particle Hilbert space, $H^{(1)}$.

$$M_{m1} : H'' \rightarrow H'' \tag{2.2}$$

$$M_{m2} : H' \rightarrow H' \tag{2.3}$$

b) For any state of distinguishable two-particles, $|\psi\rangle \in H^{(2)}$, it has an orthogonal sum

$$|\psi\rangle = \frac{1}{\sqrt{2}}\left[|\psi\rangle_{fermions} + |\psi\rangle_{bosons}\right]. \tag{2.4}$$

The operator M_m that distinguishes identical particles (fermions or bosons) does not generate entanglement by measurement (EbM) if $M_m|\psi\rangle = 0$ or $M_m P|\psi\rangle = 0$ but not both; where P is the permutation operator.

Any operator that acts on a pair of identical particles and satisfies the above two conditions is a local operation (LO). The operator of selective measure M_m is not symmetrized, see equation (2.1), then it doesn´t commute with the permutation operator P.

The proof of the above proposition is simple; both the Hilbert subspace associated with fermions or bosons decomposes in the same way

$$|\psi\rangle_{fermions} = \frac{1}{\sqrt{2}}[1-P]|\psi\rangle \tag{2.5}$$

$$|\psi\rangle_{bosons} = \frac{1}{\sqrt{2}}[1+P]|\psi\rangle \tag{2.6}$$

where **1** is the identity operator and P is the permutation operator. The ket $|\psi\rangle$ is the quantum state that characterizes two distinguishable particles, with $|\psi\rangle$ orthogonal to $P|\psi\rangle$.

We distinguish two fermions (or bosons) when measured with M_m

$$M_m|\psi\rangle_{fermions} = \frac{1}{\sqrt{2}}\left[M_m|\psi\rangle - M_m P|\psi\rangle\right] \tag{2.7}$$

$$M_m|\psi\rangle_{bosons} = \frac{1}{\sqrt{2}}\left[M_m|\psi\rangle + M_m P|\psi\rangle\right] \tag{2.8}$$

and two events are defined as: $M_m|\psi\rangle$ and $M_m P|\psi\rangle$. If one of the two events $M_m|\psi\rangle$ or $M_m P|\psi\rangle$ is zero, but not both; then M_m distinguishes particles without increasing entanglement and is called local operation (LO). Otherwise, M_m distinguish particles, and contemplates the two events $M_m|\psi\rangle$ and $M_m P|\psi\rangle$ with nonzero probability and $\langle\psi|M_m P|\psi\rangle = 0$. (That is, the kets $M_m|\psi\rangle$ and $M_m P|\psi\rangle$ are geometrically orthogonal); then the ket $|\psi\rangle_{fermions}$ (or $|\psi\rangle_{bosons}$) collapses to a entangled quantum state if the initial state is not entangled, see equation (2.7) or (2.8). We will call this procedure entanglement by measurement (EbM).

2.1. Example of "Local Operation"

Given the non-entangled two particle, [5].

$$|\psi\rangle_{\substack{fermions \\ or\ \ bosons}} = \frac{1}{\sqrt{2}}\left[|k_a k_b\rangle|\uparrow\downarrow\rangle \mp |k_b k_a\rangle|\downarrow\uparrow\rangle\right] \tag{2.9}$$

and the selective measure operator

$$M_m = |k_a k_b\rangle\langle k_a k_b| \tag{2.10}$$

Here the state of two distinguishable particles that builds the states of indistinguishable particle $|\psi\rangle_{fermions}$ (or $|\psi\rangle_{bosons}$) is

$$|\psi\rangle = |k_a k_b\rangle|\uparrow\downarrow\rangle \tag{2.11}$$

and

$$P|\psi\rangle = |k_b k_a\rangle|\downarrow\uparrow\rangle \tag{2.12}$$

with $k_a \neq k_b$.

In this projection, we have used momentum variables of each particle as labels to distinguish particles. We can see that $M_m P|\psi\rangle = 0$ and then $|\psi\rangle_{\substack{fermions \\ or\ bosons}}$ collapses to:

$$M_m|\psi\rangle_{\substack{fermions \\ or\ bosons}} = \frac{1}{\sqrt{2}}|k_a k_b\rangle|\uparrow\downarrow\rangle \tag{2.13}$$

and this selective measure does not increase entanglement in the initial state, see equation (2.9) and (2.13).

The stat $|\psi\rangle_{\substack{fermions \\ or\ bosons}}$, (2.9), describes a particle with momentum k_a and spin up, and the other with momentum k_b and spin down and $M_m|\psi\rangle_{\substack{fermions \\ or\ bosons}}$, (2.13), characterizes the particle k_a with spin up and the particle k_b with spin down. Note that the momentum variables are used as labels. Within the distinguishable approach the identical two-particle state (2.9) can be expressed [15], as:

$$|\psi\rangle_{dist} = |\uparrow_{k_a}\downarrow_{k_b}\rangle \tag{2.14}$$

2.2. An example with operation no-local.
Again, we begin with the state $|\psi\rangle_{\substack{fermions \\ or\ bosons}}$, (2.9), but now it describes an one-dimensional system of two identical particles. We assume, for the sake of the argument, that the momentum $k_a = k$ y $k_b = -k$ have the same magnitude k, but opposite directions.

The state $|\psi\rangle_{\substack{fermions \\ or\ bosons}}$, (2.9), is a momentum self-ket, then the position of each particle is no well-defined and the two events can happen with the same probability a) particles approach each other, or b) away from one another, see figure 1.

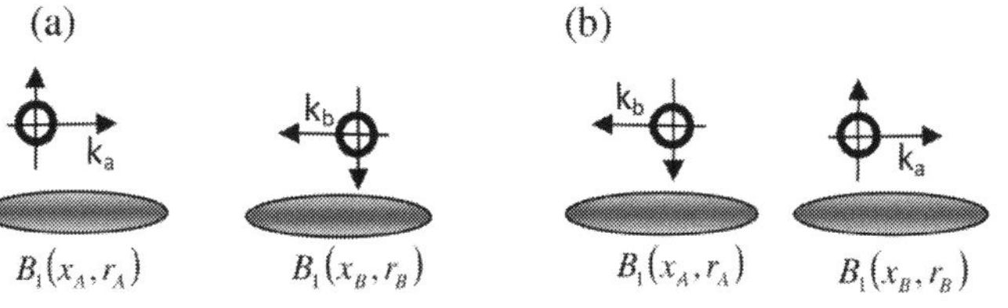

Figure 1. The state $|\psi\rangle_{\substack{fermions \\ or\ bosons}}$, (2.9) describes two possible events (a) particles approach each other, or (b) away from one another.

Lets take an operator of selective measure M_m, projector of the position self-ket.

$$M_m = \int_{B_1(x_A,r_A)} |x_1\rangle dx_1 \langle x_1| \otimes \int_{B_1(x_B,r_B)} |x_2\rangle dx_2 \langle x_2|, \quad (2.15)$$

with

$$B_1(x_A,r_A) = \left(x_A - {r_A}/{2}, x_A + {r_A}/{2} \right), \quad (2.16)$$

and

$$B_1(x_B,r_B) = \left(x_B - {r_B}/{2}, x_B + {r_B}/{2} \right). \quad (2.17)$$

In this projection, we have used position variables for each particle as labels to distinguish particles. If $r_A = r_B \approx \Delta x << 1$, development to the lowest order in Δx, so the ket $|\psi\rangle_{\substack{fermions \\ or\ bosons}}$ collapses to:

$$M_m |\psi\rangle_{\substack{fermions \\ or\ bosons}} = \Delta x^2 \langle x_A | k_A \rangle \langle x_B | k_B \rangle \left\{ |x_A \uparrow\rangle |x_B \downarrow\rangle \pm e^{-i\Delta k \Delta x} |x_A \downarrow\rangle |x_B \uparrow\rangle \right\} \quad (2.18)$$

with $\Delta k = k_B - k_A$ and $\Delta x = x_B - x_A$.

The ket $M_m |\psi\rangle_{\substack{fermions \\ or\ bosons}}$, (2.18), describes two events and the information of the whole systems is that the particles located in different regions $B_1(x_A,r_A)$ and $B_1(x_B,r_B)$have opposite spin component, but we do not know the orientation of the spin in each region $B_1(x_A,r_A)$ and $B_1(x_B,r_B)$, see figure 1. Here, the location of the particles has been entangled in the internal degrees of freedom (spin) and its origin is based on M_m is not operation local.

3. Conclusions

One of the goals of quantum computation is the systematic manipulation of physical objects with quantum properties. In the context of solid state physics, a common task in quantum computing is to use electrons in a material and separate them to two different users, Alice and Bob. The assignment of a quantum system of identical particles to each user, Alice and Bob, is a process of distinguishing identical particles.

The process of associating an identical particle to each user is to give an electron to Alice and one to Bob. In this paper, we have observed that the process of assigning identical particles to different users can be done in different ways.

If the operator of measure selective M_m is local operation (LO), the whole quantum system has the same information to each of its parts, one located in Alice's lab and another in the Bob´s lab. Conversely, if the operator selective measurement M_m is not a local operation, it increases entanglement between parts and the process is called entanglement by measurement (EBM).

In future work [13] we will discuss a more complete theory that contains both distinguishable and indistinguishable approaches to a system of identical particles, this will be done to satisfy a more complete theory in the context of the quantum information.

References

[1] Baym G 1969 *Lectures on Quantum Mechanics* (New York: W.A. Benjamin)
[2] Amico L, Fazo R, Osterloh A, and Vedral V 2008 *Rev. of Mod. Phys.* **80** 517
[3] Horodecki R, Horodecki P, Horodecki M, and Horodecki K 2009 *Rev. of Mod. Phys.* **81** 865
[4] Li Y S, Zeng B, Liu X S, and Long G L 2001 *Phys. Rev.* A **64** 0543022
[5] Schliemann J, Cirac J I, Kus M, Lewenstein M, and Loss D 2001 *Phys. Rev.* A **64** 0223035
[6] Shi Y 2003 *Phys. Rev.* A **67** 0243017
[7] Zanardi P 2002 *Phys. Rev. A* **65** 042101
[8] Zanardi P, and Wang X 2002 *J. Phys.* A **35** 7947
[9] Wiseman H M, and Vaccaro J A 2003 *Phys. Rev. Lett.* **91** 010404
[10] Vedral V 2003 *Cent. Eur. J. Phys.* **1** 197
[11] Cavalcanti D, França Santos M, Terra Cunha M O, Lunkes C, and Vedral V 2005 *Phys. Rev.* A **72** 062307
[12] Herbut F 2001 *Am. J. Phys.* **69** 207
[13] To be submitted.
[14] Nielsen M A, and 2000 *Quantum Computation and Quantum Information* (Cambridge: Cambridge University Press)
[15] Bose S, and Home D 2002 *Phys. Rev. Lett.* **88** 050401

Suppression of Superfluidity upon Overflow of Trapped Fermions. Quantal and Thomas-Fermi Studies

P. Schuck[1,2], **X. Viñas**[3]
[1]Institut de Physique Nucléaire, IN2P3-CNRS, Université Paris-Sud, F-91406 Orsay-Cédex, France
[2] Laboratoire de Physique et Modélisation des Milieux Condensés, CNRS and Université Joseph Fourier, 25 Avenue des Martyrs, Boîte Postale 166, F-38042 Grenoble Cedex 9, France
[3]Departament d'Estructura i Constituents de la Matèria and Institut de Ciències del Cosmos, Facultat de Física, Universitat de Barcelona, Diagonal *647*, *E-08028* Barcelona, Spain

E-mail: xavier@ecm.ub.es

Abstract. Two issues will be discussed in this contribution: (i) the generic fact that if a fermionic superfluid in the BCS regime overflows from a narrow container into a much wider one, pairing is much suppressed at the overflow point. Two examples pertinent to the physics of the outer crust of neutron stars and superfluid fermionic atoms in traps will be presented. (ii) A novel Thomas-Fermi approach to inhomogeneous superfluid Fermi-systems is presented and shown that it works well in cases where the local density approximation breaks down.

1. Introduction

Superfluid fermions in finite systems can exist in traps of cold atoms, in nuclear systems, in small metallic clusters, etc. An interesting question arises what happens to the superfluid if its Fermi level reaches the edge of a finite container, i.e. either the fluid overflows into the continuum or it pours into another container of much larger dimension. A trapping potential of this type has experimentally already been generated for the study of cold bosonic atoms [1]. It should also be possible to use it for fermionic atoms [2].

In the inner crust of neutron stars, there also may occur the situation where coexists a superfluid neutron gas of variable density in between the lattice of (superfluid) nuclei [3, 4]. This situation often is mimicked by a Wigner Seitz cell, where the single particle (self-consistent) potential has a pocket, representing the nucleus, embedded in a large container. Of course, finite nuclei at the neutron drip line also constitute cases of this category but, as we will see, the situation there is not so clear due to strong shell effects. Still other systems may exist with similar situations.

The purpose of this work is to study superfluidity of fermions at the overflow (drip) in the BCS regime. Since the quantal solution of BCS equations in geometries with rapidly varying single particle potentials with a large number of particles is numerically difficult, we will present, as a second objective of this work, a Thomas-Fermi (TF) approach to inhomogeneous superfluidity which shows good performance in situations where the local density approximation (LDA) fails.

Journal of Physics: Conference Series **338** (2012) 012016 doi:10.1088/1742-6596/338/1/012016

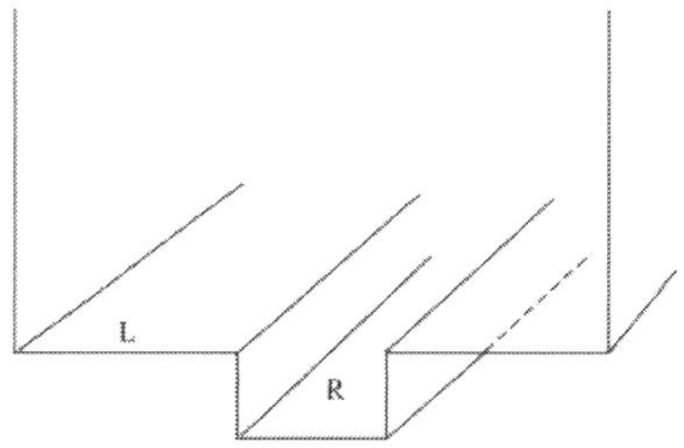

Figure 1. Schematic view of the slab used in this work with a perspective view of the potential which is translationally invariant in x, y direction.

For our study we first will consider a schematic model of slab geometry with a transverse potential of large extension L posessing at the origin a 'pocket' of variable depth and size R much smaller than the outer container. Schematically such a potential is shown in Fig. 1. This slab configuration may roughly mock up one sheet of a so-called Lasagne configuration in the inner crust of neutron stars [5]. We, therefore, will use nuclear dimensions for the slab model but they can easily be replaced by dimensions relevant for other systems. Our model and the ensuing generalisations treated below, therefore, are believed to be generic. We will study the slab configuration also because the quantal solution of the gap equation is evaluated relatively directly and the quality of the TF approach can thus be established. Once this is achieved, we also will go over to other geometrical configurations. We, for instance, will treat a second potential, shown below in Fig. 4, left panel, with spherical symmetry, a kind of which, as already mentioned, has been used for bosonic atoms in [1].

2. The formalism

Let us start out with writing down the usual BCS equations in three dimensions.

$$\Delta_\nu = -\sum_{\nu'} V_{\nu,\nu'} \frac{\Delta_{\nu'}}{2\sqrt{(\varepsilon_{\nu'} - \mu)^2 + \Delta_{\nu'}^2}}, \tag{1}$$

where the ε_ν's are the single particle energies and μ the chemical potential which can be used to fix the particle number $N = \sum_\nu n_\nu$ with

$$n_\nu = \frac{1}{2}[1 - \frac{\varepsilon_\nu - \mu}{\sqrt{(\varepsilon_\nu - \mu)^2 + \Delta_\nu^2}}]. \tag{2}$$

The wave functions and eigenenergies of a box as shown in Fig. 1a with a potential-hole are given in [6]. For pairing, we use a contact force with a cut off Λ, to make things simple. The single particle states in a slab configuration then become $|\nu\rangle = |n, \mathbf{p} >$ where n are the discrete quantum numbers in transverse direction and $\mathbf{p}$ the momentum quantum numbers in slab direction. Integrating over momenta in slab direction the above gap equation

$$\Delta_n = -\sum_{n'} \int \frac{d^2p}{(2\pi\hbar)^2} V_{nn'} \Theta(\Lambda - \varepsilon_{n'} - \varepsilon_p) \frac{\Delta_{n'}}{2E_{n'}(p)}, \tag{3}$$

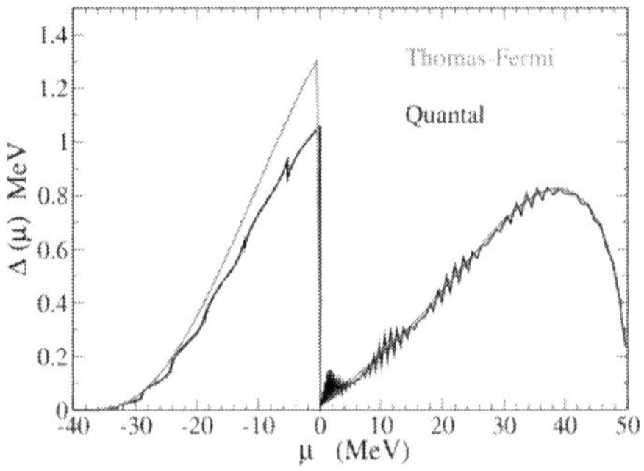

Figure 2. (Coloronline) Quantal and TF pairing gap in the slab geometry as a function of the chemical potential.

with $E_n(p) = \sqrt{(\varepsilon_n + \varepsilon_p - \mu)^2 + \Delta_n^2}$ the quasiparticle energy, $\Theta(x)$ the step function, and $\varepsilon_n, \varepsilon_p$ being the discrete single particle energies in transverse direction and kinetic energies in slab direction, respectively. After performing the integral, one arrives at the following gap equation

$$\Delta_n = -\sum_{n'} \Theta(\Lambda - \varepsilon_{n'}) V_{nn'} K_{n'}. \tag{4}$$

The pairing tensor in equation (4) is then given by

$$K_n = \frac{m}{4\pi\hbar^2} \Delta_n \ln \frac{\Lambda - \mu + \sqrt{(\Lambda - \mu)^2 + \Delta_n^2}}{\varepsilon_n - \mu + \sqrt{(\varepsilon_n - \mu)^2 + \Delta_n^2}} \tag{5}$$

where m is the particle mass and the indices n stand for the level quantum numbers in the confining potential of Fig. 1a. The matrix elements $V_{nn'} = -g \int_{-L}^{+L} |\varphi_n(z)|^2 |\varphi_{n'}(z)|^2 dz$ of the contact force $-g\delta(\mathbf{r} - \mathbf{r}')$ can be evaluated straightforwardly from the wave functions $\varphi_n(z)$ given in [6].

For an example we take as cut off $\Lambda = 50$ MeV counted from the edge of the pocket potential whose depth be $V_0 = -40$ MeV. Its extension ranges from $-R$ to $+R$ with $R = 10$ fm. The wide potential with infinite walls has extension from $-L$ to $+L$ with $L = 100$ fm. For the coupling strength we take $g = 150$ MeV fm^3. The numerical values for Λ and g yield maximum gap values around 1 MeV, typical for nuclear systems.

Before we show the results, let us explain our Thomas-Fermi (TF) approach for this problem. In the weak coupling regime, we have $\Delta/\mu << 1$. In this case the canonical basis can be replaced by the Hartree-Fock or mean-field one:

$$H|n\rangle = \epsilon_n |n\rangle. \tag{6}$$

At equilibrium and for time reversal invariant systems canonical conjugation and time reversal operation are related by

$$\langle \mathbf{r}|\bar{n}\rangle = \langle n|\mathbf{r}\rangle \Rightarrow \langle \mathbf{r}_1\mathbf{r}_2|n\bar{n}\rangle = \langle \mathbf{r}_1|\hat{\rho}_n|\mathbf{r}_2\rangle, \tag{7}$$

with $\hat{\rho}_n = |n\rangle\langle n|$. For the pairing matrix element, we, therefore, can write

$$V_{nn'} = \langle n\bar{n}|v|n'\bar{n}'\rangle = \int \langle \mathbf{r}_2|\hat{\rho}_n|\mathbf{r}_1\rangle\langle \mathbf{r_1 r_2}|v|\mathbf{r}'_1\mathbf{r}'_2\rangle\langle \mathbf{r'}_1|\hat{\rho}_{n'}|\mathbf{r'}_2\rangle d\mathbf{r_1}d\mathbf{r_2}d\mathbf{r}'_1 d\mathbf{r}'_2. \tag{8}$$

The Schroedinger equation (6) can be writeen in terms of $\hat{\rho}_n$ as

$$(H - \epsilon_n)\hat{\rho}_n = 0. \tag{9}$$

Taking the Wigner transform of this latter equation, we obtain in the $\hbar \to 0$ limit the following c-number equation: $(H_{cl.} - \epsilon)f_\epsilon(\mathbf{R}, \mathbf{p}) = 0$. The solution of this equation in the sense of distribution theory is with $x\delta(x) = 0$ given by

$$f_E(\mathbf{R}, \mathbf{p}) = \frac{1}{g^{TF}(E)}\delta(E - H_{cl.}) + O(\hbar^2). \tag{10}$$

with

$$H_{cl.} = \frac{p^2}{2m^*(\mathbf{R})} + V(\mathbf{R}) \quad \text{and} \quad g^{TF}(E) = \frac{1}{(2\pi\hbar)^3}\int d\mathbf{R}d\mathbf{p}\delta(E - H_{cl.}). \tag{11}$$

Equation (10) means that the phase space distribution corresponding to a state $|n\rangle$ at high energy is concentrated around the classical energy shell what, indeed, is a well known fact.
The TF version of the gap equation (4) then reads

$$\Delta(E) = -\int_{V_0}^{\Lambda} dE' g(E')V(E, E')K(E') \tag{12}$$

with $K(E)$ an obvious generalisation of K_n. The matrix elements $V(E, E')$ can be evaluated in replacing $|\varphi_n(z)|^2$ by [7]

$$\rho_E^{TF}(z) = \int \frac{dp}{2\pi\hbar} f_E(z, p) = \frac{1}{g^{TF}(E)}\frac{1}{2\pi}(\frac{2m}{\hbar^2})^{1/2}[E - V(z)]^{-1/2}, \tag{13}$$

which is the on-shell TF density in transverse direction (please note that we are in a 1D case here, contrary to what is treated above where it is 3D). As the reader will easily realise, the way of proceeding is very different from usual LDA where the finite size dependence is put into the (local) chemical potential, $\mu(z) = \mu - V(z)$, whereas here it is put into the matrix elements of the pairing force.

3. Results

We are now in a position to solve the quantal and TF gap equations for the above mentioned parameter values of our model. The result for the gap at the chemical potential μ is shown in Fig. 2 as a function of μ. We start with μ from the bottom of the pocket well, i.e. with zero density. We then increase μ, i.e. the density. We see that once the fill up of the pocket reaches its top, the values of the gap sharply drop and practically reach zero. In the continuum the gaps slowly rise again. We see that quantal and TF values are in close agreement. The overshoot of the TF solution for negative μ very likely is due to the smallness of the pocket which only can accomodate nine bound levels. Before we come to an explanation of the drop of the gaps at overflow (drip), let us study the gaps as a function of position in transverse direction: $\Delta(z) = -gK(z)$ with $K(z) = \sum K_n|\varphi_n(z)|^2$. Semiclassically, this expression becomes:

$$K(z) = \int_{V_0}^{\Lambda} dE g^{TF}(E)K(E)\rho_E^{TF}(z). \tag{14}$$

In Fig. 3, left panel, we show the gap profiles for three values of μ: $\mu = 40$, 0.5, and - 5 MeV. We see that quantal and TF results agree, up to shell fluctuations, very well. We also show the

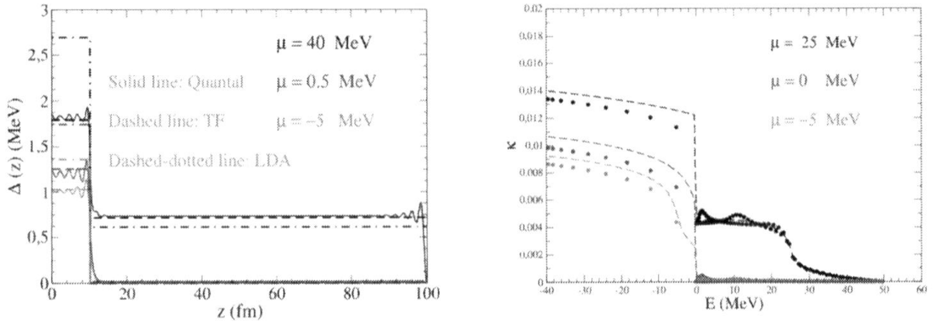

Figure 3. (Coloronline) Left panel: Position dependence of the gap in the slab geometry for different values of the chemical potential. Quantal, TF, and LDA results are shown. Notice that Δ for $\mu = 0.5$ and -5.0 MeV is practically zero in the gas region. Right panel: Comparison of quantal (dots) and TF (broken lines) values of the pairing tensor K.

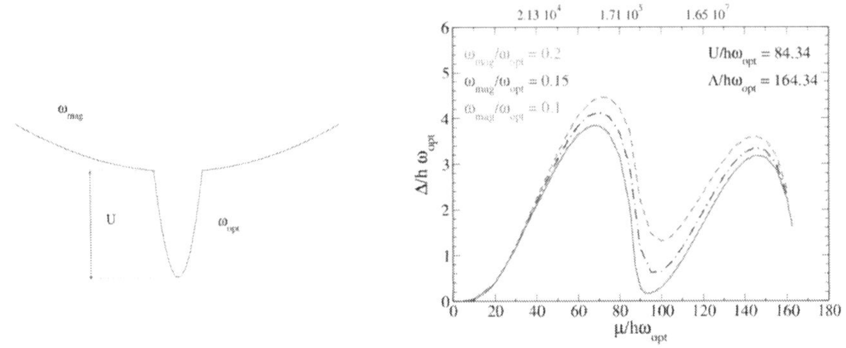

Figure 4. (Coloronline) Average TF gaps at the Fermi energy as a function of the chemical potential (right panel) for the potential shown in the left panel. In the completely filled optical trap ($\mu = U$) we accomodate 10^5 atoms in each spin state. The total number of atoms in the trap with $\mu/\hbar\omega_{opt}$=40, 80 and 120 are indicated in the upper horizontal axis.

LDA results. They can be as wrong as by 50 percent. For other choices of system parameters the LDA error may even be worse. This stems from the fact that in TF (and, of course, also quantally), there is coupling between inside and outside the pocket, i.e. the Cooper pair wave function extends into both regions what tends to equilibrate the values of the gaps. In LDA the contrast is much too strong. The drop of the gaps when crossing the threshold can be explained by the fact that the single particle states are strongly delocalised in the outer container and, thus, their contribution to the pairing matrix element $V_{n,n'}$ becomes very small. In right panel of Fig. 3, we show the quantal and TF pairing tensors, K_n and $K(E)$, defined in (4), (5), and (12), respectively. We emphasize the close agreement between quantal and TF results.

Having gained faith into our TF approach, we now can explore other geometries and other systems, which are more difficult for quantal solutions. In Fig. 4 we show the result for Δ in the spherical double harmonic oscillator potential which may be realised with cold fermionic atoms

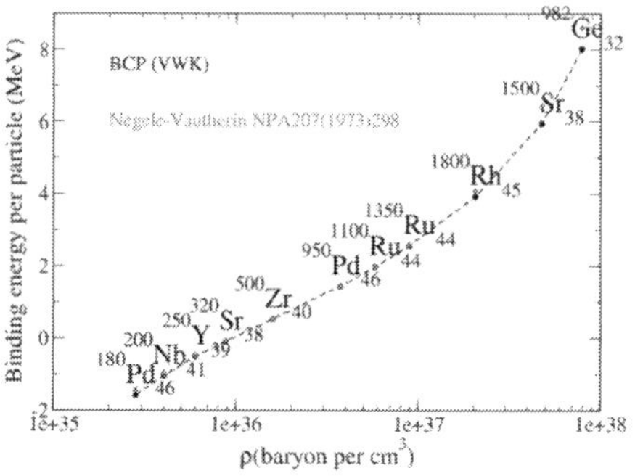

Figure 5. (Coloronline) Binding energies per particle as a function of average density in a Wigner-Seitz cell. Red dots indicate quantal Skyrme HF calculations by Negele-Vautherin and black dots correspond to semiclassical results with the variational Wigner-Kirkwood (VWK) method and the Gogny D1S force.

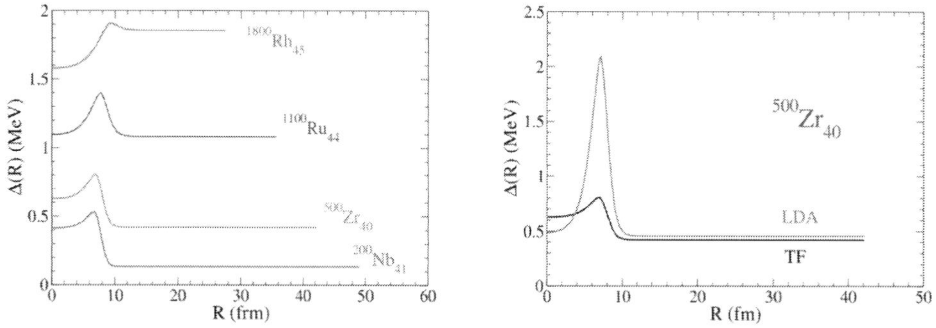

Figure 6. (Coloronline) Left panel: Radial dependence of the TF gap in the considered WS cells. The end points indicate the radius of the WS cells. Right panel: Comparison between TF and LDA gaps as a function of the position in a WS cell containing a single $^{500}_{40}$Zr nucleus.

to study the overflow situation. A zero range force with strength g=-1.0 and cut off Λ=164.34 (in the corresponding optical trap units with $\omega_{opt} = 2\pi \times 1000$ Hz taken from [2]) is used. We see that the result is qualitatively similar to the slab case, though in this spherical geometry the dip does not quite reach zero and also is shifted slightly to an energy above the break of the potential. Note that this depends strongly on the choice of the ratio $\omega_{mag}/\omega_{opt}$ as it can be seen in the figure. Also the gap starts to decrease towards the minimum quite early. This is contrary to what happens in the slab case where the change is very abrupt, the reason probably lying in the spherical symmetry of the considered system. It could be interesting to see whether our prediction can be verified experimentally. We should emphasise that the spherical case has been chosen here for convenience. Nothing essential will, however, change with other geometries, like they are more often realised for trapping potentials with cold atoms.

Let us now make a more realistic study of Wigner-Seitz (WS) cells including electrons in β equilibrium to simulate the inner crust of neutron stars [5]. The mean-field, as explained in [11], is computed selfconsistently using the BCP energy density functional [10] together with the TF approach.

The semiclassical description of the WS cells including pairing correlations at TF level is obtained from this mean-field using the finite range part of the Gogny D1S force [12] in the pairing channel [13]. It must be pointed out that the total energy per baryon obtained with our TF approach is in very good agreement with the old quantal calculation of Negele and Vautherin [14], as it is explicitly discussed in Ref.[11] and again shown here in Fig. 5. In the left panel of Fig. 6, we display the radial dependences of the gaps in some selected Wigner-Seitz cells It is seen that when the gap is small outside the region of the nucleus, then the gap also is small inside the nucleus. This stems from the very large coherence length where one neutron of a Cooper pair can be in the huge volume of the gas and the other inside the small volume of the nucleus. In this way the gas imprints its behavior for the gap also inside the nucleus. Such a conclusion was also given in a quantal Hartree-Fock-Bogoliubov (HFB) calculation by Grasso et al. in [15] what shows that the here employed BCS approximation apparently yields very similar answers as a full HFB calculation for WS cells [3, 16]. Finally, in the right panel of Fig. 6, we show a comparison of LDA and present TF results for the gaps in a particular WS cell. We see a huge difference in the surface region of the nucleus. This simply stems from the fact that in this case of the $^{500}_{40}$Zr nucleus in the WS cell the gap is very small and, therefore, the coherence length very large invalidating LDA. A study with examples a little less unfavorable for LDA is given in [17]. We also show in Fig. 7 the gaps for several WS cells including some which correspond to the outer crust where all neutrons are bound. Again we see that the gap practically vanishes at zero chemical potential when the neutrons start to drip.

For isolated nuclei at the neutron drip the difference between HFB and BCS approaches may be more significant. Also strong shell fluctuations surely play an important role. Somewhat conflicting results in this respect exist in the literature. In ref [18] very similar results to ours are found for S-wave pairing, see Fig. 8 . On the other hand in [19] the gap seems to rise towards the drip before it bends down. Similar results have recently been found in [20]. However, very small Oxygen isotopes have been considered in [19] and [20] where shell fluctuations may be dominant. The HFB calculation of Hamamoto has recently been repeated and extended passing from negative chemical potentials to positive ones and it was found that the S-wave gap of Fig. 8 clearly continues down to zero , touching zero at a slightly positive value of μ [21]. In explaining why in other works the gap is rising towards the drip, one has to keep in mind that an average gap should be calculated with the pairing tensor and *not* with the density matrix. The latter picks up the gaps at all energies which may not be small at all, even though the gap at the Fermi level is very small, see Fig. 2. On the other hand, an average with the pairing tensor only picks up the gaps around the Fermi level. It also is intuitively clear that for other than S-wave (i.e. $l > 0$) gaps the situation will be different, see Fig. 8 and [18]. This is due to resonant states below the centrifugal barrier with wave function concentrated on the domain of the nucleus. In situations where the resonant states are sufficiently close to the Fermi level, so that they can contribute to pairing, $l > 0$ gaps will exist also at positive energies. This remark also holds for spherical nuclei in WS cells, since, as we have seen, the gas couples very little with states belonging to the nucleus. Therefore, our results of above for nuclei in WS cells are only valid for S-wave pairing or for cases where the resonance states under the centrifugal barrier are almost not contributing to pairing. For deformed nuclei, the situation of resonant states in the continuum is less clear and depends on the magnitude of the deformation. For one dimensional systems, as the slab considered above, the situation will be like with S-waves, since in 1D there is no centrifugal barrier.

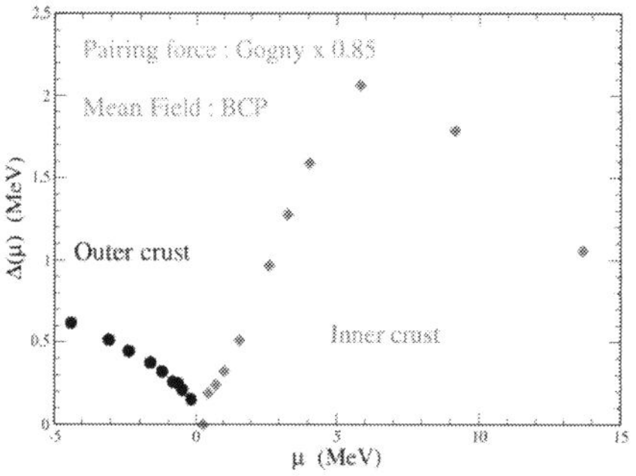

Figure 7. (Coloronline) Gap values for the Wigner-Seitz cells corresponding to Fig. 5. The zero point indicates the transition from isolated nuclei (black dots) to the dripped case (red squares).

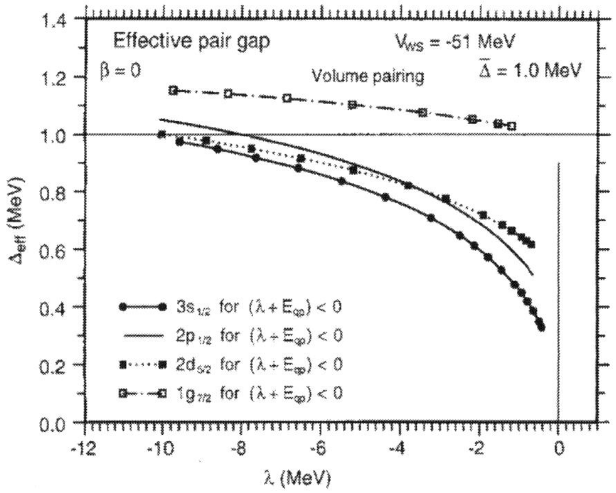

Figure 8. (Coloronline) *S*-wave gap values from schematic HFB calculation as a function of the distance of the *S*-level from the continuum [18].

4. Summary

Summarizing, we have studied superfluid fermions in a large container, either external (cold atoms) or created self consistently (nuclei) for situations where the top of the fluid reaches the edge of a small pocket located at the origin of the wide confining potential. The gap drops to zero at the edge before rising again when the density fills up the outer container. This at first somewhat surprising phenomenon can be explained quite straightforwardly. Such situations, as already mentioned, can exist in cold atoms and nuclei in the inner crust of neutron stars, two examples treated here with their specific form of containers. For small systems, like isolated nuclei at the neutron drip, the situation may be blurred by shell effects.

As an important second aspect of this work, we showed that a novel Thomas-Fermi approach to inhomogeneous situations can cope with situations where LDA fails. This means that our TF approach is free of the restrictive condition, prevailing for LDA, that the Cooper pair coherence length must be shorter than a typical length l (the oscillator length in the case of a harmonic container) over which the mean field varies appreciably. On the contrary, our TF theory has the usual TF validity criterion, namely that local wavelengths must be shorter than l.

The accuracy of our TF approach opens wide perspectives for a treatment of inhomogeneous superfluid Fermi-systems with a great number of particles not accessible for a quantal solution of the BCS (HFB) equations. Such systems may be cold atoms in deformed containers (eventually reaching millions of particles), superfluid-normal fluid (SN) interfaces, vortex profiles, etc. As a matter of fact, as is well known [9], the TF approach becomes the more accurate, the larger the system. Thus the TF approximation is complementary to the quantal one in the sense that the former works where the latter is difficult or even impossible to be obtained numerically.
Part of this work has in the meantime been published in [22].

We thank W. Ketterle and A. Minguzzi for their interest in this work and for pointing out refs. [1, 2]. We are greatful to Michel Farine for contributions and to Michael Urban for useful discussions and a critical reading of the manuscript. We also thank K. Hagino for pointing to Ref.[19] and sending their own results prior to publication. Discussions with J.Margueron on the role of resonant states under the centrifugal barrier have been appreciated. This work has been partially supported by the IN2P3-CAICYT collaboration (ACI-10-000592). One of us (X.V.) acknowledges grants FIS2008-01661 (Spain and FEDER), 2009SGR-1289 (Spain) and Consolider Ingenio Programme CSD2007-00042 for financial support.

[1] D.M. Stamper-Kurn, H.-J. Miesner, A.P. Chikkatur,S. Inouye, J. Stenger and W. Ketterle, Phys. Rev. Lett. **81**, 2194 (1998).
[2] L. Viverit, S. Giorgini, L.P. Pitaevskii and S. Stringari, Phys. Rev. **A63**, 033603 (2001).
[3] M. Baldo, U. Lombardo, E.E. Saperstein and S.V. Tolokonnikov, Nucl. Phys. **A750**, 409 (2005); M. Baldo, E.E. Saperstein and S.V. Tolokonnikov, Eur. Phys. J. **A32**, 97 (2007).
[4] N. Chamel, S. Goriely, J.M. Pearson and M. Onsi, Phys. Rev. **C81**, 045804 (2010).
[5] P. Haensel, *Neutron Stars 1: Equation of State and Structure*, (Springer-Verlag, New York, 2007).
[6] S. Flügge, *Practical Quantum Mechanics*, (Springer-Verlag, Berlin, 1974).
[7] X. Viñas, P. Schuck, M. Farine and M. Centelles, Phys. Rev. **C67**, 054307 (2003).
[8] X. Viñas, P. Schuck and M. Farine, Int. J. Mod. Phys. **E20** 399 (2011)
[9] P. Ring and P. Schuck, *The Nuclear Many-Body Problem*, (Springer-Verlag, Berlin, 1980).
[10] M. Baldo, P. Schuck and X. Viñas Phys. Lett. **B663**, 390 (2008).
[11] X. Viñas, P. Schuck and M. Farine, arXiv:1106.0187, J. of Phys.:Conf.Ser. **321**, 012024 (2011).
[12] J. Dechargé and D. Gogny, Phys. Rev. **C21**, 1568 (1980); J.-F. Berger, M. Girod and D. Gogny, Comp. Phys. Comm. **63**, 365 (1991).
[13] The D1S force for pairing has been multiplied with a factor 0.85 to compensate for the use of the bare mass in the calculation.
[14] J.W. Negele and D. Vautherin, Nucl. Phys. **207**, 298 (1973).
[15] M. Grasso, E. Khan, J. Margueron and N. Van Giai, Nucl. Phys. **A807**, 1 (2008).
[16] Our TF theory can be generalised to the HFB case what shall be done in future work.
[17] A. Pastore, F. Barranco, R.A. Broglia, and E. Vigezzi, Phys. Rev. **C78**, 024315 (2008).
[18] I. Hamamoto, Phys. Rev. **C71**, 037302 (2005).
[19] N. Tajima, Eur. Phys. J. **25**, 571 (2005).
[20] K. Hagino and H. Sagawa, arXiv:1105.5469
[21] We thank K. Hagino for sending us his results.
[22] P. Schuck, X. Viñas, Phys. Rev. Lett. 107,205301 (2011)

Journal of Physics: Conference Series **338** (2012) 012017 doi:10.1088/1742-6596/338/1/012017

Generalized Master equation approach to mesoscopic time-dependent transport

Kristinn Torfason[1,2], Andrei Manolescu[1,*], Valeriu Molodoveanu[3], Vidar Gudmundsson[2]

[1] School of Science and Engineering, Reykjavik University, Menntavegur 1, IS-101 Reykjavik, Iceland

[2] Science Institute, University of Iceland, Dunhaga 3, IS-107 Reykjavik, Iceland

[3] National Institute of Materials Physics, P. O. Box MG-7, Bucharest-Magurele, Romania

[*] E-mail: manoles@ru.is

Abstract. We use a generalized Master equation (GME) formalism to describe the non-equilibrium time-dependent transport through a short quantum wire connected to semi-infinite biased leads. The contact strength between the leads and the wire are modulated by out-of-phase time-dependent functions which simulate a turnstile device. One lead is fixed at one end of the sample whereas the other lead has a variable placement. The system is described by a lattice model. We find that the currents in both leads depend on the placement of the second lead. In the rather small bias regime we obtain transient currents flowing against the bias for short time intervals. The GME is solved numerically in small time steps without resorting to the traditional Markov and rotating wave approximations. The Coulomb interaction between the electrons in the sample is included via the exact diagonalization method.

1. Introduction

The theoretical description of time-dependent transport in semiconductor nanostructures has received considerable attention in the last few years. Non-equilibrium Greens' function techniques and density-functional methods were developed for transient current calculations in interacting and non-interacting structures (see [1–4] and references therein). These methods were employed to study the response of a mesoscopic sample to a time-dependent (possibly pulsed) voltage applied on the leads and/or to check the expected crossover to a steady-state.

A strong motivation behind these studies is the need to model and predict the transient response of open and interacting nanodevices subjected to time-dependent signals. Since in real systems both the contacts and sample geometry as well as the charging and correlation effects are important, the numerical implementation of the various formal methods requires extensive and costly computational work.

Recently we reported transport calculations for a two-dimensional parabolic quantum wire in the turnstile setup [5], where the Coulomb interaction between electrons was neglected. The wire was in contact with external leads, seen as particle reservoirs, and the contact regions were described by a phenomenological ansatz. We provided a careful analysis of the electronic propagation on the edge states that develop in the presence of a strong perpendicular magnetic field. Let us remind here that the turnstile setup was experimentally realized by Kouwenhoven *et al.* [6]. It essentially involves a time-dependent modulation (pumping) of the tunneling barriers between the finite sample and drain and source leads, respectively. During the first half of the pumping cycle the system opens only to the source lead whereas during the second half of the cycle the drain contact opens. At certain parameters an integer number of electrons is transferred across the sample in a complete cycle. Similar transient current measurements were performed in a pump-and-probe configuration [7–9]. More complex turnstile pumps have been studied by numerical simulations, like one-dimensional arrays of junctions [10] or two-dimensional multidot systems [11]. The turnstile regime differs from the adiabatic quantum pumping where charge is transferred along a sample even in the absence of a bias.

In this paper we describe the turnstile regime of a one-dimensional (1D) quantum wire ("the sample"). The effect of the electron-electron interaction is included in the sample via the exact diagonalization method while the time-dependent transport is performed within the generalized Master equation (GME) formalism as it is described in Ref. [12]. The implemented GME formalism can be used to describe both the initial transient regime immediately after the coupling of the leads to the sample and the evolution towards a steady state achieved in the long time limit. To the best of our knowledge these are the first numerical simulations of electronic transport through a Coulomb interacting and spatially extended quantum turnstile. We discuss for the first time the effect of contacts' location on the transient currents. More precisely, we show that if the drain lead is attached to different regions of the quantum wire the currents in both leads are considerably affected.

The paper is organized as follows: The model and the methodology are described in Section 2, the numerical results are presented in Section 3, and the conclusions in Section 4.

2. The physical model and the GME

2.1. Setup

The physical system consists in a sample connected to two leads acting as particle reservoirs. We shall adopt a tight-binding description of the system: the sample is a 10-site quantum wire and the leads are 1D and semi-infinite. A sketch is given in Fig. 1. The left lead (or the source, marked as L) is contacted at one end of the sample and the right lead (or the drain, marked as R) may be contacted on any other site. The Hamiltonian of the coupled and electrically biased system reads as

$$H(t) = \sum_{\ell} H_\ell + H_S + H_T(t) = H_0 + H_T(t)\,, \tag{1}$$

where H_S is the Hamiltonian of the isolated sample, including the electron-electron interaction, and H_ℓ, with $\{\ell\} = (L, R)$, corresponds to the left and the right leads. H_T describes the time-dependent coupling between the single-particle basis states $\{|\phi_n\rangle\}$ of the isolated sample and the states $\{\psi_{q\ell}\}$ of the leads:

$$H_T(t) = \sum_n \sum_\ell \int \mathrm{d}q\, \chi_\ell(t)(T^\ell_{qn} c^\dagger_{q\ell} d_n + h.c.)\,. \tag{2}$$

The function $\chi_\ell(t)$ describes the time-dependent switching of the sample-lead contacts, while $d^\dagger_n$ and $c_{q\ell}$ create/annihilate electrons in the corresponding single-particle states of the sample or leads, respectively. The coupling coefficient $T^\ell_{qn} = V_0 \psi^*_{q\ell}(0)\phi_n(i_\ell)$ involves the two eigenfunctions evaluated at the contact sites $(0, i_\ell)$, 0 being the site of the lead ℓ and i_ℓ the site in the sample [12]. In our present calculations we keep the left lead connected to the site $i_L = 1$, while the position of the right lead is $i_R = 10$ or $i_R = 3$. The parameter V_0 plays the role of a coupling constant between the sample and the leads.

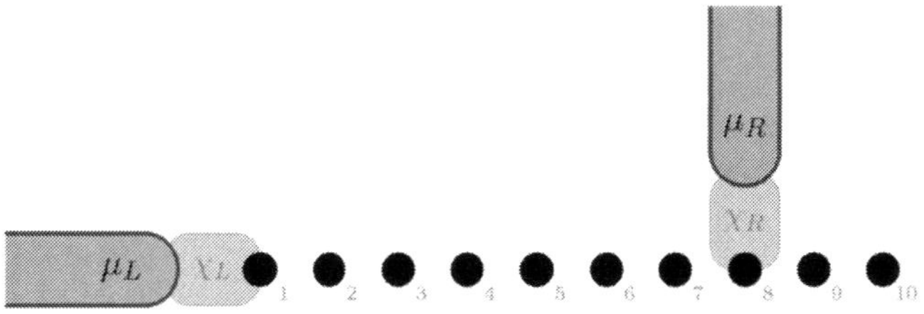

Figure 1. A sketch of the system under study. A 1D lattice with 10 sites ("the sample") is connected to two semi-infinite leads via tunneling. The left lead is connected to the left end of lattice, while the position of the right lead can be changed. The contacts are modulated in time with the functions $\chi_{L,R}$ given in Eq. 5.

2.2. GME

Following the Nakajima-Zwanzig technique [13] we define the reduced density operator (RDO), $\rho(t)$, by tracing out the degrees of freedom of the environment, the leads in our case, over the statistical operator of the entire system, $W(t)$:

$$\rho(t) = \mathrm{tr}_L\, \mathrm{tr}_R\, W(t)\,, \quad \rho(0) = \rho_S\,. \tag{3}$$

The initial condition corresponds to a decoupled sample and leads when the RDO is just the statistical operator of the isolated sample ρ_S. For a sufficiently weak coupling strength (V_0) one obtains the non-Markovian integro-differential Master equation for the RDO:

$$\dot{\rho}(t) = -\frac{i}{\hbar}[H_S, \rho(t)] - \frac{1}{\hbar^2}\sum_{\ell}\int \mathrm{d}q\, \chi_\ell(t)([\mathcal{T}_{q\ell}, \Omega_{q\ell}(t)] + h.c.)\,,$$

where the operators $\Omega_{q\ell}$ and $\Pi_{q\ell}$ are defined as

$$\Omega_{q\ell}(t) = e^{-itH_S}\int_0^t \mathrm{d}s\, \chi_\ell(s)\Pi_{q\ell}(s)e^{i(s-t)\varepsilon_{q\ell}}e^{itH_S}\,,$$
$$\Pi_{q\ell}(s) = e^{isH_S}\left(\mathcal{T}_{q\ell}^\dagger\rho(s)(1-f_\ell) - \rho(s)\mathcal{T}_{q\ell}^\dagger f_\ell\right)e^{-isH_S}\,,$$

and f_ℓ is the Fermi function of the lead ℓ. The operators $\mathcal{T}_{q\ell}$ and $\mathcal{T}_{q\ell}^\dagger$ describe the 'transitions' between two many-electron states (MES) $|\alpha\rangle$ and $|\beta\rangle$ when one electron enters the sample or leaves it:

$$(\mathcal{T}_{q\ell})_{\alpha\beta} = \sum_n T_{qn}^\ell \langle\alpha|d_n^\dagger|\beta\rangle\,. \tag{4}$$

The GME is solved numerically by calculating the matrix elements of the RDO in the basis of the interacting MES, in small time steps, following a Crank-Nicolson algorithm. Using the RDO we obtain the time dependent charge and currents in the system. More details can be found in Ref. [12].

2.3. Coulomb interaction

We will ignore the Coulomb effects in the leads, where we assume a high concentration of electrons and thus strong screening and fast particle rearrangements. The Coulomb electron-electron interaction is considered in detail only in the sample, where Coulomb blocking effects may occur. We calculate the MES in the sample following the exact diagonalization method, i. e. without any mean field approximation. The interacting MES are calculated in the Fock

Journal of Physics: Conference Series **338** (2012) 012017 doi:10.1088/1742-6596/338/1/012017

space as superpositions of non-interacting MES derived from Slater determinants [12]. Since the sample is open the number of electrons is not fixed, but the Coulomb interaction conserves the number of electrons. With 10 lattice sites we obtain 10 single-particle eigenstates and thus $2^{10} = 1024$ elements in the Fock space spanned by the occupation numbers. The Coulomb effects are measured by the ratio of a characteristic Coulomb energy $U_C = e^2/(\kappa a)$ and the hopping energy $t_s = \hbar^2/(2m_{eff}a^2)$. Here a denotes the inter-site distance (the lattice constant of the discretized system), while κ and m_{eff} are material parameters, the dielectric constant and the electron effective mass, respectively. In our calculations the *relative* strength of the Coulomb interaction, $u_C = U_C/t_s$, is considered a free parameter. We will use $u_C = 0.5$. For a material like GaAs this value would correspond to a sample length of $9a \approx 22$ nm. This length may not look very realistic, but the choice of the parameters was determined by the computational time spent in solving the GME which grows very fast with the number of MES.

The chemical potentials in the leads create a bias window $\Delta\mu = \mu_L - \mu_R$. The MES of the sample which participate in the transport correspond to chemical potentials $\mu_N^{(i)} := \mathcal{E}_N^{(i)} - \mathcal{E}_{N-1}^{(0)}$ situated within the bias window, or maybe only slightly outside, depending on the sample-leads coupling constant V_0 [12]. Here $\mathcal{E}_N^{(i)}$ is an energy of the sample spectrum containing N particles, $i = 0$ indicating to the ground state and $i > 0$ the excited states. A diagram of the chemical potentials is shown in Fig. 2.

2.4. *Time-dependent switching*

The switching-functions in Eq. 2 act on the contact regions shaded blue in Fig. 1 and are used to mimic potential barriers with time dependent height. In the present study they are made by combining two quasi Fermi functions that are shifted relatively to each other,

$$\chi_\ell(t) = 1 - \frac{1}{e^{t-\gamma_s^\ell-\delta}+1} - \frac{1}{e^{-(t-\gamma_s^\ell)+(T_p^\ell+\delta)}+1}, \qquad t \in [0,\, 2T_p^\ell], \tag{5}$$

where $\gamma_s^\ell = \{0,\, T_p^L\}$ defines the phase shift between the leads ($\ell = L, R$) and $T_p^\ell = 30$ is the pulse length, the same in the two leads. The parameter δ controls the shape of the pulse and is fixed at the value $\delta = 10$. The time unit used is $\hbar/t_s$. The time dependent contact functions are graphed at the bottom of Fig. 3. The initial values are $\chi_{L,R}(0) = 0$, i. e. the leads and the sample are initially disconnected.

3. Results

We chose the bias window $\Delta\mu = \mu_L - \mu_R$ to include the ground state with $N = 3$ electrons. We performed transport calculations for two bias windows: a larger one, with $\mu_L = 3.30$ and $\mu_R = 2.90$, and a narrower one, with $\mu_L = 3.20$ and $\mu_R = 2.98$, as indicated also in Fig. 2. In addition, we also collected the results at vanishing bias $\mu_L = \mu_R = 3.20$.

We start the time-dependent calculations with $N = 3$ electrons in the sample, initially in the ground state. We place the left lead (L) in contact with site 1 of the sample (the left end) and the right lead (R) in two different locations: first on site 10 (the right end) and then on site 3. In each case we turn on the turnstile, i. e. $\chi_{L,R}(t)$ follow Eq. (5), and calculate the time dependent charge in the sample and the currents in the left and right leads. In Fig. 3 we show the evolution of the total charge in the system for the two contact configurations mentioned above and the narrower bias window. The modulating signals impose charge oscillations, which after some time become periodic. Due to the choice of the bias window the dominant populations correspond to the three particle ground state and to a two-particle excited state. Single particle states are practically unpopulated and inactive in this case. We see that the charge accumulated in the sample 'feels' the different placements of the right lead. Indeed, when the drain lead is coupled to the site 3 the oscillations of the total charge diminish compared to the case when the drain is

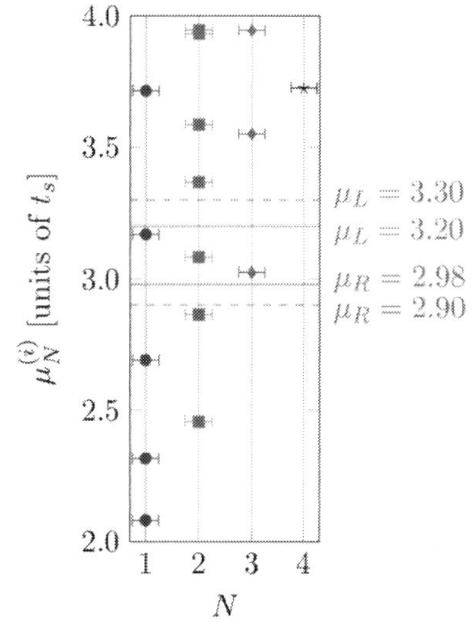

Figure 2. The μ-diagram vs. the number of particles N for $u_C = 0.5$. The blue points correspond to single-particle states, the red points to two-particle states, the brown points to three-particle states, and the only black point to four particles. The bias window is $\Delta\mu = \mu_L - \mu_R$. The small bias window is $\Delta\mu = 3.20 - 2.98 = 0.22$ (green). The large one is $\Delta\mu = 3.30 - 2.90 = 0.40$ (violet). Both bias windows include the ground state of three particles, but also excited one and two-particle states. So the expected number of electrons in the steady state is slightly below three.

on site 10. One can say that part of the charge located beyond the right contact, i. e. between sites 3 and 10, is somehow frozen and does not contribute to transport. Note however than in both configurations the populations of the two-particle and three-particle states have opposite variations in time, i. e. the gain of one is partly compensated by the loss of the other one.

The same behavior can be observed in the charge distribution along the sample, which is shown in Fig. 4 at a particular time moment. The charge distribution is far from being homogeneous, less charge being localized at the edges of the wire than in the center. The charge distribution is also not symmetric along the sample, and at each site it oscillates in time with frequencies related to the charging times of the active MES. Consequently the currents in the leads also oscillate in time within each pumping period.

In the next figures we show the calculated currents in the two leads. In Fig. 5 we show the time dependent currents for zero bias. The oscillations of the currents within each cycle average practically to zero, so we can say that no net charge is transferred through the sample. The positive values of the currents correspond to the direction $L \to R$ and the negative values to the opposite direction, $R \to L$. One can notice some differences in the current oscillations between the two placements of the right contacts, on site 10 and on site 3, respectively. More and sharper oscillations of the currents occur in the later case, but still one can say that qualitatively the currents look similar in the two cases.

This situation changes totally with increasing the bias window. In the next figure, 6, we show the currents for the small bias window, i. e. $\mu_L = 3.20$ and $\mu_R = 2.98$. Now the current profiles for the two placements of the right contact are qualitatively different. With the right lead on site 10 we obtain positive currents in both leads, describing charge flow in the direction imposed by the the bias, i. e. from the left to the right of the system. Moreover, the current in the left leads partially resembles the pulse shape. But for the other placement, i. e. with the right

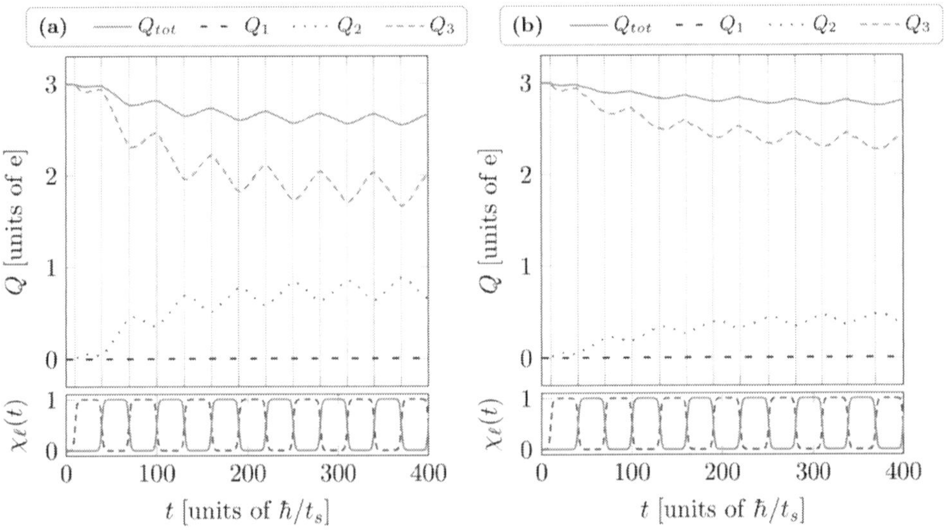

Figure 3. The time dependent charge for $\mu_L = 3.20$, $\mu_R = 2.98$. The contact functions $\chi_{L,R}$, given by Eq. 5, are shown at the bottom. (a) Left lead at site 1 (blue), right lead at site 10 (red). (b) Left lead at site 1, right lead at site 3. The total charge is shown by the continuous brown line. The dashed lines show the population of the states with 1, 2, and 3 electrons (indicated as Q_1, Q_2 and Q_3 on the top of the figures).

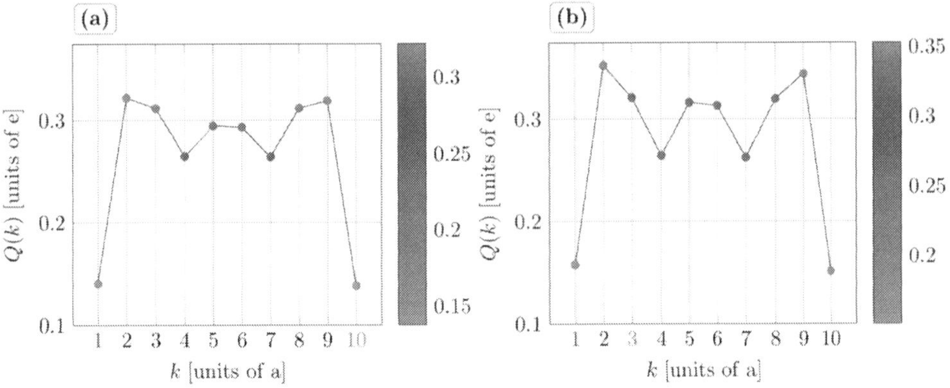

Figure 4. A snapshot of the charge distribution along the 10 sites, for $\mu_L = 3.20$, $\mu_R = 2.98$. The time is $t = 400$. The position of the right contact is shown on the horizontal axis in red. (a) Left 1 Right 10. (b) Left 1 Right 3.

contact at site 3, we see much stronger oscillations, and even negative currents in both leads for short times. These negative currents are actually *against* the bias. In spite on their rather small amplitude one should note that these negative currents do not vanish in the long time limit, that is when the evolution of the system is periodic in time. Another particular feature of the asymmetric contact geometry is that it leads to pronounced spikes in the left lead current.

We then increase the bias window and obtain the results shown in Fig. 7. The current pulses in the contact placement $L1 - R10$ do not change qualitatively from the previous case of the smaller bias window, but those for the placement $L1 - R3$ do change: the negative currents

Journal of Physics: Conference Series **338** (2012) 012017 doi:10.1088/1742-6596/338/1/012017

occur only in the right lead (the red solid line), but not in the left lead (the blue dashed line). In this case some charge bounces back an forth between the sites 3 and 10, and enter or exit the sample through the R lead, but it does not reach the L lead at site 1. In other words the placement of the right contact is qualitatively important for the current profiles for a finite bias window. As we have checked, the negative currents survive for a longer period of the pulses.

It is interesting to compare the current in one lead to the time derivative of the charge, Figures 6 and 3. In the simplest interpretation, when the left contact is open and the right contact is closed the charge in the sample increases, and so the left current is positive and the right current is zero. In the next phase, when the left contact closes and the right one opens, the charge decreases, and the right current is positive, the left current being now zero. The charging or discharging are actually complex processes, because different states are occupied with different time constants, and thus the charging is not constant in time, and the currents have oscillations, more or less following the time derivative of the charge. The fine structure of the currents is however more complicated, also related to the charge oscillations on the particular site where the contact is attached.

Finally, we want to comment on the Coulomb effects. The Coulomb interaction which is included in the present calculations has an important role in the charge distribution, through Coulomb blocking and correlation effects, and so the currents are also affected. However, it is quite difficult to compare the results without and with the Coulomb interaction included. If Coulomb effects are neglected ($u_C = 0$) the whole energy spectrum changes and new states are present within the bias window. Then the chemical potentials in the leads have to be shifted accordingly in order to capture in the bias window states with similar number of electrons as in the interacting case. This means one cannot compare the two situations just by changing only one parameter.

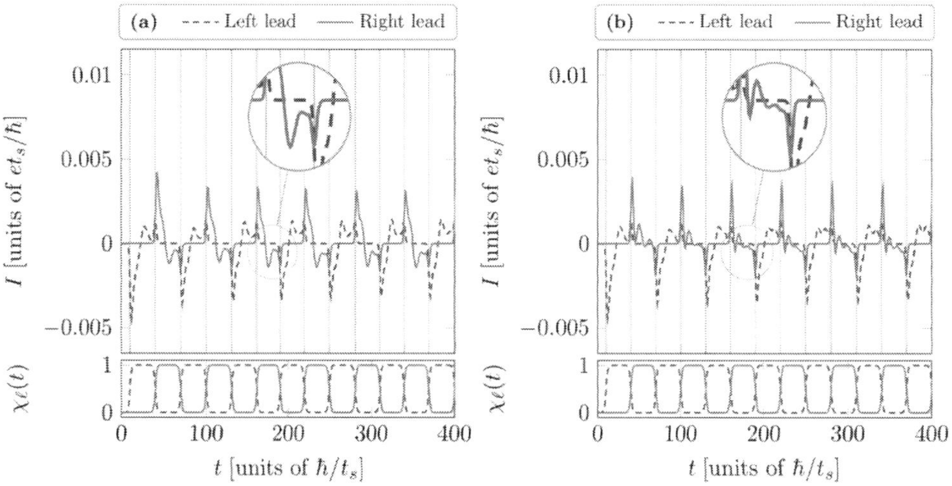

Figure 5. The currents in the two leads at zero bias, $\mu_L = \mu_R = 3.20$: (a) Left 1 Right 10. (b) Left 1 Right 3. At $t = 0$ the sample contains three electrons in the ground state.

4. Concluding remarks

Using a lattice model we have simulated the time dependent transport through a one-dimensional finite quantum wire attached to two leads. Out-of-phase time-dependent signals are applied at the contact regions, generating a turnstile device. The calculations are performed by solving

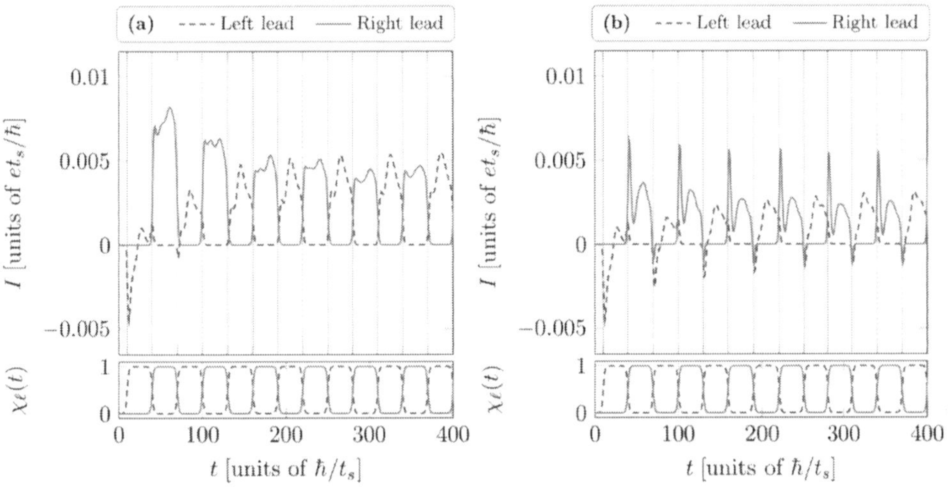

Figure 6. $\mu_L = 3.20$, $\mu_R = 2.98$. The currents in the two leads: (a) Left 1 Right 10. (b) Left 1 Right 3. The initial condition corresponds to three electrons in the ground state.

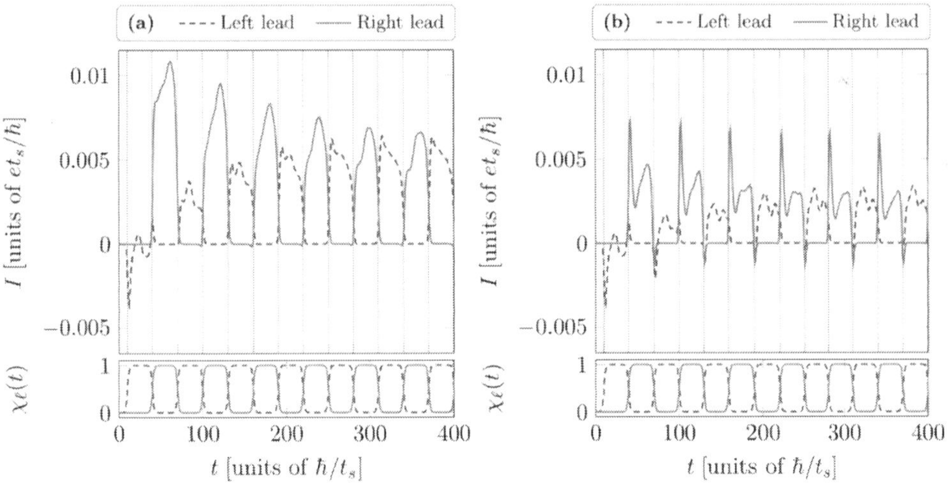

Figure 7. $\mu_L = 3.30$, $\mu_R = 2.90$. The currents in the two leads: (a) Left 1 Right 10. (b) Left 1 Right 3. The initial condition corresponds to three electrons in the ground state.

the generalized Master equation of the reduced density operator. The latter acts in the Fock space of the many-body states of the electrons in the sample, which are calculated via exact diagonalization. We show that the location of the contacts with the leads along the sample leaves clear fingerprints on the transient currents. In particular, when changing the location of the drain lead the currents may flow against the bias for a short time. Even though the currents calculated in these examples are small and we are limited here only to the qualitative effects, the observed counterflow may be seen in future experiments. One can also expect that a suitable placement of the source and drain leads along the sample would be a way to deliver modulated output currents with a desired shape and period.

Acknowledgments
This work was financially supported by the Icelandic Research Fund.

References

[1] Kurth S, Stefanucci G, Almbladh C O, Rubio A and Gross E K U 2005 *Phys. Rev. B* **72**(3) 035308

[2] Stefanucci G and Almbladh C O 2004 *Phys. Rev. B* **69**(19) 195318

[3] Myöhänen P, Stan A, Stefanucci G and van Leeuwen R 2009 *Phys. Rev. B* **80**(11) 115107

[4] Myöhänen P, Stan A, Stefanucci G and van Leeuwen R 2010 *J. Phys.: Conf. Series* **220** 012017

[5] Gainar C M, Moldoveanu V, Manolescu A and Gudmundsson V 2011 *New. J. Phys.* **13** 013014

[6] Kouwenhoven L P, Johnson A T, van der Vaart N C, Harmans C J P M and Foxon C T 1991 *Phys. Rev. Lett.* **67**(12) 1626–1629

[7] Fujisawa T, Austing D G, Tokura Y, Hirayama Y and Tarucha S 2003 *J. Phys. Cond. Mat.* **15** R1395

[8] Lai W T, Kuo D M and Li P W 2009 *Physica E* **41** 886 – 889

[9] Naser B, Ferry D K, Heeren J, Reno J L and Bird J P 2007 *Appl. Phys. Lett.* **90** 043103

[10] Mizugaki Y 2003 *J. Appl. Phys* **94** 4480–4484

[11] Ikeda H and Tabe M 2006 *J. Appl. Phys* **99** 073705

[12] Moldoveanu V, Manolescu A, Tang C S and Gudmundsson V 2010 *Phys. Rev. B* **81**(15) 155442

[13] Timm C 2008 *Phys. Rev. B* **77**(19) 195416

Advanced many-body and statistical methods in mesoscopic systems IOP Publishing
Journal of Physics: Conference Series **338** (2012) 012018 doi:10.1088/1742-6596/338/1/012018

Off-diagonal carrier-phonon coupling and polaron transport

Jun Ye[1], **Dongmeng Chen**[1,2], **and Yang Zhao**[1]

[1] *School of Materials Science and Engineering, Nanyang Technological University, 639798, Singapore*

[2] *College of Physics Science and Technology, China University of Petroleum, Dongying 257061, China*

E-mail: YZhao@ntu.edu.sg

Abstract. A method extending the Munn-Silbey approach has been applied to obtain the temperature dependence of transport properties of a generalized Holstein model incorporating simultaneous diagonal and off-diagonal carrier-phonon coupling. The Hamiltonian is partially diagonalized by a canonical transformation, and optimal transformation coefficients are determined in a self-consistent manner. Effects of off-diagonal coupling on the optimal transformation coefficients and diffusion coefficients have been discussed in details. The off-diagonal coupling has been revealed as a localization factor as well as a transport mechanism. Moreover, momentum-space variation of the transformation coefficients are found to be responsible for enhanced transport due to off-diagonal coupling.

1. Introduction

Theory of organic molecular crystals was established to deal with the complexity of charge transport from as early as 1950s [1, 2], when the Holstein molecular crystal model was introduced to account for carrier-phonon coupling. Charge transport in organic semiconductors, often described by polaron and disorder models as reviewed recently in Ref. [3], is governed by a number of factors such as the electronic coupling and carrier-phonon interactions. Early theories on polaron transport include the phenomenological model [4] and the polaron effective mass approach [5]. Microscopic models for polaron transport were later developed by Yarkony, Munn and Silbey[6, 7, 8, 9, 10] in 1980s using a density matrix approach, capable to account for the effects of electronic coupling, diagonal and off-diagonal carrier-phonon interaction of arbitrary strength over a full range of temperatures. More recently, models using the generalized master equation approach[11, 12, 13] and dynamical mean-field theory[14] have also been developed.

The effect of off-diagonal coupling is more difficult to tackle than its diagonal counterpart, and as a result, off-diagonal coupling has not been extensively treated before[15]. Following up on the Munn-Silbey transformation, Zhao *et al.* [16] devised a self-consistent routine to determine the transformation coefficients, demonstrating substantial corrections of the polaron band structure thanks to a built-in momentum dependence of the transformation coefficients. In our recent work[17], a combination of the Munn-Silbey transformation and Zhao's self-consistent routine has been shown to be a robust approach to provide new insights into charge transport in the presence of off-diagonal carrier-phonon coupling. To further account for the effect of off-diagonal coupling, a more systematic study will be performed in this paper.

Journal of Physics: Conference Series **338** (2012) 012018 doi:10.1088/1742-6596/338/1/012018

The rest of the paper is organized as follows. Theoretical formulations for charge transport are first introduced in Section II. In Section III numerical results on transport properties obtained with our approach are displayed and discussed in detail with a focus on the effect of off-diagonal coupling on polaron transport. Finally, the conclusions are drawn in Section IV.

2. Methodology

The generalized Holstein Hamiltonian and canonical transformation to partially diagonalize it have been described in detail previously by Munn and Silbey[8, 9, 10], and more detailed work combining the self-consistent routine devised by Zhao *et al.* [16] and the Munn-Silbey approach [8, 9, 10] has been documented elsewhere. Thus, in this Section we will only briefly introduce the formalism first constructed in Ref. [17]. The generalized Holstein Hamiltonian in the momentum-space representation looks more compact than its site-space counterpart due to translational symmetry

$$H = \sum_k \epsilon_k a_k^\dagger a_k + \sum_q \omega_q (b_q^\dagger b_q + 1/2) + N^{-1/2} \sum_{kq} \omega_q f_{-k}^q a_{k+q}^\dagger a_k (b_q + b_{-q}^\dagger), \tag{1}$$

with $\epsilon_k = \epsilon + J_k$, and

$$J_k = \sum_m e^{ik\cdot(\mathbf{R}_n - \mathbf{R}_m)} J_{nm}, \tag{2}$$

where the nearest-neighbor transfer integral $J_{n,m} = J(\delta_{n,m+1} + \delta_{n,m-1})$ is taken, and ω_q is the phonon frequencies.

The carrier-phonon coupling coefficients can also be written as:

$$f_k^q = \sum_m e^{-ik\cdot(\mathbf{R}_n - \mathbf{R}_m)} f_{n-m}^q, \tag{3}$$

with following coupling geometry:

$$f_k^q = g - i\phi[\sin(k) - \sin(k-q)], \tag{4}$$

which indicates that the antisymmetric type of coupling is used throughout the paper.

Now the transformation takes the following form in the momentum-space representation:

$$U = e^{N^{-1/2} \sum_{kq} A_{-k}^q (b_{-q}^\dagger - b_q) a_{k+q}^\dagger a_k}, \tag{5}$$

and

$$A_k^q = \sum_m e^{-i(\mathbf{k}-\mathbf{q})\cdot\mathbf{R}_n} e^{i\mathbf{k}\cdot\mathbf{R}_m} A_{nm}^q = (A_{k-q}^{-q})^*. \tag{6}$$

The transformation can then also be written as:

$$U = e^{\sum_{kk'} a_k^\dagger \mathbf{S}_{kk'} a_{k'}}, \tag{7}$$

with

$$\mathbf{S}_{kk'} = N^{-1/2} A_{-k}^{k-k'} (b_{k'-k}^\dagger - b_{k-k'}). \tag{8}$$

where $\mathbf{S}_{kk'}$ is an operator creating a net phonon with momentum $k'-k$. This operator is essential towards the evaluation of a series of important matrices, where

$$a_k \rightarrow \sum_{k'} \theta_{kk'} a_{k'}, a_k^\dagger \rightarrow \sum_{k'} \theta_{kk'}^\dagger a_{k'}^\dagger \tag{9}$$

with

$$\theta_{kk'} = [\exp(-\mathbf{S})]_{kk'}, \theta^{\dagger}_{kk'} = [\exp(\mathbf{S})]_{kk'} \tag{10}$$

where $\mathbf{S}$ stands for matrix $\mathbf{S}_{kk'}$. The transformed Hamiltonian can be partitioned into H_0 terms (zeroth order part) and the perturbative V terms. Thermal average of the $\theta_{kk'}$ operator has the property:

$$\langle\theta^{\dagger}_{k,k+q}\theta_{k',k'+q'}\rangle = \langle\theta^{\dagger}_{k,k+q}\theta_{k',k'+q}\rangle\delta_{qq'}, \tag{11}$$

which ensures the correct thermal equilibrium behavior of H_0 by keeping it diagonal in excitation wave vector.

By applying the thermal average routine shown in Ref. [9], the transformed Hamiltonian can be written in a simpler form, which facilitates the numerical calculations. The zeroth order Hamiltonian H_0 now can be written as:

$$\tilde{H}_0 = \sum_k \tilde{\epsilon}_k a_k a_k + \sum_q \omega_q(b^{\dagger}_q b_q + \frac{1}{2}), \tag{12}$$

and

$$\tilde{\epsilon}_k = \epsilon + \tilde{J}_k - N^{-1}\sum_q |A^q_k|^2\omega_q. \tag{13}$$

The renormalized transfer integral in Eq. 13 is given by:

$$\tilde{J}_k = \sum_{k'} \langle\theta_k\rangle^2(\exp E^0)_{kk'}J_{k'}, \tag{14}$$

where the triadic $E^q_{kk'}$ is a quantity introduced to yield a simpler form of the transformed Hamiltonian, with following relationship:

$$E^q_{kk'} = N^{-1}(2n_{k-k'}+1)(A^{k-k'}_{k-q})^* A^{k-k'}_k, \tag{15}$$

here n_q is the Bose-Einstein distribution, with $2n_q + 1 = \coth(\frac{1}{2}\beta\hbar\omega)$.

With the form of A^q_k shown in Eq. 18, the perturbation part V of $\tilde{H}$ can be written as:

$$V = \sum_{kk'k''}[J_k T_{kk';kk''} - 2N^{-1}\sum_q \omega_q f^q_{-k}A^{-q}_{-k''}T_{k+q,k';k,k''-q} + N^{-1/2}\sum_q \omega_q f^q_{-k}T_{k+q,k';kk''}(b_q + b^{\dagger}_{-q})a^{\dagger}_{k'}a_{k''}]. \tag{16}$$

where

$$T_{kk';uu'} = \theta^{\dagger}_{kk'}\theta_{uu'} - \langle\theta^{\dagger}_{kk'}\theta_{uu'}\rangle, \tag{17}$$

and A^q_k takes following form:

$$A^q_{-k} = \sum_{k'} f^q_{-k'}\langle\theta^{\dagger}_{k'+q,k}\theta_{k',k-q}\rangle, \tag{18}$$

to curb this possible uncontrolled growth of $\theta_{kk'}$.

In this study, the transformation coefficients A^q_k (A-matrix) will be obtained numerically prior to the calculations of the transport properties. The detailed numerical variational procedures can be found in Ref. [16]. In performing the variational calculations, the transformation coefficients must satisfy following self-consistency equations together with Eq. 15:

$$\langle\theta_k\rangle = \exp[-\frac{1}{2}\sum_{k'} E^0_{kk'}], \tag{19}$$

$$A_k^q = \langle\theta_{k-q}\rangle\langle\theta_k\rangle \sum_{k'} f_{k'}^q [\exp(E^q)]_{kk'}. \tag{20}$$

The self-consistency equations above follow the Munn-Silbey secular elimination scheme. In Ref. [9], this set of equations was solved upon further approximates on A_k^q, whereas in this paper, it is solved numerically with an accuracy unavailable previously.

In the Munn-Silbey approach[9], the transformation coefficients A_k^q was at first approximated by the scaling parameters ξ and η as:

$$A_k^q = g\xi - i\phi\eta[\sin(k) - \sin(k-q)]. \tag{21}$$

Numerical evaluation of the Eqs. 15-20 is facilitated by rewriting the Eq. 21 with real matrices ξ_k^q and η_k^q in the following form:

$$A_k^q = g\xi_k^q - i\phi\eta_k^q[\sin(k) - \sin(k-q)]. \tag{22}$$

In the equation above, the sine function leads to zero imaginary part along the line with $q = 0, 2k \pm \pi$. In the numerical calculation, the lines with $q = 0$ and $q = 2k \pm \pi$ are treated as removable singularities. Moreover, the values of η_k^q along these lines are chosen to be analytically connect with neighboring values[16]. The variational matrices ξ_k^q and η_k^q chosen can preserve the symmetry properties of A_k^q in a way such that $\xi_k^q = \xi_{k-q}^{-q}$ and $\eta_k^q = \eta_{k-q}^{-q}$.

The diffusion coefficient is closely related to the mean square displacement of an excitation in a given period of time. For the transformation scheme shown in this paper, the mean square displacement can be calculated from the excitation density matrix. The diffusion coefficient can then be expressed as[8, 10]:

$$D = a^2\langle\langle \nu_k^2/\Gamma_{kk} + \gamma_{kk}\rangle\rangle, \tag{23}$$

where a is the nearest neighbor distance, $\mathbf{k}$ is the wave vector, and $\nu_k = \nabla_k E_k$. The double bracket denotes the thermal average over the polaron states of energy E_k, where Γ_{kk} and γ_{kk} are the scattering and hopping rates expressed as

$$\Gamma_{k'k'} = N^{-1} \sum_{k \neq k'} W_{k,k;k',k'}, \tag{24}$$

$$\gamma_{kk} = \frac{1}{2}\nabla_q^2 \sum_k \mathrm{Re}(\frac{1}{2} W_{k,k;k'+q,k'+q} - W_{k,k+q;k',k'+q})|_{q=0}. \tag{25}$$

where $W_{k,k+q;k',k'+q}$, crucial to the calculation of diffusion coefficient, are given by

$$\begin{aligned} W_{k,k+q;k',k'+q} &= \int_0^\infty dt \{\langle V_{k'+q,k+q} V_{k,k'}(t)\rangle \exp[-i(E_{k'+q} - E_{k+q})t] \\ &+ \langle V_{k'+q,k+q}(t) V_{kk'}\rangle \exp[-i(E_k - E_{k'})t]\}, \end{aligned} \tag{26}$$

here the single angular bracket denotes the average of the residual interaction $V_{kk'}(t)$ over phonon states. Using the formulism above, the evaluation of D will be done numerically here without introducing further approximations as in Ref. [8, 9]. With the diffusion coefficient D, the mobility can be easily obtained with the Einstein relation $\mu = e\beta D$. As this work focus more on the numerical studies of off-diagonal coupling effect, one can refer to Ref. [17] for the details of correlation functions.

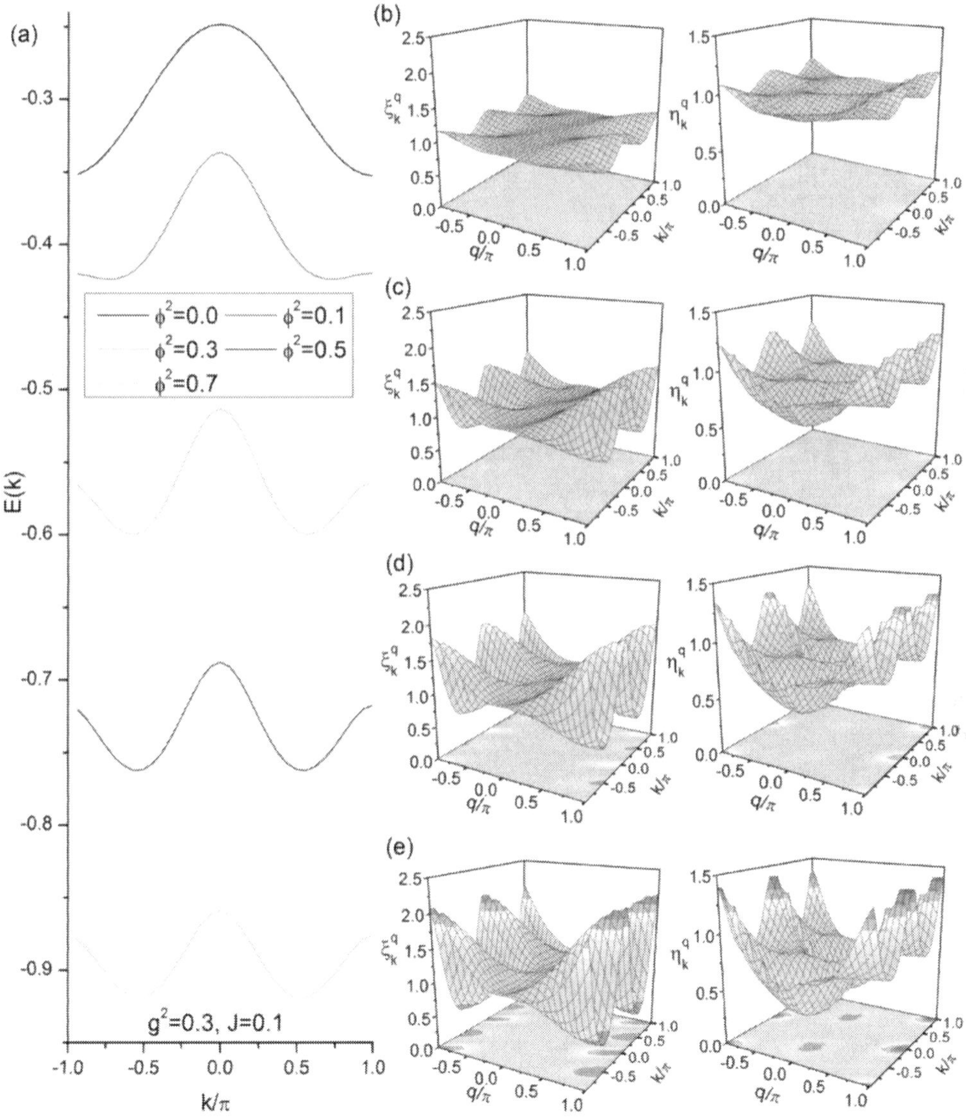

Figure 1. (a) Polaron band structure with different values of off-diagonal coupling strength ϕ^2 ranging from 0.0 to 0.7. A-matrix parameters ξ_k^q and η_k^q as functions of increasing off-diagonal coupling strength ϕ^2 of (b)0.1, (c) 0.3, (d) 0.5 and (e) 0.7. The other parameters are: J=0.1, g^2=0.3 and T=1.0.

3. Results and Discussion

The numerical evaluation involves the solution of the self-consistency equations Eqs. 15, 19 and 20. Since the procedures for the evaluation of transformation coefficients A_k^q have been detailed in Ref. [16, 17], here, we only concentrate on some of the important issues related to the transport properties, especially the role of off-diagonal coupling.

As demonstrated in Figs. 1(b)-(e), the A-matrix parameters ξ_k^q and η_k^q in Eq. 22 obtained with

Journal of Physics: Conference Series **338** (2012) 012018 doi:10.1088/1742-6596/338/1/012018

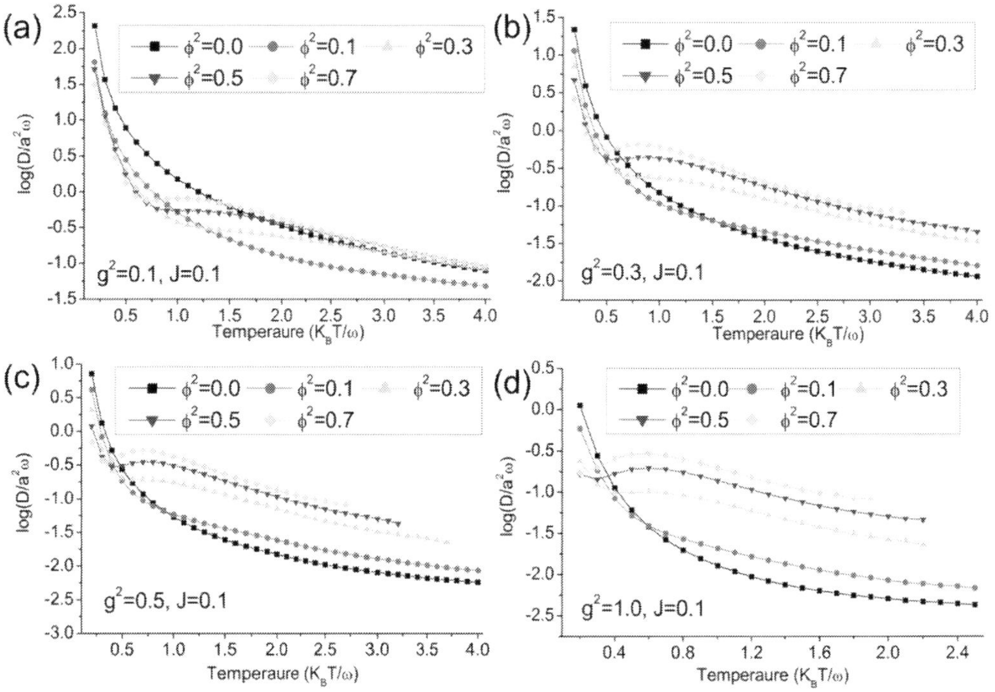

Figure 2. Diffusion coefficients D versus scaled temperature $k_B T/\omega$ for a 1 dimensional chain with 6 sites, where the diagonal coupling strength g^2 takes (a) 0.1, (b) 0.3, (c) 0.5 and (d) 1.0. The off-diagonal coupling strength ϕ^2 ranges from 0.0 to 0.7 for all figures. Finally, the transfer integral J=0.1 for all cases.

the self-consistency procedure have a significant wave vector dependence. More importantly, the wave vector dependence of ξ_k^q and η_k^q leads to significant changes in the structure of the polaron energy band $E(k)$ as exhibited in Fig. 1(a). It can be observed that with an increasing value of ϕ^2, $E(k)$ clearly shows bimodal variation. Critical modifications of the transport properties that can arise due to such alterations in the structure of the polaron states and the energy band will be explicitly examined in the rest of this section.

In the vicinity of $q = 0$, ξ_k^q is weakly structured and relatively flat with respect to k, while the wave vector dependence of η_k^q is much more pronounced than that of ξ_k^q. Regardless of the coupling strength, the wave vector dependence of ξ_k^q and η_k^q has a characteristic "manta-ray" shape. As revealed from Figs. 1(b)-(e), the A-matrix parameters tend to have stronger wave vector dependence for larger ϕ. It is also interesting to look into the change in the polaron band structure as a function of ϕ. Greater wave vector dependence found for stronger off-diagonal coupling cases indicates that this parameter is responsible for the formation of bimodal shape of the polaron bands. It is however counterintuitive to see that the greater momentum-space variability of ξ_k^q and η_k^q due to larger ϕ translates into a narrower polaron band. The key towards a possible explanation lies in the binding energy being proportional to the average of $|A_k^q|^2$. The average binding energy obtained in the current approach is significantly larger while the average Debye-Waller factor is much smaller, leading eventually to the narrowing of the band. Further, the ϕ dependence in Fig. 1 (a) can also be comprehended since a larger wave vector variation in ξ_k^q and η_k^q gets transformed into a weaker distortion of the energy band, an effect explained in detail in Ref. [16, 17].

Stronger ϕ^2 leads to a smaller diffusion coefficient at very low temperatures as shown in Fig. 2. However, it is more interesting to note the nonmonotonic changes in the diffusion coefficient (*i.e.*, a "hump") as a function of the temperature in Fig. 2. This nonmonotonic behavior becomes even more pronounced when g^2 and ϕ^2 are both large as indicated in Fig. 2(d), suggesting that the diffusion coefficient reaches its peak value at $T \approx 0.5$. Such behavior is a strong evidence that the off-diagonal carrier-phonon coupling can be a transport mechanism, which in turn suggests a competitive effect of the diagonal and the off-diagonal coupling by hindering and promoting transport respectively[10]. For a fixed value of g^2, the "hump" becomes more pronounced with an increase in ϕ^2, which further supports the role of the off-diagonal coupling in enhancing transport. It has been shown[3, 10, 17] that the band-like transport dominates in the low temperature regime. Inclusion of carrier-phonon coupling will introduce phonon scattering that is deleterious to the band-like transport through the first term in Eq. 23. A decrease in the diffusion coefficients at low temperature regime with increasing ϕ^2 is thus an evidence of localization role of the off-diagonal coupling via increasing phonon scattering. Although an increase in ϕ^2 will lead to an increase in hopping contribution of the diffusion coefficient, the number of phonons that is greatly reduced at ultra-low temperature regime largely restricts the contribution from hopping transport. A combined effect of reduced band-like transport and slightly enhanced hopping transport at low temperature regime thus results in a decrease in the total diffusion coefficient.

While considering both Fig. 1 and Fig. 2(b) together, it needs be noted carefully that the first figure directly corresponds to the band structures and the A-matrix parameters for diffusion coefficients shown in the latter figure correspond to $T = 1$. An increase in ϕ^2 leads to a greater momentum space variation of the A-matrix parameters, resulting in a higher value of D at $T = 1$. One may thus suspect that the greater momentum space variation in the A-matrix parameters should be responsible for larger value of D at the concerned temperature. More importantly, if one refers to Eq. 25 which corresponds to hopping rates, one may observe a $q = 0$ constraint when performing summation of the W quantities. This suggests that the variation of the A-matrix parameters along $q = 0$ line is important for those temperature regimes where hopping diffusion dominates. Further examination of Figs. 1(b)-(f) reveals an interesting k-dependence of ξ_k^q and η_k^q along the $q = 0$ line. It is also found that ξ_k^q are nearly k-independent and η_k^q have much greater variation along the $q = 0$ line, which is eventually reflected in the increased value of D for an increase in ϕ^2 at a given T. Such distinct k-dependence bolsters the role of off-diagonal coupling strength ϕ in promoting hopping transport. Even more interestingly, the temperature regime for the nonmonotonic variations in the diffusion coefficient points to possible activated behavior of hopping transport owing to the simultaneous inclusion of the diagonal and the off-diagonal coupling, where the activation energy should be related to the diagonal polaron binding energy[10].

4. Conclusions

We have numerically computed the diffusion coefficients while taking into account the diagonal and the off-diagonal carrier-phonon coupling simultaneously for a wide range of temperatures following the Munn-Silbey approach. The structured transformation coefficients determined self-consistently are found to impose substantial influence on the transport properties. Effects of the off-diagonal carrier-phonon coupling strength have been discussed in detail via examination of the momentum space dependencies of the transformation coefficients and nonmonotonic variations in the temperature dependence of the diffusion coefficients. Importantly, off-diagonal coupling is revealed as a transport mechanism while also being an agent for polaronic localization. Inclusion of off-diagonal coupling in the current approach allows us to see the critical role it plays in determining both the polaron structures and the transport properties over a wide range of temperatures. A greater momentum-space variability of ξ_k^q and η_k^q due to larger ϕ is shown to lead to a narrower polaron band and an obvious bimodal variation of polaron band structure $E(k)$. Moreover, strong k-dependence of η_k^q compared to ξ_k^q along the $q = 0$ line at given temperature can be regarded as a source to promote hopping transport. The nonmonotonic behavior in diffusion coefficient observed for an increase in temperature, is explained as a result of the competitive effect of diagonal coupling in hindering and off-diagonal coupling in promoting transport. Such an effect also suggests activated behavior of hopping transport with the activation energy being related to the diagonal polaron binding energy. At low temperatures, total diffusion coefficient is reduced owing to the phonon scattering that increases with off-diagonal coupling, and remains a major source of reduction in dominant band-like transport.

Acknowledgments
Support from the Singapore National Research Foundation through the Competitive Research Programme (CRP) under Project No. NRF-CRP5-2009-04 is gratefully acknowledged.

References
[1] T Holstein 1959 *Ann. Phys.* **8** 325
[2] T Holstein 1959 *Ann. Phys.* **8** 343
[3] V Coropceanu, *et al.* 2007 *Chem. Rev.* **107** 926
[4] E Silinsh, V Cápek, Organic Molecular Crystals: Interaction, Localization, and Transport Phenomena, American Institute of Physics: New York, 1994
[5] R M Glaeser and R S Berry 1966 *J. Chem. Phys.* **44** 3797
[6] D R Yarkony and R J Silbey 1976 *J. Chem. Phys.* **65** 1042
[7] D R Yarkony and R J Silbey 1977 *J. Chem. Phys.* **67** 5818
[8] R Silbey and R W Munn 1980 *J. Chem. Phys.* **72** 2763
[9] R W Munn and R Silbey 1985 *J. Chem. Phys.* **83** 1843
[10] R Silbey and R W Munn 1985 *J. Chem. Phys.* **83** 1854
[11] V M Kenkre, *et al.* 1989 *Phys. Rev. Lett.* **62** 1165
[12] P E Parris and V M Kenkre 2004 *Phys. Rev. B* **70** 064304
[13] L Giuggioli, J D Andersen, and V M Kenkre 2003 *Phys. Rev. B* **67** 045110
[14] S Fratini and S Ciuchi 2003 *Phys. Rev. Lett.* **91** 256403
[15] G D Mahan, *Many-Particle Physics* (Kluwer Academic/Plenum Publishers, New York, 2010)
[16] Y Zhao, D W Brown, and K Lindenberg 1994 *J. Chem. Phys.* **100** 2335
[17] D M Chen, J Ye, H J Zhang, and Y Zhao, 2011 *J. Phys. Chem. B* **115** 5312

Journal of Physics: Conference Series **338** (2012) 012019 doi:10.1088/1742-6596/338/1/012019

Isovector pairing effect on the moments of inertia of proton-rich heated nuclei.

I Ami [1,2], M Fellah [2,3], N Benhamouda [2], N H Allal [2,3] and M Belabbas [4]

[1] Faculté des Sciences et de la Technologie, Université Dr. Yahia Fares de Médéa, Quartier Ain D'heb, 26000 Médéa, Algeria
[2] Laboratoire de Physique Théorique, Faculté de Physique,USTHB, BP 32, El Alia, 16111 Bab Ezzouar, Algiers, Algeria
[3] Centre de Recherche Nucléaire d'Alger, 2 Bd. Frantz Fanon, BP. 399 Alger-Gare, Algiers, Algeria
[4] Faculté des Sciences et Sciences de l'Ingénieur, UHBB, BP151, Chlef, Algeria.

Abstract. The perpendicular and parallel moments of inertia are calculated as a function of the temperature by taking into account the isovector pairing. The used single-particles energies are those of a deformed Woods-Saxon mean-field. The obtained results are compared to their homologues of the conventional Finite Temperature BCS (FTBCS) theory. With this aim, the generalized gap equations have been solved for even-even heated deformed nuclei such as $Z = 30 - 38$ and $N - Z = 0, 2, 4$. The isovector pairing effect leads to a change in the behavior of the perpendicular and parallel moments of inertia. Moreover, there is a non-negligible discrepancy between the perpendicular and parallel moments of inertia values calculated within the two models when $T < T_{cnp}$ (T_{cnp} being the critical temperature beyond which the neutron-proton (np) gap parameter vanishes). Beyond this temperature, a discrepancy between the two models persists. It is due to the shift of the critical temperatures of the proton (T_{cpp}) and neutron (T_{cnn}) systems when evaluated with and without inclusion of the isovector pairing effect.

1. Introduction

The neutron-proton (np) pairing correlations [1, 2, 3, 4] in proton-rich nuclei plays a very significant role in the evaluation of physical observables. Indeed, it is already well established that in $N \simeq Z$ nuclei, the neutron and proton Fermi levels are close to each other and therefore pairing, and in particular np pairing, correlations are expected to play an important role in their structure. One also expects that they play a significant role in the evaluation of the moment of inertia. Indeed, this observable has been the subject of many studies because it is involved in the description of rotation motion of nuclei. The nuclear moment of inertia has been evaluated by taking into account pairing between like-particles correlations at zero as well as at finite temperature [5, 6, 7, 8]. It has also been calculated at zero temperature including isovector pairing correlations [9, 10, 11, 12]. However, only a few studies of the np pairing correlations effect on the moment of inertia have been performed at finite temperature. Recently, expressions of temperature-dependent perpendicular and parallel moments of inertia including np pairing effects [13] have been derived using the cranking method as well as isovector temperature-dependent gap equations [14, 15]. Numerical calculations have been carried out within the framework of the schematic Richardson model [16] as well as for even-even $N = Z$ nuclei such as $30 \leq Z \leq 38$, using the single-particles energies and eigen-states of a deformed Woods-Saxon mean field [13]. The aim of the present work is to complete this study by considering even-even nuclei such as $N - Z = 0, 2, 4$. The formalism is briefly recalled in next section. Numerical results are presented and discussed in section 3.

2. Formalism

Let us consider a system constituted by N neutrons and Z protons which is cranked around the Oi ($i = x, z$) axis (Oz being the symmetry axis) of a rotating frame. The grand-partition function of a such a hot nuclear system is given by:

$$Z = Tr\left\{\exp(-\beta[H - \sum_t \lambda_t N_t - \hbar\omega J_i])\right\} \qquad i = x, z \tag{1}$$

where β is the inverse of the temperature T and H is the Hamiltonian of the system given, in the second quantization and isotopic spin formalism, in the isovector pairing case, by [11, 12, 13, 14, 15, 16]:

$$H = \sum_{\nu>0,t} \varepsilon_{\nu t}(a^{\dagger}_{\nu t}a_{\nu t} + a^{\dagger}_{\tilde{\nu}t}a_{\tilde{\nu}t}) - \sum_{tt'} G_{tt'} \sum_{\nu,\mu>0} \left(a^{\dagger}_{\nu t}a^{\dagger}_{\tilde{\nu}t'}a_{\tilde{\mu}t'}a_{\mu t} + a^{\dagger}_{\nu t}a^{\dagger}_{\tilde{\nu}t'}a_{\tilde{\mu}t}a_{\mu t'}\right) \tag{2}$$

The subscript t corresponds to the isotopic spin component ($t = n, p$), $a^{\dagger}_{\nu t}$ and $a_{\nu t}$ respectively represent the creation and annihilation operators of the particle in the state $|\nu t\rangle$, of energy $\varepsilon_{\nu t}$; $|\tilde{\nu}t\rangle$ is the time-reverse of $|\nu t\rangle$, $G_{tt'}$ characterizes the pairing- strength. The neutron and proton are supposed to occupy the same energy levels. In all that follows, it is assumed that the single-particles energies are independent from the temperature [17]. $\lambda_t(t = n, p)$ are Lagrange parameters which represent the Fermi level energies and the N_t are the particle-number operators given by:

$$N_t = \sum_{\nu>0} \left(a^{\dagger}_{\nu t}a_{\nu t} + a^{\dagger}_{\tilde{\nu}t}a_{\tilde{\nu}t}\right) \qquad t = n, p \tag{3}$$

ω is the rotation frequency and J_i is the i projection of the angular momentum.

The usual Inglis [18] expression of the energy may be easily generalized in order to include the temperature effects [5] as well as the np pairing correlations, that is [13, 16]:

$$E = \left(-\frac{\partial \ln Z}{\partial \beta}\right)_{\lambda_t\beta=cte} \tag{4}$$

Its expansion to the second order in ω is given by:

$$E \simeq E_0 - \omega^2\hbar^2 \int_0^{\beta} \langle J_i(\beta)\, J_i(\chi)\rangle\, d\chi, \qquad with \qquad i = x, z \tag{5}$$

where

$$J_i(\kappa) = e^{\kappa H'} J_i e^{-\kappa H'} \; and \quad \langle J_i(\beta)\, J_i(\chi)\rangle = \frac{Tre^{-\beta H'} J_i(\beta)\, J_i(\chi)}{Tre^{-\beta H'}} \qquad \kappa = \beta, \chi \qquad i = x, z \tag{6}$$

$J_i(\kappa)$ is the Heisenberg transform of J_i and the thermal average in Eq. (5) is evaluated using the grand-canonical ensemble associated to the Hamiltonian without rotation H' given by:

$$H' = H - \sum_t \lambda_t N_t \tag{7}$$

This thermal average value may easily be determined using the quasiparticle representation. In the latter, the auxiliary Hamiltonian has been approximately diagonalized by means of the Feynman path integral technique [14] and using the Hubbard-Stratonovich transformation [19]. In fact, the diagonal form of the Hamiltonian H' becomes:

$$H' = \sum_{\nu>0,\ \tau=1,2} E_{\nu\tau}\left(\alpha^{\dagger}_{\nu\tau}\alpha_{\nu\tau} - \alpha_{\tilde{\nu}\tau}\alpha^{\dagger}_{\tilde{\nu}\tau}\right) + \sum_{\nu>0,t} \tilde{\varepsilon}_{\nu t} \tag{8}$$

where we set

$$\tilde{\varepsilon}_{\nu t} = \left(\varepsilon_{\nu t} - \lambda_t\right) - \left(G_{tt} + G_{np}\right)/2 \qquad t = p, n \tag{9}$$

$\alpha^{\dagger}_{\nu\tau}$ (respectively $\alpha_{\nu\tau}$) is the creation (respectively annihilation) operator of a quasiparticle of $\tau\,(\tau = 1,2)$ type, given by the generalized Bogoliubov-Valatin transformation:

$$\alpha^{\dagger}_{\nu\tau} = \sum_{\nu>0,\ t} \left(u_{\nu\tau t}a^{\dagger}_{\nu t} + v_{\nu\tau t}a_{\tilde{\nu}t}\right) \qquad \tau = 1,2 \tag{10}$$

The corresponding energies $E_{\nu\tau}$ are defined in Ref. [14, 15]. As $E = E_0 - \frac{1}{2}\Im\omega^2$ ($\Im$ being the moment of inertia), the perpendicular and parallel moments of inertia, are then respectively defined by:

$$\Im^{np}_{\perp(\parallel)} = 2\hbar^2 \int_0^{\beta} \left\langle J_{x(z)}(\beta)\, J_{x(z)}(\chi)\right\rangle_0 d\chi \tag{11}$$

One has, after some algebra, for the perpendicular moment of inertia:

$$\begin{aligned}\Im^{np}_{\perp} &= \hbar^2 \sum_{\nu\mu\tau\tau' tt'} \langle \nu t\,|J_x|\,\mu t\rangle \langle \mu t'\,|J_x|\,\nu t'\rangle \\ &\left\{ \left[v_{\nu\tau t}u_{\mu\tau' t} - v_{\mu\tau' t}u_{\nu\tau t}\right]\left[v_{\nu\tau t'}u_{\mu\tau' t'} - v_{\mu\tau' t'}u_{\nu\tau t'}\right] \left(\frac{\tanh\left(\frac{\beta E_{\nu\tau}}{2}\right) + \tanh\left(\frac{\beta E_{\mu\tau'}}{2}\right)}{E_{\nu\tau} + E_{\mu\tau'}}\right)\right. \\ &\left. + \left[u_{\nu\tau t}u_{\mu\tau' t} + v_{\nu\tau t}v_{\mu\tau' t}\right]\left[u_{\nu\tau t'}u_{\mu\tau' t'} + v_{\nu\tau t'}v_{\mu\tau' t'}\right] \left(\frac{\tanh\left(\frac{\beta E_{\mu\tau'}}{2}\right) - \tanh\left(\frac{\beta E_{\nu\tau}}{2}\right)}{E_{\mu\tau'} - E_{\nu\tau}}\right)\right\}\end{aligned} \tag{12}$$

In the same way, the parallel moment of inertia $\Im^{np}_{\parallel}$ takes the following form:

$$\begin{aligned}\Im^{np}_{\parallel} &= \hbar^2 \sum_{\nu\tau\tau' tt'} \langle \nu t\,|J_z|\,\nu t\rangle \langle \nu t'\,|J_z|\,\nu t'\rangle \\ &\left\{ \left[u_{\nu\tau t}u_{\nu\tau' t} + v_{\nu\tau t}v_{\nu\tau' t}\right]\left[u_{\nu\tau t'}u_{\nu\tau' t'} + v_{\nu\tau t'}v_{\nu\tau' t'}\right] \left(\frac{\tanh\left(\frac{\beta E_{\nu\tau}}{2}\right) - \tanh\left(\frac{\beta E_{\nu\tau'}}{2}\right)}{E_{\nu\tau} - E_{\nu\tau'}}\right)\right. \\ &\left. + \left[u_{\nu\tau t}v_{\nu\tau' t} - u_{\nu\tau' t}v_{\nu\tau t}\right]\left[u_{\nu\tau t'}v_{\nu\tau' t'} - u_{\nu\tau' t'}v_{\nu\tau t'}\right] \left(\frac{\tanh\left(\frac{\beta E_{\nu\tau}}{2}\right) + \tanh\left(\frac{\beta E_{\nu\tau'}}{2}\right)}{E_{\nu\tau} + E_{\nu\tau'}}\right)\right\}\end{aligned} \tag{13}$$

If the np pairing effects vanish, Eqs. (12) and (13) reduce to the sum of the moments of inertia of the neutron and proton systems considered separately in the framework of the pairing between like-particles [5, 6].

Journal of Physics: Conference Series **338** (2012) 012019 doi:10.1088/1742-6596/338/1/012019

3. Numerical results-Discussion

3.1. Pairing gap parameters

The previously described formalism has been numerically applied using the single-particles and eigen-states of a deformed Woods-Saxon mean-field. We used the parameters of Ref. [20] with a maximum number of shells $N_{\max} = 12$. The ground state deformations are those of the Möller Table [21]. We considered even-even nuclei such as $N - Z = 0, 2, 4$. Indeed, in such nuclei, it had been shown that np pairing effect is non-negligible [3, 4, 12] at zero temperature. In the present work, we considered only nuclei with a deformed ground-state. In what follows, we present the results that correspond to Zirconium and Strontium nuclei chosen as an example. The pairing-strength values G_{pp}, G_{nn} and G_{np} have been deduced from the Δ_{pp}, Δ_{nn} and Δ_{np} values at zero temperature. The latter are obtained using the odd-even mass differences [1].

Figures 1 and 2 show the variations of the Δ_{pp}, Δ_{np} and Δ_{np} gap parameters as a function of the temperature as well as those of the perpendicular and parallel moments of inertia. The values obtained within the framework of the conventional $FTBCS$ method are also given in the same figures. In the latter case, the pairing-strength values G_{pp} and G_{nn} have also been chosen in such a way to reproduce the Δ_{pp} and Δ_{nn} values at zero temperature.

As it is pointed out in Refs. [13, 15], the $\Delta_{np}(T)$ parameter has a behavior which is similar to that of the conventional $FTBCS$ Δ_{pp}, Δ_{nn} gap parameters, i.e. a quasi-constant value on a given interval, and then a decreasing until it vanishes at $T = T_{cnp}$ (T_{cnp} being the critical value of the temperature).

Dealing with Δ_{pp} and Δ_{nn} of the present model, one observes a plateau in the region where Δ_{np} is constant and then a sudden variation in the vicinity of T_{cnp} which is a consequence of the sudden increasing of Δ_{np}. This behavior is also observed for all studied nuclei. The phase transition when $\Delta_{np} = 0$ (i.e. $T = T_{cnp}$) is sharp. As in the pairing between like-particles case, this abrupt phase transition could be smoothed out if one includes the quantum fluctuations that have been neglected in the partition function [14]. It could be also smoothed out by including a particle number-projection [23, 24, 25].

When $T = T_{cnp}$ (i.e. $\Delta_{np} = 0$), the behavior of Δ_{pp} and Δ_{nn} is similar to that of the $FTBCS$ method. However, there exists a shift between the proton and neutron critical temperature values T_{cpp} and T_{cnn} of the present model and those of the $FTBCS$ one. The discrepancy varies from about 34 % (32%) for T_{cpp} (T_{cnn}) in the ^{60}Zn case to about 13.5% (5%) for T_{cpp} (T_{cnn}) in the ^{76}Sr case.

Moreover, one notices that for each studied element (for a given Z), the critical temperature T_{cnp} is as weak as the $(N - Z)$ value is important. Indeed, for the Strontium for example, the critical temperatures T_{cnp} varies between $0.156MeV$ for the ^{76}Sr and $0.070MeV$ for the ^{80}Sr. Its value is lowered by about 55.13%. It is well established that at zero temperature Δ_{np} rapidly decreases as a function of $(N - Z)$ and practically vanishes when $(N - Z) > 8$ [3, 4, 12]. At finite temperature, the present study shows that for a given Z, the critical temperatures T_{cnp} are as weak as $(N - Z)$ is important and the np pairing effect practically disappears with the increasing values of N for a given Z.

3.2. Moments of inertia

Panels (b) and (c) in Figs. 1 and 2 show respectively the variations of the perpendicular and parallel moments of inertia as a function of temperature for the previously cited nuclei ($Z = 30$ and $Z = 38$ with $N - Z = 0, 2, 4$). The values obtained using the conventional $FTBCS$ method (i.e. within the framework of pairing between like-particles) are also given in the same figures.

With regard to the perpendicular moment of inertia $\Im_{\perp}$, one observes a plateau in the region where Δ_{np} is constant (i.e. $0 \leq T \leq T_{cnp}$) followed by a sudden variation in the vicinity of T_{cnp} which is a consequence of the sudden decreasing of Δ_{np}. This behavior is also observed for all studied nuclei. Besides, the relative discrepancy between the present model values and the

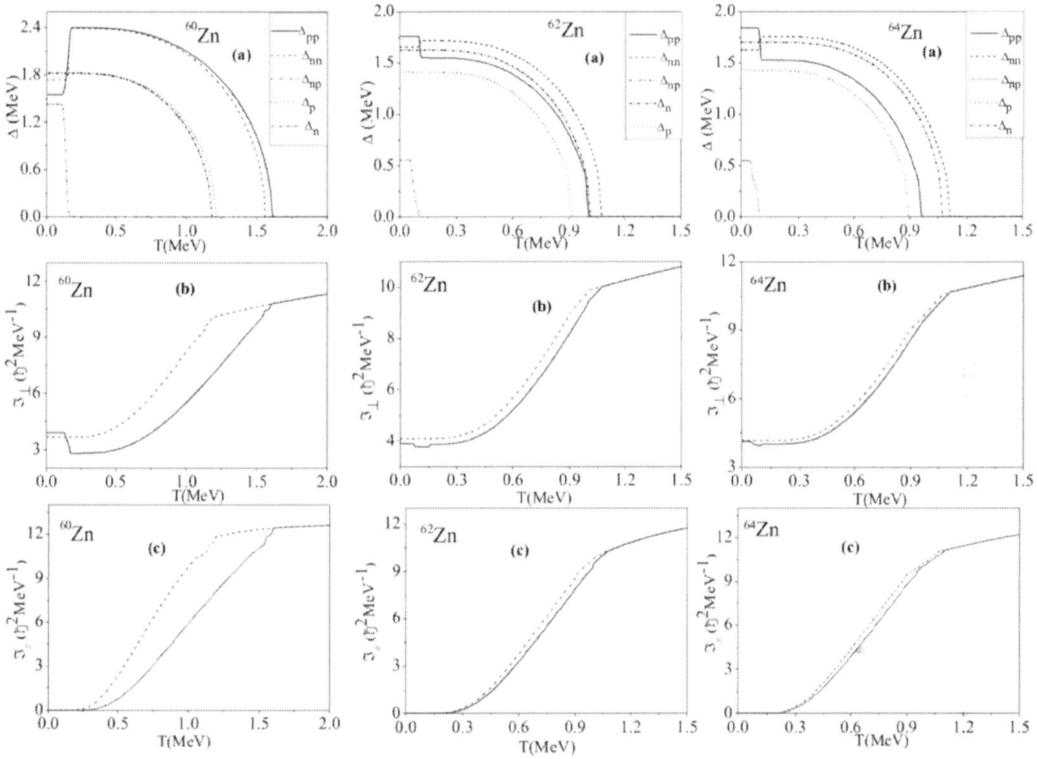

Figure 1. Variation of the various gap parameters (a), the perpendicular moment of inertia (b) and the parallel moment of inertia (c) as a function of the temperature for the nucleus ^{60}Zn, ^{62}Zn and ^{64}Zn. Straight lines refer to the present model and dashed lines to the conventional $FTBCS$ model.

$FTBCS$ ones, defined by $\left|\Im_{\perp}^{np} - \Im_{\perp}^{FTBCS}\right| / \Im_{\perp}^{FTBCS}$ varies from about 32% in the ^{78}Sr case to 1.14% in the ^{64}Zn case. Its average value for all considered nuclei is of the order of 12.23% in this region.

When $T > T_{cnp}$, the present model curve and the $FTBCS$ one behave likely. However, there is a quasi-constant shift between the two curves. Indeed, although the gap parameters [13, 14, 16, 15] as well as the expression (12) of $\Im_{\perp}^{np}$ reduce to their FTBCS homologues when Δ_{np} vanishes, the pairing-strength values are different in each case. This is the reason why Δ_{pp}, Δ_{nn}, T_{cpp}, T_{cnn} and $\Im_{\perp}$ are different from their homologues of the $FTBCS$ model. One notes that the maximum value of the relative discrepancy varies from one nucleus to another. It varies from 33.72% in the ^{60}Zn case to 1.91% in the ^{80}Sr case. It is about 11.46% on average in this region for all studied nuclei.

When there is no more pairing effect, i.e. when $\Delta_{pp} = \Delta_{nn} = 0$, the two curves join and then correspond to the rigid body values.

Dealing with the parallel moment of inertia $\Im_{\parallel}$, the np pairing effect is not visible in the

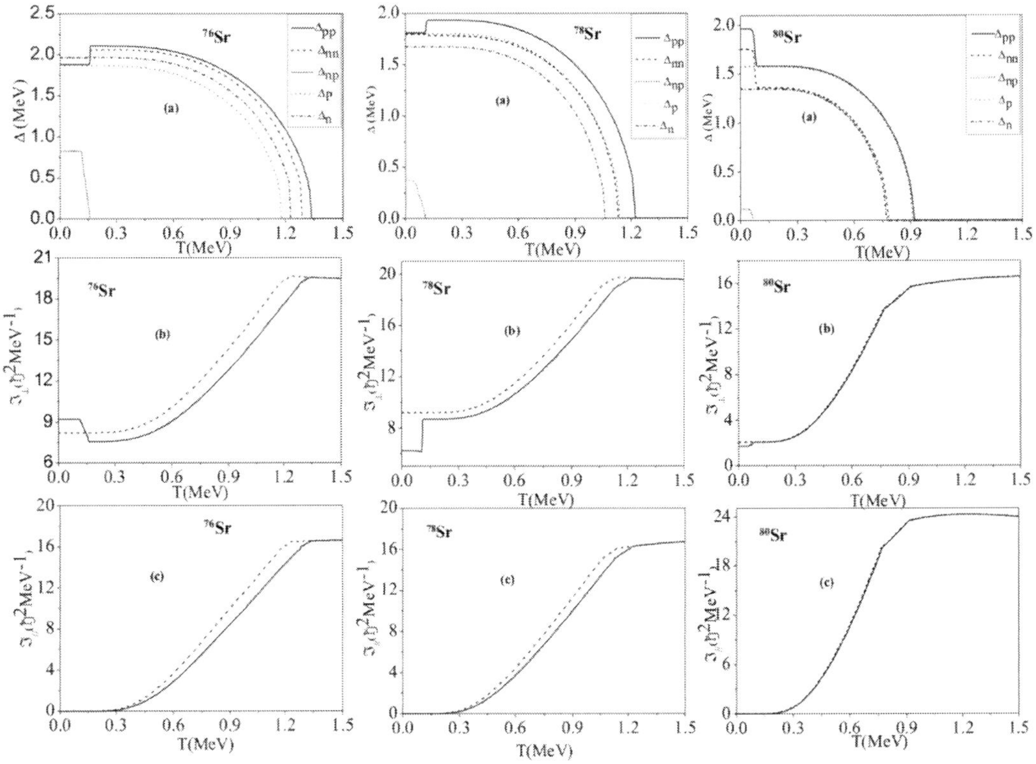

Figure 2. Same as Fig. 1 for the nucleus ^{76}Sr, ^{78}Sr and ^{80}Sr.

region $0 \leq T \leq T_{cnp}$ for all the studied nuclei. In fact, the parallel moment of inertia is nil in this region and hence there is no difference between the present model previsions and those of the conventional $FTBCS$ method. This fact is due to the weak values of T_{cnp}.

When $T > T_{cnp}$, the behavior of the parallel moment of inertia $\Im_{\parallel}$ is similar to that of the $FTBCS$ method. However, one observes, as in the perpendicular moment of inertia case, a shift between the two curves which is due to the previously discussed shift between the critical temperatures. The maximum value of the relative discrepancy between the present model and the $FTBCS$ one, defined by $\left|\Im_{\parallel}^{np} - \Im_{\parallel}^{FTBCS}\right| / \Im_{\parallel}^{FTBCS}$, differs from one nucleus to another. It is about 38.84% in the ^{60}Zn case and about 2.16% in the ^{80}Sr one. It is of the order of 13.51% on average for all the studied nuclei.

Naturally, both models join when $T > T_{cnp}$ since, in this case, there is no more pairing effect ($\Delta_{pp} = \Delta_{nn} = 0$) and the obtained values are those of the rigid body.

4. Conclusion

The temperature-dependent perpendicular and parallel moments of inertia including isovector pairing effect have been calculated as a function of the temperature. The used single-particles energies are those of a deformed Woods-Saxon mean-field. The obtained results have been compared to their homologues of the conventional *FTBCS* theory. It has been shown that:

(i) In the region where $0 \leq T \leq T_{cnp}$, regarding the perpendicular moment of inertia $\Im_{\perp}$, the study shows that the np pairing effect leads to a maximum discrepancy of 12.23% on average for all the considered nuclei. Dealing with the parallel moment of inertia $\Im_{\parallel}$, in this region, it has been shown that there is no difference between the previsions of both models since for all the considered nuclei $\Im_{\parallel}$ is nil. In fact, T_{cnp} is small and the np pairing effect on $\Im_{\parallel}$ is not visible.

(ii) In the region where $T > T_{cnp}$, in spite of the fact that there is no more np pairing, there exists a discrepancy between the present model values and those of the FTBCS method, which is due to the shift between the critical temperatures. The maximum relative discrepancy between the present model results and the *FTBCS* ones is of the order of 11.46% on average for the perpendicular moment of inertia and about 13.51% on average for the parallel moment of inertia for the six studied nuclei.

Finally, it has been noticed that, at finite temperature, for a given Z, the critical temperatures T_{cnp} are as weak as $(N - Z)$ is important and the np pairing effect practically disappears with increasing values of N.

As a conclusion, the isovector pairing effect on both perpendicular and parallel moments of inertia is non negligible at finite temperature. These correlations must be taken into account in study of warm rotating nuclei in the $N \simeq Z$ region.

References

[1] Simkovic F, Moustakidis Ch. C, Pacearescu L and Faessler A, 2003 *Phys. Rev* C **68** 054319.
[2] Perlinska E, Rohozinski S. G, Dobaczewski J and Nazarewicz W, 2004 *Phys. Rev.* C **69** 014316.
[3] Mokhtari D, Allal N. H and Fellah M, 2004 *Acta. Phys. Hung. A : Heavy Ion Phys.* **19** 187.
[4] Civitarese O and Reboiro M, 1997 *Phys. Rev.* C **56** 1179.
[5] Allal N.H, Fellah M , Benhamouda N and Oudih M.R, 2008 *Phys. Rev.* C **77** 054310.
[6] Brack M, Damgaard J, Jensen A. S, Pauli H. C, Strutinsky V. M and Wong C. Y, 1972 *Rev. Mod. Phys.* **44** N2 320.
[7] Allal N.H and Fellah M, 1991 *Phys. Rev.* C **43** 2648.
[8] Hasegawa M and Tazaki S, 1993 *Phys. Rev.* C **47** 188.
[9] Meftunoglu E. et al., 1998 *J. Phy. G : Nucl. Part. Phys.* **24** 107.
[10] Gerceklioglu M and Calik A. E, 2005 *Acta Phys. Slo.* **55** 197.
[11] Mokhtari D, Ami I, Fellah M and Allal N.H, in *Proc. Int. Conf. On Nuclear Data for Science and Technology*, April 22-27, 2007, Nice, France; Bersillon ed. O, Gunsing F, Bauge E, Jacqmin R and Leray S , 159 (EDP Sciences, 2008).
[12] Mokhtari D, Ami I, Fellah M and Allal N.H, 2008 *Int. J. Mod. Phys.*E **17** 655.
[13] Ami I, Fellah M, Allal N.H, Benhamouda N, Belabbas M and Oudih M.R, 2011 *Int. J. Mod. Phys.* E. **20** 1947.
[14] Fellah M, Allal N.H, Belabbas M, Oudih M.R and Benhamouda N, 2007 *Phys. Rev.* C **76** 047306.
[15] Belabbas M, Fellah M, Allal N. H, Benhamouda N, Ami I and Oudih M. R., 2010 *Int. J. Mod. Phys.* E **19** 1973.
[16] Ami I, Belabbas M, Benhamouda N, Fellah M and Allal N.H, *European Nuclear Physics Conference 2009*, Bochum, Germany, March 2009.
[17] Brack M and Quentin P, 1981 *Nucl. Phys.* A **361** 35.
[18] Inglis D.R, 1954 *Phys. Rev.* **96** 1059.
[19] Hubbard J, 1959 *Phys. Rev. Lett.* **3** 77 ; Stratonovich R.L, 1958 *Sov. Phys. Dokl.* **2** 416.
[20] Allal N.H and Fellah M, 1994 *Phys. Rev.* C **50** 1404.
[21] Möller P, Nix J. R, Myers W. D and Swiatecki W. J, 1995 *Atomic Data and Nuclear Data Tables* **59** 185.
[22] Rossignoli R, Canosa N and Egido J. L, 1996 *Nucl. Phys.* A **605** 1.
[23] Kargar Z, 2007 *Phys. Rev.*C **75** 064319.
[24] Dinh Dang N., 2007 *Phys. Rev.* C **76** 064320.
[25] Puddu G, 1993 *Phys. Rev.* C **47** 1067.

Journal of Physics: Conference Series **338** (2012) 012020 doi:10.1088/1742-6596/338/1/012020

Kinetic approaches to phase transitions in strongly interacting matter

V Baran[1], **M. Colonna**[2], **V. Greco**[2,3], **M. Di Toro**[2,3], **S. Plumari**[2] **and R. Zus**[1]

[1] Physics Faculty, University of Bucharest, Romania
[2] Laboratori Nazionali del Sud INFN, I-95123, Catania, Italy
[3] Physics and Astronomy Dept., University of Catania, Italy

E-mail: `baran@nipne.ro`

Abstract. Strongly interacting matter manifests a very rich dynamical behavior and the associated phase diagram was intensively explored through heavy ions collisions during the last years. Around Fermi energies the nuclear multifragmentation shows up analogies with the liquid-gas phase transition. Within a microscopic transport model based on Boltzmann-Nordheim-Vlasov equation we investigate the kinetics of this process as well as the features related to the two-component character of the nuclear matter. The evolution of the fragmentation mechanism with the centrality and the role of various instabilities are also discussed. At ultra-relativistic energies recent results from RHIC and LHC experiments evidenced the manifestation of quarks and gluons degrees of freedom during the evolution of the hot fireball created in these collisions. Based on a relativistic transport model we inquire upon the role of chiral symmetry breaking on the collective features of expanding quark-gluon plasma.

1. Introduction

The fragment production in heavy ions reactions at intermediate energies represents an important dissipation mechanism. Very often, in discussing the nuclear multifragmentation, the analogy with the liquid-gas phase transition is invoked. However, in this comparison one should be aware of the differences determined by the Coulomb, finite-size or quantum effects. Moreover, at variance to the situation regarding the phase transitions in macroscopic systems, in heavy ions collisions the reaction times are comparable to the fragments growth time. The fast expansion may quench the system inside the instability region of the phase diagram and the spinodal decomposition will represent a significant kinetic mechanism for the fragment formation during the first stages of the nuclear fragmentation at Fermi energies. A binary system, including asymmetric nuclear matter (ANM), manifests a richer thermodynamical behavior since has to accommodate one more conservation law.

At ultra-relativistic energies, with the advance of new experimental programs ending with Relativistic Heavy Ion Collider (RHIC) at BNL and Large Hadron Collider (LHC) at CERN, the heavy ions collisions are probing new phases of strongly interacting matter. These investigations provide the means to test under extreme conditions of temperature and/or density the predictions of fundamental theory of strong interaction, Quantum Chromodynamics (QCD), by spanning different regions of the QCD phase diagram [1]. QCD is a non-Abelian gauge theory based on color SU(3) group and describes the interaction of quarks and gluons. It dictates the structure of hadrons and nuclei and the interaction among them and has a very rich dynamical content. In particular it manifest the so called asymptotic freedom which means a decrease of the interaction strength with energy scale. At low energies, of the order of $\Lambda_{QCD} = 300MeV$, the coupling becomes large and a perturbation approach cannot be applied. Two important non-pertubative features of QCD are the color confinement, related to the fact that quarks and gluons are locally confined to color singlet bound states and spontaneous chiral symmetry breaking. This symmetry, which is exact in the limit of massless quarks, expresses the invariance of QCD Lagrangian density to global transformations $SU_L(N_F) \times SU_R(N_F)$, where N_F is the number of flavors, and is equivalent to invariance under $SU_V(N_F) \times SU_A(N_F)$. For one flavor case this leads to conservation of quark vector current $V^\mu(x) = \bar{q}(x)\gamma^\mu q(x)$ and of axial vector current $A^\mu(x) = \bar{q}(x)\gamma^\mu\gamma_5 q(x)$ respectively. Several experimental indications, including the masses of mesons and baryons of opposite parity and pion properties suggests that the chiral symmetry is realized in Nambu-Goldstone mode, i.e. the symmetry of the Lagrangian is not realized in the ground state. So the full chiral symmetry mentioned above is spontaneously broken down to the diagonal vector symmetry $SU_V(N_F)$. As a consequence it is expected the appearance in the spectrum of pseudoscalar mesons of mass zero, the Goldstone bosons. For two flavors case they are identified with the pions. Their mass small but finite, indicates an explicit breaking of the chiral symmetry as a consequence of non-zero bare (current) masses of up and down quarks (5 MeV and 10 MeV respectively). The pseudoscalar meson octet is associated to the Goldstone bosons in the three-flavor case but in this situation the explicit breaking symmetry is more evident since the mass of strange quark is greater, approx. 200 MeV.

In the first part of this work we discuss briefly the nature of the instabilities and of the related density fluctuations in binary systems. Then we describe some aspects of the kinetics of phase transition in ANM in the linear regime. For heavy ion collisions at intermediate energies, within a microscopic transport model, we explore the relevance of these results for nuclear multifragmentation and show the evolution of the fragmentation mechanisms with the centrality of the reaction.

At ultra-relativistic energies one of the important questions regards the the role of chiral symmetry breaking on quark-gluon plasma dynamics. Therefore in the second part of the paper, based on a relativistic transport model including a Nambu-Jona-Lasinio quark interaction, we investigate the influence of chiral symmetry breaking on the properties of the elliptic flow at RHIC energies.

2. Nuclear fragmentation at Fermi energies

The spinodal decomposition mechanism in the liquid-gas phase transition is related to the fact that dynamically the system can enter in the instability region of the phase diagram and is associated to the exponential amplification of density fluctuations.

An appropriate framework for the study of instabilities in two-components systems is provided by the Fermi liquid theory [2], which has been applied, for instance, to symmetric binary systems as SNM (the two components being protons and neutrons) [3] and the liquid ^3He (spin-up and spin-down components) [4, 5].

One-component systems may become unstable against density fluctuations as the result of the attractive interaction between constituents. In symmetric binary systems, like SNM, one may encounter two kinds of density fluctuations: isoscalar, when the densities of the two components fluctuate in phase with equal amplitude, and isovector when the two densities fluctuate still with equal amplitude but out of phase. Thermodynamically, the mechanical instability is associated to the instability against isoscalar fluctuations, leading to cluster formation, while chemical instability is related to instability against isovector fluctuations, leading to species separation.

An extension to asymmetric binary systems is obtained by a generalization of the formalism introduced in [4] to the asymmetric case. At $T = 0$ the thermodynamical stability requires the energy of the system to be an absolute minimum for the undistorted distribution functions, so that the relation:

$$\delta H - \mu_p \delta\rho_p - \mu_n \delta\rho_n > 0 \tag{1}$$

is satisfied when we deform proton and neutron Fermi seas, determined by the distribution functions $f_q(\epsilon_p^q) = \Theta(\mu_q - \epsilon_p^q)$, $q = n, p$. Here H is the energy density, μ_q, ρ_q are the corresponding chemical potentials and densities.

The nucleons interaction is characterized by the Landau parameters:

$$F^{q_1 q_2} = N_{q_1} V^2 \frac{\delta^2 H}{\delta f_{q_1} \delta f_{q_2}} = N_{q_1} \frac{\delta^2 H}{\delta\rho_{q_1} \delta\rho_{q_2}} \tag{2}$$

$$N_q(T) = \int \frac{-2\ d\mathbf{p}}{(2\pi\hbar)^3} \frac{\partial f_q(T)}{\partial \epsilon_p^q} \tag{3}$$

where V is the volume and N_q is the single-particle level density at the Fermi energy. At $T = 0$ $N_q(0) = m p_{F,q}/(\pi^2 \hbar^3) = 3\rho_q/(2\epsilon_{F,q})$, where $p_{F,q}$ and $\epsilon_{F,q}$ are Fermi momentum and Fermi energy of the q-component. Only monopolar deformations will be considered here and $F_{l=0}^{q_1 q_2}$ are the only non-zero Landau parameters. Eq.(1) becomes

$$\delta H - \mu_p \delta\rho_p - \mu_n \delta\rho_n = \frac{1}{2}(a \delta{\rho_p}^2 + b \delta{\rho_n}^2 + c \delta\rho_p \delta\rho_n) > 0 \tag{4}$$

where

$$\begin{aligned} a &= N_p(0)(1 + F_0^{pp}) \ ; \ \ b = N_n(0)(1 + F_0^{nn}) \ ; \\ c &= N_p(0) F_0^{pn} + N_n(0) F_0^{np} = 2 N_p(0) F_0^{pn}. \end{aligned} \tag{5}$$

The r.h.s. of Eq.(4) is diagonalized by the following transformation:

$$\begin{aligned} u &= cos\beta\ \delta\rho_p + sin\beta\ \delta\rho_n, \\ v &= -sin\beta\ \delta\rho_p + cos\beta\ \delta\rho_n, \end{aligned} \tag{6}$$

where the *mixing* angle $0 \leq \beta \leq \pi/2$ is given by

$$tg\ 2\beta = \frac{c}{a-b} = \frac{N_p(0) F_0^{pn} + N_n(0) F_0^{np}}{N_p(0)(1 + F_0^{pp}) - N_n(0)(1 + F_0^{nn})}. \tag{7}$$

Then the condition (4) takes the form

$$\delta H - \mu_p \delta\rho_p - \mu_n \delta\rho_n = X u^2 + Y v^2 > 0 \ , \tag{8}$$

where

$$\begin{aligned} X &= \frac{1}{2}(\ a + b + sign(c)\sqrt{(a-b)^2 + c^2}\ ) \\ &\equiv \frac{(N_p(0) + N_n(0))}{2}(\ 1 + F_{0g}^s\ ) \end{aligned} \tag{9}$$

and

$$\begin{aligned} Y &= \frac{1}{2}(\ a + b - sign(c)\sqrt{(a-b)^2 + c^2}\ ) \\ &\equiv \frac{(N_p(0) + N_n(0))}{2}(\ 1 + F_{0g}^a\ ), \end{aligned} \tag{10}$$

are defining the new generalized Landau parameters $F_{0g}^{s,a}$.

Hence, thanks to the rotation Eq.(6), it is possible to separate the total variation (see Eq.(1) into two independent contributions, the "normal" modes, characterized by the "mixing angle" β, which depends on the density of states and the details of the interaction. The thermodynamical stability requires $X > 0$ *and* $Y > 0$. Equivalently, the following conditions have to be fulfilled:

$$1 + F_{0g}^{s} > 0 \quad and \quad 1 + F_{0g}^{a} > 0, \tag{11}$$

representing the Pomeranchuk stability conditions extended to asymmetric binary systems. It can be proved that [6]:

$$\begin{aligned} XY &= N_p(0)N_n(0)[(1+F_0^{nn})(1+F_0^{pp}) - F_0^{np}F_0^{pn}] \\ &= \frac{[N_p(0)N_n(0)]^2}{(1-y)\rho^2}\left(\frac{\partial P}{\partial \rho}\right)_{T,y}\left(\frac{\partial \mu_p}{\partial y}\right)_{T,P} \end{aligned} \tag{12}$$

and:

$$\begin{aligned} \left(\frac{\partial P}{\partial \rho}\right)_{T,y} &= \frac{\rho y(1-y)}{N_p(0)N_n(0)}(\ t\ a + \frac{1}{t}\ b + c\ ) \\ \propto X(\sqrt{t}cos\beta &+ \frac{1}{\sqrt{t}}sin\beta)^2 + Y(\sqrt{t}sin\beta - \frac{1}{\sqrt{t}}cos\beta)^2 \\ with \quad t &= \frac{y}{1-y}\frac{N_n(0)}{N_p(0)}. \end{aligned} \tag{13}$$

Consequently the new stability conditions, Eq.(11), can be related to mechanical and chemical stability of the thermodynamical state, [7], i.e.

$$\left(\frac{\partial P}{\partial \rho}\right)_{T,y} > 0 \quad and \quad \left(\frac{\partial \mu_p}{\partial y}\right)_{T,P} > 0 \tag{14}$$

where P is the pressure and y the proton fraction. Indeed, let us assume that in the density range we are considering the quantities a and b remain positive. Then one can study the effect of the interaction between the two components, given by c, on the instabilities of the mixture. If $c < 0$, i.e. for an attractive interaction between the two components, the system is always stable against isovector-like fluctuations (see Eq.(10)) However it becomes unstable against isoscalar-like fluctuations if $c < -2\sqrt{ab}$ (see Eq.(9)). Thermodynamically this instability will show up as a chemical instability if $(-ta - b/t) < c < -2\sqrt{ab}$ or as a mechanical instability if $c < (-ta - b/t) < -2\sqrt{ab}$ (see Eq.(13)). We conclude that the relative strength of the interactions among the various species will determine the kind of thermodynamical instability which shows up. As we already mentioned, in the symmetric two-components systems it is a unique correspondence between the nature of thermodynamical instability and the nature of the unstable fluctuation. However we see that for asymmetric binary systems this correspondence is lost. The relevant instability region is defined only in terms of the instabilities against isoscalar fluctuations and therefore we can speak about a unique spinodal region. But inside the spinodal the system can be mechanically or chemically unstable. This is the case of asymmetric nuclear matter (ANM). For a Skyrme-like interaction, selected to reproduce the saturation properties of nuclear matter, $a, b > 0$ and $c < 0$, we deal only with isoscalar-like unstable fluctuations. We show in Fig.1, for three values of the proton fraction, the spinodal line corresponding to isoscalar-like instability (circles).

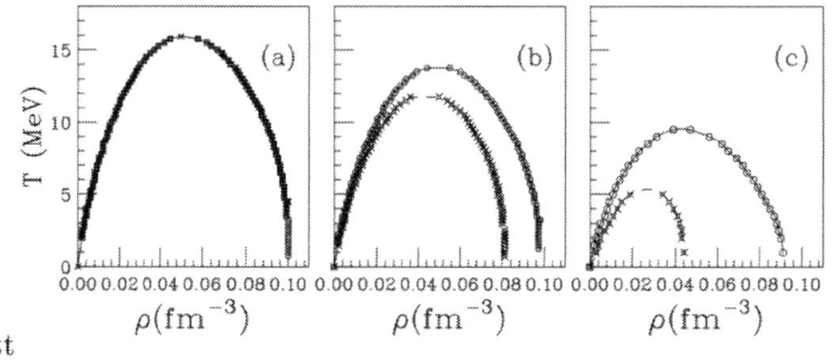

tt

Figure 1. Spinodal boundary corresponding to insoscalar-like instability (circles) for three proton fractions: (a) $y = 0.5$, (b) $y = 0.25$, (c) $y = 0.1$. The mechanical instability line, inside spinodal region, is indicated by crosses.

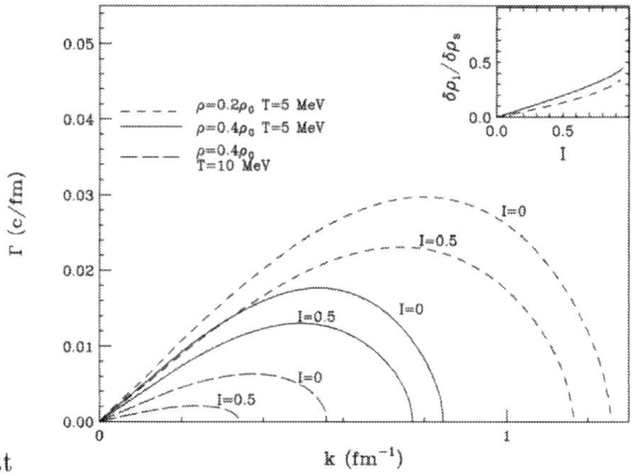

tt

Figure 2. Growth rates of instabilities as a function of the wave vector, as calculated from the dispersion relation for three situations inside the spinodal region. Lines are labeled with the asymmetry value I. The insert shows the asymmetry of the perturbation $\delta\rho_I/\delta\rho_S$, as a function of the asymmetry I of the initially uniform system, for the most unstable mode, in the case $\rho = 0.4\rho_0$, $T = 5\ MeV$.

For asymmetric matter, $y < 0.5$, under this border one encounters either chemical instability, in the region between the two lines, or mechanical instability, under the inner line (crosses). The latter is defined by the set of values (ρ, T) for which $\left(\frac{\partial P}{\partial \rho}\right)_{T,y} = 0$.

To complete our analysis we consider now the case $c > 0$, i.e. when the interaction between the two components is repulsive. Now the thermodynamical state is always stable against isoscalar-like fluctuation. In can becpme unstable to isovector fluctuations if $c > 2\sqrt{ab}$. With our choices of the parameters, $a, b, c > 0$, the isovector instability is always associated with chemical instability. Such situation will lead to a component separation of the liquid mixture while the total density does not change. Summarizing, a complete analysis of the instabilities of any binary system can be performed, in connection to signs, strengths and density dependence of the interactions among the components.

Once the stability conditions of ANM were investigated, the linear response will provide informations about the growing rate of the unstable modes, relevant for the fragment formation

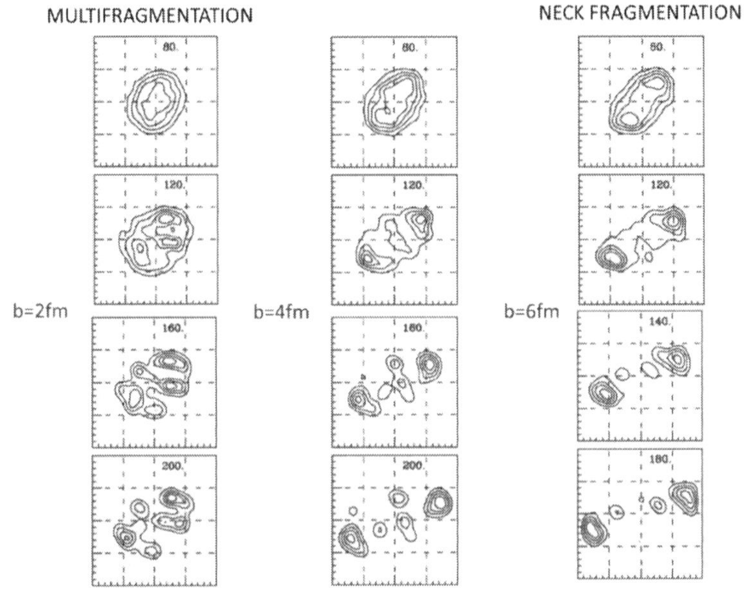

Figure 3. Density contour plots for ^{124}Sn + ^{124}Sn collision at 50AMeV for three impact parameters. From the left b=2 fm, b=4 fm and b=6fm.

time scales in multifragmentation. The approach is based on Vlasov equation,

$$\frac{\partial f_q(\mathbf{r},\mathbf{p},t)}{\partial t} + \frac{\mathbf{p}}{m}\frac{\partial f_q}{\partial \mathbf{r}} - \frac{\partial U_q(\mathbf{r},t)}{\partial \mathbf{r}}\frac{\partial f_q}{\partial \mathbf{p}} = 0 \,, \qquad q = n, p. \tag{15}$$

where $U_q(\mathbf{r},t)$ is the self-consistent mean field potential in a Skyrme-like form [8, 9]. By linearization, corresponding to a semiclassical RPA approach, it is obtained a dispersion relation quadratic in the frequency ω of the mode $\delta f_q(\mathbf{r},\mathbf{p},t) \sim \exp(-i\omega t)$. One finds two independent solutions (isoscalar-like and isovector-like solutions): ω_s^2 and ω_i^2. For the unstable isoscalar-like mode we report in Fig.2 the growth rate $\Gamma = \text{Im}\ \omega(k)$ as a function of the wave vector k. Results are shown for symmetric ($I = 0$) and asymmetric ($I = 0.5$) nuclear matter.

The growth rate has a maximum $\Gamma_0 = 0.01 \div 0.03$ c/fm corresponding to a wave-vector value around $k_0 = 0.5 \div 1\ \text{fm}^{-1}$ and becomes equal to zero at $k \simeq 1.5k_0$, due to k-dependence of the Landau parameters. The instabilities are reduced when increasing the temperature, an effect also present in the symmetric N = Z case [10, 11, 12]. At larger initial charge asymmetry the development of the spinodal instabilities is slower. The conclusion is that the fast spinodal decomposition mechanism in neutron-rich matter will dynamically leads to more symmetric fragments surrounded by a more netron rich gas. Some recent experimental observations from fragmentation reactions with neutron rich nuclei at the Fermi energies seem to be in agreement with this result: nearly charge symmetric Intermediate Mass Fragments (IMF) have been detected together with very neutron-rich light ions [13]. All these considerations provide a firm ground for numerical simulations performed in the nonlinear regime, corresponding to realistic initial conditions of heavy-ions collisions at Fermi energies. In Fig. 3 we show the evolution of the reaction mechanism with the impact parameter. A transition from multifragmentation, at $b = 2fm$, to neck fragmentation, at $b = 6fm$, is observed. These results were obtained within a microscopic transport model, Stochastic Mean Field (SMF)[14]. Fragment formation and growth is quite consistent with spinodal decomposition mechanism. The velocity distributions of the fragments are in agreement with several experimental observations. Based on these approaches the correlations between fragments isopin content, velocity distributions and IMF flows open the possibility to learn more about the density dependence of the symmetry energy in the equation of state, of interest in the study of exotic nuclei or astrophysical investigations [14].

3. Transport approach to QGP expansion including chiral symmetry breaking

The difficulties encountered in the basic theory in the nonperturbative regime motivates to look at simpler models which are capturing some essential features of QCD. One of the models appropriate to study the chiral symmetry breaking is the Nambu-Jona-Lasinio (NJL) model. The NJL Lagrangian density contains only quark degrees of freedom and the effects of gluons are included in a four fermion point interaction. Hence we deal with an effective field theory with an attractive force between quarks and antiquarks whose Lagrangian density (for one-flavor case) is

$$\mathcal{L}_{NJL} = \bar{q}i\gamma^{\mu}\partial_{\mu}q + G[(\bar{q}q)^2 + (\bar{q}i\gamma_5 q)^2] \tag{16}$$

where $(\bar{q}q)^2 \equiv \left(\sum_{a=1}^{N_c} \bar{q}^a q^a\right)^2$ and N_c is the number of color degrees of freedom. The coupling constant G is assumed to be positive so that the forces between quarks and antiquarks are attractive. In analogy with the superconductivity, where an attractive interaction between electrons with opposite momenta at Fermi surface leads to a non-pertubative change of ground state, the quark-antiquark interaction generate a similar ground state built as a coherent superposition of quark-antiquark pairs with zero total momentum and zero total helicity. A BCS approach is then possible with a variational ground state [15]:

$$|vac\rangle = \prod_{\mathbf{p},s=\pm1} [\,\cos\theta(p) - s\sin\theta(p) b^{\dagger}(\mathbf{p},s) d^{\dagger}(-\mathbf{p},s)]|0\rangle \tag{17}$$

The annihilation operators for quarks and antiquarks with momentum p and helicity s define the normal (perturbative) vacuum, i.e. $b(\mathbf{p},s)|0\rangle = d(\mathbf{p},s)|0\rangle = 0$. The variational principle leads to an equation analogous to the gap equation in superconductivity:

$$m^* = 4GN_cN_f \int_0^{\Lambda} \frac{d^3q}{(2\pi)^3}\frac{m^*}{E_q} \tag{18}$$

where $E_q = \sqrt{q^2 + m^{*2}}$ and $\sin\theta(p) = m^*/\sqrt{p^2 + m^{*2}}$. The three momentum cut-off parameter Λ was introduced since the point-like interaction in the NJL Lagrangian renders the model nonrenormalizable. For $N_F = 2$ the model contains three free parameters: the coupling constant G, the current quark mass m and the cutoff Λ which is used to make theory finite. These are fixed to reproduce the physical pion mass, $m_\pi = 135MeV$, the physical pion decay constant $f_\pi = 93MeV$ and the chiral condensate value $\langle\bar{\psi}\psi\rangle = \dfrac{m^* - m}{2G}$. The latter, which is the order parameter of the phase transition, having zero value in the chiral restored phase, has larger uncertainties and therefore different sets of parameters are possible. For example in [16] with the set $\Lambda = 820MeV, G\Lambda^2 = 1.99, m = 4.0MeV$ is obtained $m^* = 323MeV$ while in [17] with $\Lambda = 588MeV, G\Lambda^2 = 2.44, m = 5.6MeV$ is obtained $m^* = 400MeV$.

The NJL model can be extended to finite density and temperature [18]. The mass gap equation, including a current finite current mass becomes:

$$M(T,\mu) = m + 4\,G\,N_c M(T,\mu)\int \frac{d^3p}{(2\pi)^3}\frac{1 - f^-(T,\mu) - f^+(T,\mu)}{E_p} \tag{19}$$

with $E_p = \sqrt{p^2 + M(T,\mu)^2}$ and $f^-(T,\mu)$ $f^+(T,\mu)$ the Fermi-Dirac distribution functions for quarks and antiquarks respectively. The solution of this equation for various temperatures as a

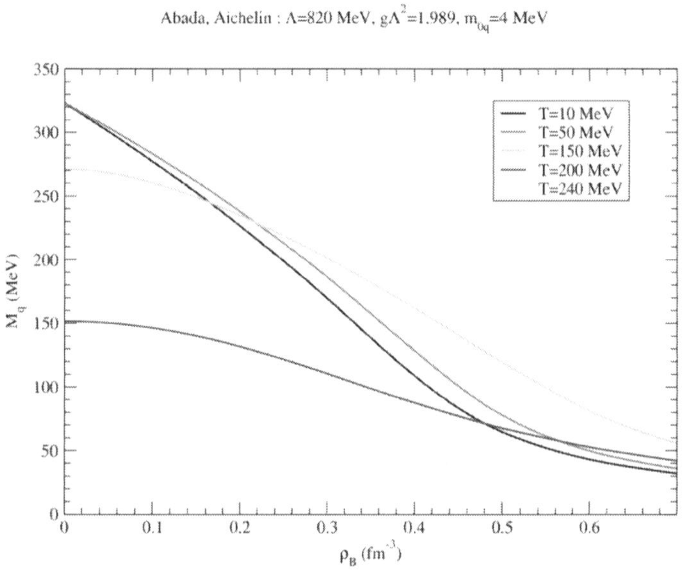

Figure 4. Quark constituent mass as a function of barionic density for various temperatures

function of barionic density with the set of parameters from [16] is plotted in Fig. 5. Thermal fluctuations as well as larger barionic densities are reducing the effect of symmetry breaking.

The theoretical views on QGP changed dramatically with the discoveries at RHIC from a system of gas of dressed quarks and gluons interacting weakly with each other to a strongly interacting matter of tightly coupled qusiparticles, an "perfect fluid (that) quenches jets almost perfectly" [19]. Within a standard scenario the evolution of the system following the heavy ions collision begin with the generation of an intense chromofields which later decay into of quarks and gluons. Then take place thermalization and entropy production, a viscous hydrodynamical expansion and hadronization. Important signals for QGP formation at RHIC includes strong azimuthal dependence of particle momenta described by the elliptic flow. Moreover the valence quark scaling of elliptic flow parameter v_2 indicates that flow pattern is developed at quark level.

One of the interesting questions is to understand the role of chiral symmetry breaking on collective features of expanding QGP. We solved numerically the microscopic transport equations of NJL model employing the test particle method [28]. One finally obtains the Boltzmann-Vlasov transport equations for the (anti-) quark phase-space distribution function $f^{\pm}$:

$$p^{\mu}\partial_{\mu}f^{\pm}(x,p) + M(x)\partial_{\mu}M(x)\partial_{p}^{\mu}f^{\pm}(x,p) = \mathcal{C}(x,p) \tag{20}$$

where $\mathcal{C}(x,p)$ is the Boltzmann-like collision integral, main ingredient of the several cascade codes already developed [20, 21, 22]. We notice that in addition to the cascade model ingredients the NJL dynamics introduces a new term associated to the mass generation. The Eq.(20) is formally the same as the widely used relativistic transport approaches for hadronic matter like RBUU,uRQMD, RLV [23, 24], but with a vanishing vector field. A key difference is however that particles do not have a fixed mass and a self-consistent approach couples the transport Eq.(20) to the mass gap equation of the NJL model that extended to the case of non-equilibrium as:

$$\frac{M(x) - m}{4GN_c} = M(x)\int \frac{d^3p}{(2\pi)^3}\frac{1 - f^{-}(x,p) - f^{+}(x,p)}{E_p(x)} \tag{21}$$

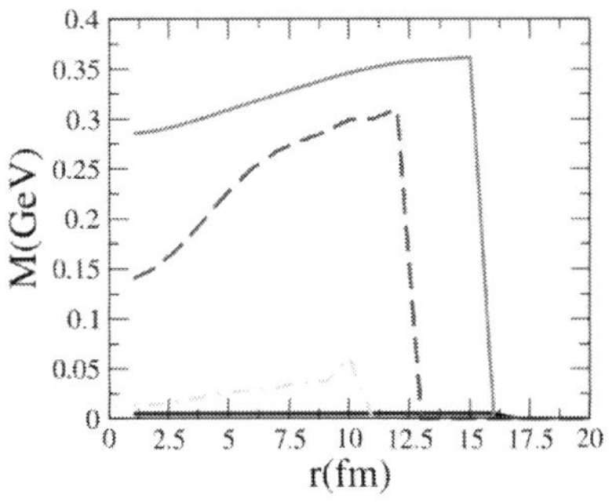

Figure 5. Quark constituent mass as a function of radial distance from the center of the expanding distribution Green dot dashed line: t=0.67 fm/c.Blue dashed line:t=3.7 fm/c. Red solid line:t=6.7 fm/c.Black line: current quark mass.

This equation determines the local mass $M(x)$ at the space-time point x in terms of the distribution functions $f^{\pm}(x,p)$. The Eqs.(20) and (21) form a closed system of equations constituting the Boltzmann-Vlasov equation associated to the NJL Lagrangian that allows to obtain self-consistently the local effective mass $M(x)$ and time dependent, nonequilibrium distribution function $f^{\pm}(x,p)$.

We have run the simulations for $Au+Au$ at $\sqrt{s_{NN}} = 200$ AGeV and $b = 6\,fm$. The density distribution in coordinate space is given by the standard Glauber model. The maximum initial temperature is $T = 340$ MeV and the initial time is $\tau_0 = 0.6$ fm/c as usually done also in hydrodynamical calculations. We follow the dynamical evolution of quarks, anti-quarks and gluons. The last has been included, even if they are not explicitly present in the NJL model, with the aim of using a realistic density for both the total and the (anti-)quark density in the simulation of the collisions. However gluons do not actively participate in the evaluation of the chiral phase transition, but they simply acquire the same mass of the quarks not contributing to its determination according to the the NJL model. The justification for this choice relies on the quasi-particle models that are fitted to lQCD thermodynamics [25, 26] where one finds a similar behavior of $M(T)$ for both gluons and quarks approximately.

In Fig. 5 plot the radial dependence of the quark constituent mass at three different instants: initial time $t = 0.6 fm/c$, $t = 3.6 fm/c$ and final time $t = 6.7 fm/c$. The mass is larger toward the frontiers of the expanding distributions where the quark densities are lower. For comparison, in the case without NJL interaction, the mass is constant, equal to the current mass.

In Fig. 6 (right) it is shown the time evolution of the average elliptic flow $\langle v_2 \rangle$ for a constant transport cross section of σ_{tr} =10 mb, a typical value that is able to reproduce the amount of v_2 observed in experiments [27]. Comparing the two lines we can see that the NJL mean field induce a decrease of $\langle v_2 \rangle$. The reduction of v_2 can be expected considering that the NJL field produce a scalar attractive field that at the phase transition results in a gas of massive particles.

In Fig.6 (left), we show the elliptic flow at freeze-out as a function of the transverse momentum p_T. One can see that the role of the mean field even increases with momentum affecting also particles at a p_T larger than the energy scale of the scalar condensate $\langle \bar{\psi}\psi \rangle \sim 300MeV$. Therefore the effect of the scalar field extends thanks to collisions into a range quite larger than one would naively think and the interplay between collisions and mean field is fundamental. We also find that at $b = 7fm$ the presence of a NJL-field that drives the chiral phase transition suppress the $v_2(p_T)$ by about 20% at $p_T > 1$ GeV [28], [29].

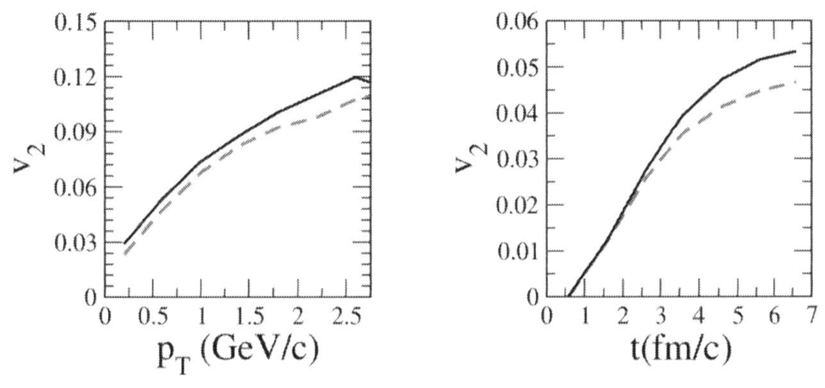

Figure 6. Left: Average elliptic flow as a function of transverse momentum for $Au + Au$ collisions in the mid-rapidity region $|y| < 1$ at b= 6 fm. Right: time evolution of the elliptic flow parameter for the same case as in the left panel. Black solid: only collisions. Red dashed:NJL mean field included.

4. Conclusions and perspectives

In this work we explored several facets of the nuclear fragmentation at Fermi energies. This can be related to a spinodal decomposition mechanism and realistic simulations based on microscopic transport models provide important informations related to the nuclear EOS. We then investigated within a relativistic transport model the effects of chiral symmetry breaking on the collective features of quark-gluon plasma at RHIC conditions. A sizable lowering of elliptic flow parameter was observed. The approach can be extended to LHC conditions and is appropriate to study the effects of fixed shear viscosity to entropy density ratio in the presence or absence of chiral dynamics. By including other interaction terms the transport model can be applied to situations corresponding to a finite barionic density.

Acknowledgments

This work for V. Baran was supported by a grant of the Romanian National Authority for Scientific Research, CNCS-UEFISCDI, project number PN-II-ID-PCE-2011-3-0972. S. Plumari and V. Greco are partially supported by the FIRB Research Grant RBFR0814TT. For R. Zus this work was supported by the strategic grant POSDRU/89/1.5/S/58852, Project "Postdoctoral programme for training scientific researchers" co-financed by the European Social Found within the Sectorial Operational Program Human resources Development 2007-2013.

References

[1] *Quark-Gluon Plasma* 2010, ed. Rudolph C Hwa and Xin-Nian Wang, Vol 4, (World Scientific)
[2] Landau L D 1957 JETP **5** 101
[3] Migdal A B 1967 *Theory of finite Fermi systems and applications to atomic nuclei*, (Wiley & Sons, N.Y.)
[4] Baym G and Pethick C J 1978 *The physics of Liquid and Solid Helium* vol 2, ed. K.H.Bennemann and J.B.Ketterson, Vol 2, (Wiley, N.Y.), p 1
[5] Pethick C J and Ravenhall D G 1988 *Ann.Phys.* **183** 131
[6] Baran V, Colonna M, Di Toro M and Greco V 2001 *Phys. Rev. Lett.* **86** 4492
[7] Landau L D and Lifshitz E M 1989 *Statistical Physics*, (Pergamonn Press), p. 288
[8] Colonna M, Di Toro M and Larionov A B 1998 *Phys.Lett.* B**428** 1
[9] Baran V, Colonna M, Di Toro M and Larionov A B 1998 *Nucl.Phys.* A **632** 287
[10] Colonna M and Chomaz P H 1994 *Phys.Rev.* C **49** 1908
[11] Colonna M, Chomaz P H and Randrup J 1994 *Nucl.Phys.* A**567** 637
[12] Chomaz P H, Colonna M and Randrup J 2004 *Phys.Rep.* **389** 263

[13] Xu H S et al. 2000 *Phys.Rev.Lett.* **85** 716
[14] Baran V, Colonna M, Di Toro M and Greco V 2005 *Phys. Rep.* **410** 335
[15] Klevansky C D 1992 *Rev. Mod. Phys.* **64** 649
[16] Abada A and Aichelin J 1995 *Phys. Rev. Lett.***74** 3130
[17] Buballa M 2005 *Phys. Rep.***407** 205
[18] Klevansky C D 1996 *Ann. Phys.***245** 445
[19] Muller B 2009 *Prog. Part. Nucl. Phys.***62** 551
[20] Molnar D and Gyulassy M 2002 *Nucl.Phys* A **697** 495
[21] Xu Z and Greiner C 2005 *Phys. Rev.* C **71** 064901
[22] Ferini G, Colonna M, Di Toro M and Greco V 2009 *Phys Lett* B **670** 325
[23] Bass S A et al. 1998 *Prog.Part. Nucl. Phys.***41** 255
[24] Fuchs C and Wolter H H 1995 *Nucl. Phys.* A**589** 732
[25] Levai P and Heinz U W 1998 *Phys. Rev* C **57** 1879
[26] Plumari S, Alberico W M, Greco V and Ratti C *preprint arXiV.hep-ph/:1103.5611*
[27] Lin Z-W and Ko C M 2002 *Phys. Rev.* C **65** 034904
[28] Plumari S, Baran V, Di Toro M, Ferini G and Greco V 2010 *Phys.Lett.* B **689** 18
[29] Plumari S 2010 *PhD thesis* (Univ. of Catania,Italy)

Neutron-Proton pairing effect on the thermodynamical quantities of even-even proton-rich nuclei

M Belabbas [1], M Fellah [2,3], N H Allal [2,3] and I Ami [4,2]

[1] Faculté des sciences et science de l'Ingénieur, UHBB, BP151, Chlef, Algeria.
[2] Laboratoire de Physique Théorique, Faculté de Physique,USTHB, BP 32, El Alia, 16111 Bab Ezzouar, Algiers, Algeria
[3] Centre de Recherche Nucléaire d'Alger, 2 Bd. Frantz Fanon, BP. 399 Alger-Gare, Algiers, Algeria
[4] Faculté des Sciences et de la Technologie, Université Dr. Yahia Fares de Médéa, Quartier Ain D'heb, 26000 Médéa, Algeria

Abstract. Expressions of the thermodynamical quantities, i.e. the energy E, the entropy S and the heat capacity C are established by including the isovector neutron-proton (np) pairing effect. They are deduced using temperature-dependent gap equations . E, S and C are numerically studied as a function of the temperature for some even-even proton-rich nuclei. The single-particle energies used are those of a Woods-Saxon deformed mean field. It is shown that the isovector pairing effect on E, S and C is non-negligible, not only in the $0 \leq T \leq T_{cnp}$ region (T_{cnp} being the critical temperature beyond which the np pairing vanishes), but also in the $T_{cnp} \leq T \leq T_{cn}$ region (T_{cn} being the neutron-system critical temperature).

1. Introduction

The experimental study of proton-rich nuclei became possible since they may be produced at radioactive nuclear beam facilities. Since in this kind of nuclei protons and neutrons Fermi levels are close to each other, it is expected that the neutron-proton (np) pairing correlations are non-negligible [1]. They have been studied in many papers at zero temperature (see e.g. [1]-[5]).

In hottest nuclei, the pairing correlations play an important role in the phase transition. They have been studied extensively in the pairing between like-particles case using for instance the finite temperature BCS (FTBCS) model [6]-[8], but for the np pairing, only a limited number of studies have been performed (see e.g. [9]-[10]). The purpose of the present work is to study the thermal effects of the np pairing correlations, in the isovector case, on some thermodynamical quantities i.e. the energy, the entropy and the heat capacity. The formalism is explicited in next section. In section 3, the model is numerically applied to $N = Z$ even-even nuclei using the single-particle energies of a deformed Woods-Saxon mean field. Main conclusions are summarized in last section.

2. Formalism

Let us start with the isovector pairing Hamiltonian, describing a system of N neutrons and Z protons [11]-[13]:

$$\mathcal{H} = \sum_{\nu>0,t} \varepsilon_{\nu t}(a^+_{\nu t}a_{\nu t} + a^+_{\tilde{\nu t}}a_{\tilde{\nu t}}) + H_{nn} + H_{pp} + H_{np} \tag{1}$$

where t corresponds to the isotopic spin component ($t = n, p$), $a^+_{\nu t}$ and $a_{\nu t}$ respectively represent the creation and annihilation operators of the particle in the state $|\nu t\rangle$, of energy $\varepsilon_{\nu t}$ (which is assumed to be independent from the temperature); $|\tilde{\nu t}\rangle$ is the time-reverse of $|\nu t\rangle$ which has the same energy and $H_{tt'}$ $(t, t\prime = n, p)$ corresponds to the pairing interaction given by:

$$H_{tt'} = -G_{tt'} \sum_{\nu,\mu>0} P^+_{\nu tt'} P_{\mu tt'} \qquad t, t' = n, p \tag{2}$$

with: $P^+_{\nu tt} = a^+_{\nu t}a^+_{\tilde{\nu t}}$, $P^+_{\nu np} = \left(a^+_{\nu p}a^+_{\tilde{\nu n}} + a^+_{\nu n}a^+_{\tilde{\nu p}}\right)$ and $G_{tt'}$ refers to the pairing-strength which is assumed to be constant.

Let us define the auxiliary Hamiltonian:

$$H = \mathcal{H} - \sum_{t=n,p} \lambda_t N_t \tag{3}$$

where λ_t are Lagrange parameters and the N_t are the particle-number operators:

$$N_t = \sum_{\nu>0}(a^+_{\nu t}a_{\nu t} + a^+_{\tilde{\nu t}}a_{\tilde{\nu t}}) \qquad t = n, p \tag{4}$$

The diagonalization of the Hamiltonian H using the Feynman path-integral technique is exposed in detail in ref. [11] and will not be recalled here. Its diagonal form is given by [11]-[13]:

$$H = \sum_{\nu>0,t} \tilde{\varepsilon}_{\nu t} + \sum_{\nu>0,\tau=1,2} E_{\nu\tau}\left(\alpha^+_{\nu\tau}\alpha_{\nu\tau} - \alpha_{\tilde{\nu}\tau}\alpha^+_{\tilde{\nu}\tau}\right) \tag{5}$$

where we set $\tilde{\varepsilon}_{\nu t} = \varepsilon_{\nu t} - \lambda_t - (G_{tt} + G_{np})/2, t = n, p$, $\alpha^+_{\nu\tau}$ is the creation operator of a quasiparticle of τ $(\tau = 1, 2)$ type, given by the generalized Bogoliubov-Valatin transformation :

$$\alpha^+_{\nu\tau} = \sum_{\nu>0}\left(u_{\nu\tau t}a^+_{\nu t} + v_{\nu\tau t}a_{\tilde{\nu t}}\right) \qquad with \quad \tau = 1, 2 and t = n, p \tag{6}$$

The quasiparticle energies $E_{\nu\tau}$ are given by:

$$E_{\nu\tau} = \sqrt{\frac{1}{2}\left\{E^2_{\nu n} + E^2_{\nu p} + 2\Delta^2_{np} + (-1)^{\tau+1} R_\nu\right\}} \tag{7}$$

with the notations:

$$\begin{aligned} R_\nu &= \sqrt{\left(E^2_{\nu n} - E^2_{\nu p}\right)^2 + 4\Delta^2_{np}\left[(\tilde{\varepsilon}_{\nu n} - \tilde{\varepsilon}_{\nu p})^2 + (\Delta_{nn} + \Delta_{pp})^2\right]} \\ E_{\nu t} &= \sqrt{\tilde{\varepsilon}^2_{\nu t} + \Delta^2_{tt}} \qquad t = n, p \end{aligned}$$

$\Delta_{tt'}$ $(t, t' = n, p)$ represent the gap parameters and are solutions of the generalized gap equations:

$$\begin{aligned} \frac{4\Delta_{nn}}{G_{nn}} &= \sum_{\nu\tau} \frac{1}{E_{\nu\tau}} \tanh\frac{1}{2}\beta E_{\nu\tau} \qquad (8) \\ &\left\{\Delta_{nn} + \frac{(-1)^{\tau+1}}{R_\nu}\left[\left(E^2_{\nu n} - E^2_{\nu p}\right)\Delta_{nn} + 2\Delta^2_{np}(\Delta_{nn} + \Delta_{pp})\right]\right\} \end{aligned}$$

(9)

$$\frac{4\Delta_{pp}}{G_{pp}} = \sum_{\nu\tau} \frac{1}{E_{\nu\tau}} \tanh \frac{1}{2}\beta E_{\nu\tau} \times \left\{ \Delta_{pp} + \frac{(-1)^{\tau+1}}{R_\nu} \left[\left(E_{\nu p}^2 - E_{\nu n}^2 \right) \Delta_{pp} + 2\Delta_{np}^2 \left(\Delta_{nn} + \Delta_{pp} \right) \right] \right\} \tag{10}$$

(11)

$$\frac{2}{G_{np}} = \sum_{\nu\tau} \frac{1}{E_{\nu\tau}} \tanh \frac{1}{2}\beta E_{\nu\tau} \times \left\{ 1 + \frac{(-1)^{\tau+1}}{R_\nu} \left[E_{\nu n}^2 + E_{\nu p}^2 + 2\left(\Delta_{nn}\Delta_{pp} - \tilde{\varepsilon}_{\nu n}\tilde{\varepsilon}_{\nu p} \right) \right] \right\} \tag{12}$$

(13)

$$\langle N_t \rangle = \sum_\nu \left\{ 1 + \sum_\tau \frac{\partial E_{\nu\tau}}{\partial \lambda_t} \tanh \frac{1}{2}\beta E_{\nu\tau} \right\} \tag{14}$$

with

$$\frac{\partial E_{\nu\tau}}{\partial \lambda_{n(p)}} = \frac{1}{2E_{\nu\tau}} \left[-\tilde{\varepsilon}_{\nu n(p)} + (-1)^{\tau+1} \frac{\tilde{\varepsilon}_{\nu n(p)} \left(E_{\nu p(n)}^2 - E_{\nu n(p)}^2 \right) + 2\Delta_{np}^2 \left(\tilde{\varepsilon}_{\nu p(n)} - \tilde{\varepsilon}_{\nu n(p)} \right)}{R_\nu} \right]$$

β being the inverse of the temperature.

The partition function of the system is then given, after some algebra, by:

$$Z = e^{\Omega} \tag{15}$$

with

$$\Omega = -\beta \left[\frac{\Delta_{nn}^2}{G_{nn}} + \frac{\Delta_{pp}^2}{G_{pp}} + \frac{\Delta_{np}^2}{G_{np}} \right] - \beta \sum_\nu \left(\tilde{\varepsilon}_{\nu n} + \tilde{\varepsilon}_{\nu p} \right) + \sum_\nu \ln \left[4^2 \cosh^2 \frac{\beta E_{\nu 1}}{2} \cosh^2 \frac{\beta E_{\nu 2}}{2} \right] \tag{16}$$

Starting from the grand potential Ω defined in (16), one may easily deduce the thermodynamical quantities. Indeed, the energy of the system E is obtained by deriving expression (16) with respect to the temperature:

$$E = -\left(\frac{\partial \Omega}{\partial \beta} \right)_{\alpha_t = cte} \qquad \alpha_t = \lambda_t.\beta \quad t = n, p \tag{17}$$

that is:

$$E = -\frac{\Delta_{nn}^2}{G_{nn}} - \frac{\Delta_{pp}^2}{G_{pp}} - \frac{\Delta_{np}^2}{G_{np}} + \sum_\nu \left(\varepsilon_{\nu n} + \varepsilon_{\nu p} \right) - \sum_{\nu\tau} \frac{\tanh \frac{\beta E_{\nu\tau}}{2}}{2E_{\nu\tau}} h_\gamma \tag{18}$$

where

$$h_\gamma = \varepsilon_{\nu n}\tilde{\varepsilon}_{\nu n} + \varepsilon_{\nu p}\tilde{\varepsilon}_{\nu p} + \frac{(-1)^{\tau+1}}{R_\nu} \times \left[2\Delta_{np}^2 \left(\varepsilon_{\nu n} - \varepsilon_{\nu p} \right) \left(\tilde{\varepsilon}_{\nu n} - \tilde{\varepsilon}_{\nu p} \right) + \left(\varepsilon_{\nu n}\tilde{\varepsilon}_{\nu n} - \varepsilon_{\nu p}\tilde{\varepsilon}_{p} \right) \left(E_{\nu n}^2 - E_{\nu p}^2 \right) \right]$$

Journal of Physics: Conference Series **338** (2012) 012021 doi:10.1088/1742-6596/338/1/012021

In the same way, using the usual definition of the entropy S:

$$S = \Omega + \beta E - \alpha_n N_n - \alpha_p N_p \tag{19}$$

one has :

$$S = \sum_{\nu\tau} \left\{ \ln\left(4\cosh^2 \frac{\beta E_{\nu\tau}}{2}\right) - \beta E_{\nu\tau} \tanh \frac{\beta E_{\nu\tau}}{2}) \right\} \tag{20}$$

Finally, the heat capacity which is proportional to $\frac{\partial S}{\partial \beta}$ may be written:

$$\begin{aligned} C &= -\beta \frac{\partial S}{\partial \beta} \\ &= \frac{1}{2} \sum_{\nu\tau} \frac{E_{\nu\tau}}{\cosh^2 \frac{\beta E_{\nu\tau}}{2}} \left(\beta^2 E_{\nu\tau} + \beta^3 \frac{\partial E_{\nu\tau}}{\partial \beta} \right) \end{aligned} \tag{21}$$

It is worth noticing that at the limit when Δ_{np} goes to zero, expressions (18), (20) and (21) reduces to their homologues of the conventional finite temperature BCS (FTBCS) method.

3. Numerical results - Discussion

We considered even-even nuclei such as $N = Z$ and of which "experimental" values of the pairing gap parameters at zero temperature are known. These values are deduced from the even-odd mass differences as defined in ref. [14] using the experimental mass values of the Möller table [15]. The pairing-strength values have been directly deduced from the experimental values of the pairing gap parameters at zero temperature. We used the single-particle energies of a Woods-Saxon mean field with the parameters defined in ref. [16] .

In the present work, we chosen four nuclei, i.e. ^{40}Ca, ^{56}Ni, ^{68}Se and ^{76}Sr as an example. The variations of the various pairing gap parameters $\Delta_{tt'}$, the energy of the system E, the entropy S as well as the heat capacity C as a function of the temperature are shown in figs. 1 to 4 and discussed in what follows. We reported in same figures the same quantities evaluated using the conventional FTBCS theory.

3.1. Gap parameters

From figs 1 to 4, one may notice that the Δ_{np} behavior is similar to that of Δ_{nn} and Δ_{pp} of the FTBCS method [7]; whereas the overall trend of the gap parameters Δ_{nn} and Δ_{pp} is different from that of the FTBCS method. Indeed, when $T = T_{cnp}$ (i.e. when Δ_{np} vanishes), the Δ_{nn} and Δ_{pp} values, which were constant, present a sudden increasing. This fact has been already underlined in refs [11]-[13]. One can also notice that the critical temperature values of the neutron and proton systems T_{cn} and T_{cp} are clearly greater than in the conventional FTBCS case. Finally, the phase transition when $\Delta_{np} = 0$ is very sharp.

3.2. Energy of the system

Figs 1 to 4 show that the energy E is quasi-constant when $T < T_{cnp}$ (i.e. when the isovector pairing effect exists) and suddenly grows when $T = T_{cnp}$, that is when Δ_{np} vanishes. The sharp transition of E is due to the sharp transition of Δ_{np}. The np pairing effect then leads to a diminution of the energy, besides the discrepancy which is due to the shift in the T_{cn} and T_{cp} values. Moreover, one notices that the increase of energy is more important at $T = T_{cnp}$ than at the critical temperatures of the neutron and proton systems.

3.3. Entropy

The isovector pairing effect on the entropy differs from one nucleus to another (see figs 1 to 4): in the $0 \leq T \leq T_{cnp}$ region, the difference between the two models only appears if the T_{cnp} value is sufficiently important, as in the ^{76}Sr case. For the ^{40}Ca, ^{56}Ni and ^{68}Se nuclei, in the interval where the np pairing exists, the entropy is still zero and there is no difference between the present model predictions and the conventional FTBCS ones.

For all studied nuclei, the discrepancy that appears between the np and FTBCS predictions in the $T_{cnp} \leq T \leq T_{cn}$ interval is due to the shift in the T_{cn} and T_{cp} values which has previously been mentioned.

3.4. Heat capacity

Variations of the heat capacity C as a function of the temperature are given in figs. 1 to 4 in the isovector pairing case, as well as in the pairing between like-particles one. As in the entropy case, the behavior of C differs from one nucleus to another. Indeed, for the nucleus ^{76}Sr, one observes three peaks which correspond to the three critical temperatures (even if those that correspond to T_{cnp} and T_{cp} are somewhat small). For nuclei ^{40}Ca, ^{56}Ni and ^{68}Se, the isovector pairing effect on the heat capacity is not visible since C is still zero in both models when T reaches its critical value T_{cnp}.

Here again, the shift between the T_{cn} and T_{cp} values of the FTBCS model and the present one explain the discrepancy between the two graphs when $T > T_{cnp}$. Obviously, when $T > T_{cn}$, the two models lead to the same predictions, since there is no more pairing in this region.

4. Conclusion

Expressions of the main nuclear thermodynamical quantities i.e. the energy E, the entropy S and the heat capacity C including the isovector np pairing correlations have been derived from recently established generalized equations. They have been numerically studied as a function of the temperature T within the framework of a deformed Woods-Saxon mean-field for $N = Z$ nuclei. The obtained results have been compared to those obtained using the conventional FTBCS method.

It has been shown that the isovector np pairing effect on E, S and C is non-negligible. Moreover, the behavior of these quantities as a function of the temperature may be explained by the behavior of the various gap parameters. Indeed, Δ_{np} behaves like Δ_{nn} and Δ_{pp} in the conventional FTBCS model; whereas the overall trend of the gap parameters Δ_{nn} and Δ_{pp} of the present model is different from that of the FTBCS method. It has also been noticed that the critical value of temperature T_{cnp} beyond which $\Delta_{np} = 0$ is smaller than T_{cp} and T_{cn} and that the critical values of the temperature T_{cp} and T_{cn} in the present model are greater than those of the FTBCS model. As a consequence, in the region $T > T_{cnp}$, although $\Delta_{np} = 0$, the np pairing effect leads to a discrepancy between the present model and the FTBCS previsions for all studied thermodynamical quantities.

In the interval where $\Delta_{np} \neq 0$, the np pairing effect on the energy is a lowering of about 1%, on average, for all considered nuclei; whereas its effect on the entropy and the heat capacity depends on the T_{cnp} value.

References

[1] Goodman A L , 1979 *Adv. Nucl. Phys.* **11** 263.
[2] Civitarese O and Reboiro M, 1997 *Phys. Rev. C* **56** 1179.
[3] Mokhtari D, Ami I, Fellah M and Allal N H, 2008 *Int. J. Mod. Phys. E* **17** 655.
[4] Kerrouchi S, Mokhtari D, Allal N H and Fellah M, 2009 *Int. J. Mod. Phys. E* **18** 141.
[5] Kerrouchi S, Allal N H, Fellah M and Douici M, 2010 *Int. J. Mod. Phys. E* **19** 1383.
[6] Sano M and Yamazaki S, 1963 *Prog. Theor. Phys.* **29** 397.
[7] Moretto L G, 1972 *Nucl. Phys.A* **185** 145.

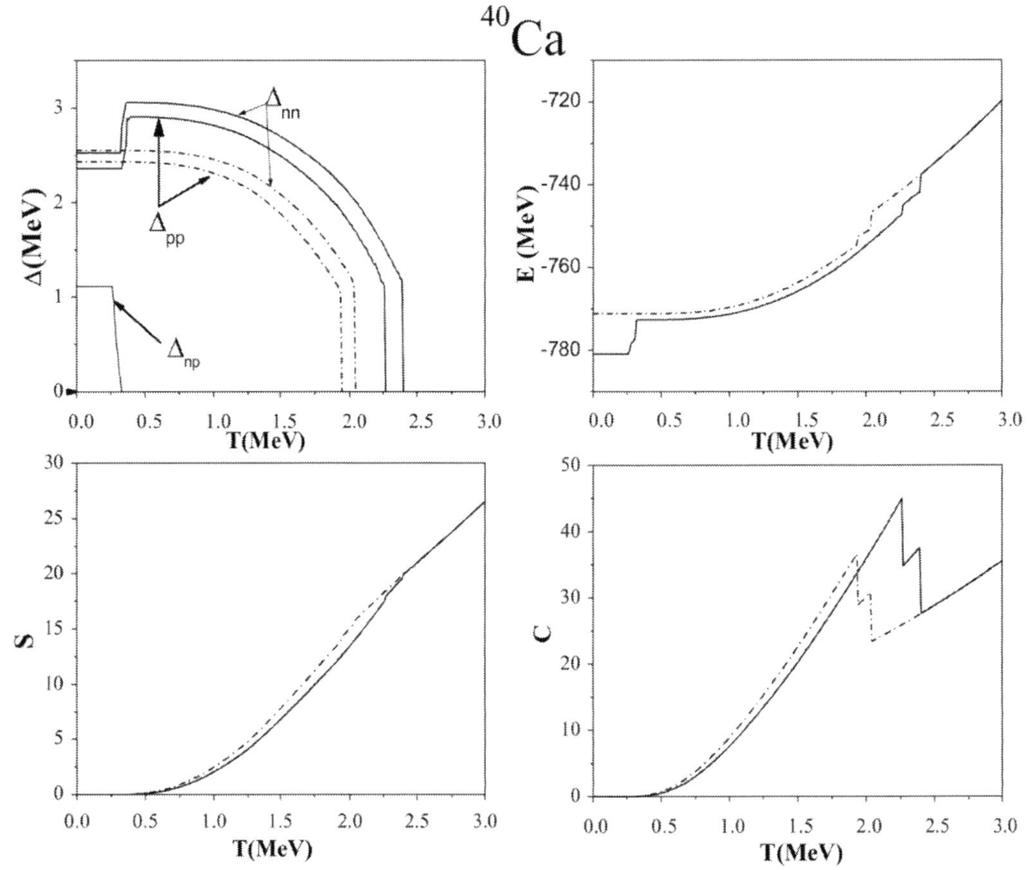

Figure 1. Variations of the various gap parameters, the energy E, the entropy S, and the heat capacity C of the system as a function of the temperature T for the nucleus ^{40}Ca. Solid lines refer to the isovector pairing and dashed lines to the pairing between like-particles.

[8] Goodman A, 1981 *Nucl.Phys.A* **352** 205.
[9] Langanke K, Dean D J, Koonin S E and Radha P B, 1997 *Nucl. Phys. A* **613** 253.
[10] Kaneko K and Hasegawa M, 2005 *Phys. Rev. C* **72** 031302(R).
[11] Fellah M, Allal N H, Belabbas M, Oudih M R and Benhamouda N, 2007 *Phys. Rev. C* **76** 047306.
[12] Belabbas M, Fellah M, Allal N H, Benhamouda N, Ami I and Oudih M R, 2010 *Int. J. Mod. Phys. E* **19** 1973.
[13] Ami I, Fellah M, Allal N H, Benhamouda N, Belabbas M and Oudih M R, 2011 *Int. J. Mod. Phys. E* **20** 1947.
[14] Simkovic F, Moustakidis Ch C, Pacearescu L and Faessler A, 2003 *Phys. Rev. C* **68** 054319.
[15] Möller P and Nix J R, 1995 *Atomic Data and Nuclear Data Tables* **59** 185.
[16] Allal N H and Fellah M, 1994 *Phys. Rev. C* **50** 1404.

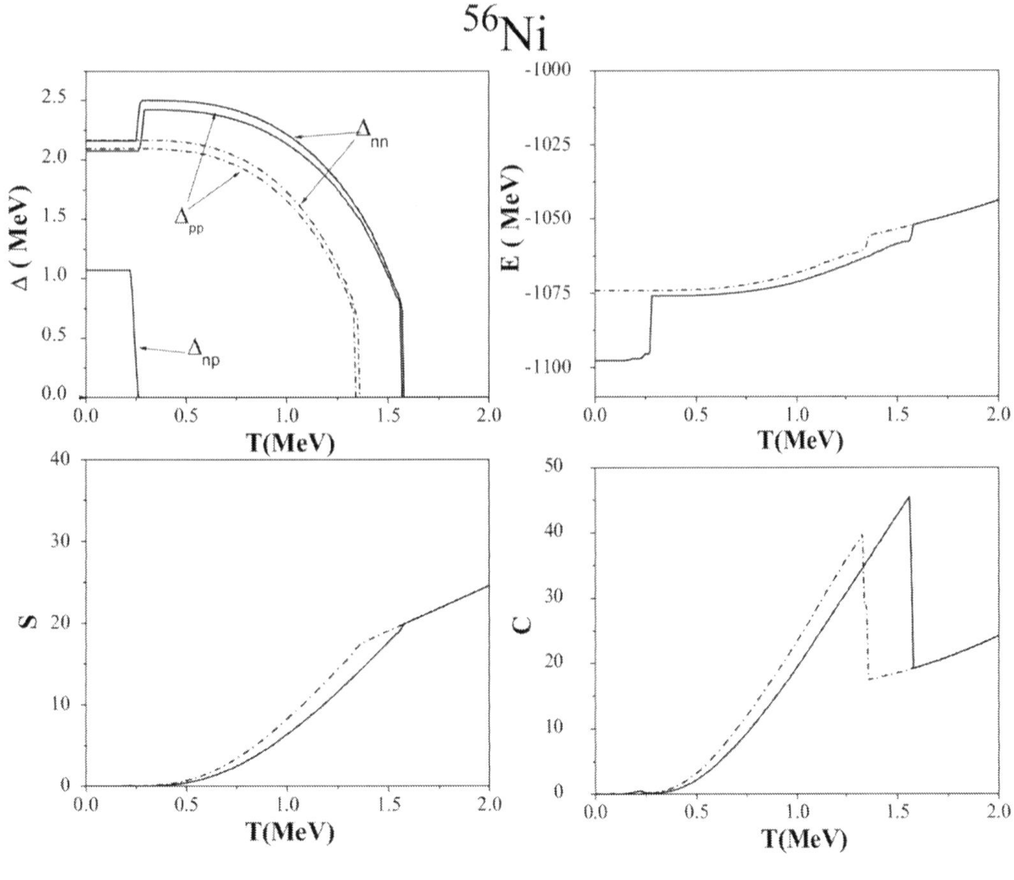

Figure 2. Same as Fig. 1 for the nucleus ^{56}Ni.

Journal of Physics: Conference Series **338** (2012) 012021 doi:10.1088/1742-6596/338/1/012021

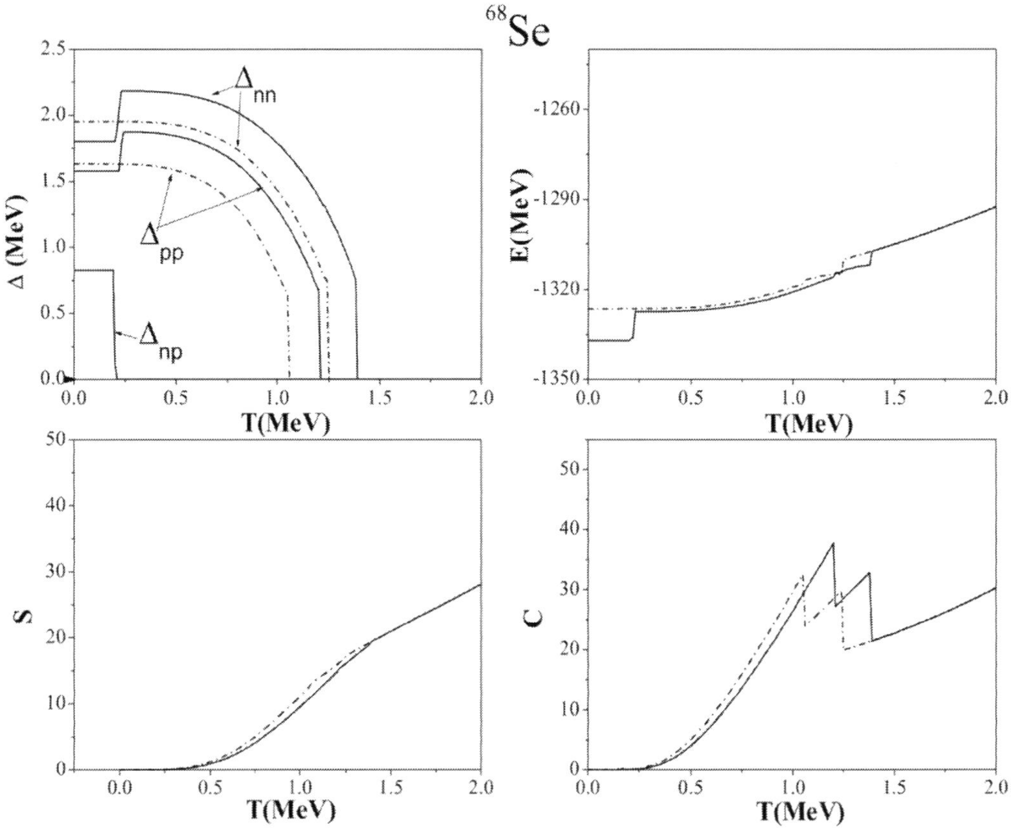

Figure 3. Same as Fig. 1 for the nucleus ^{68}Se.

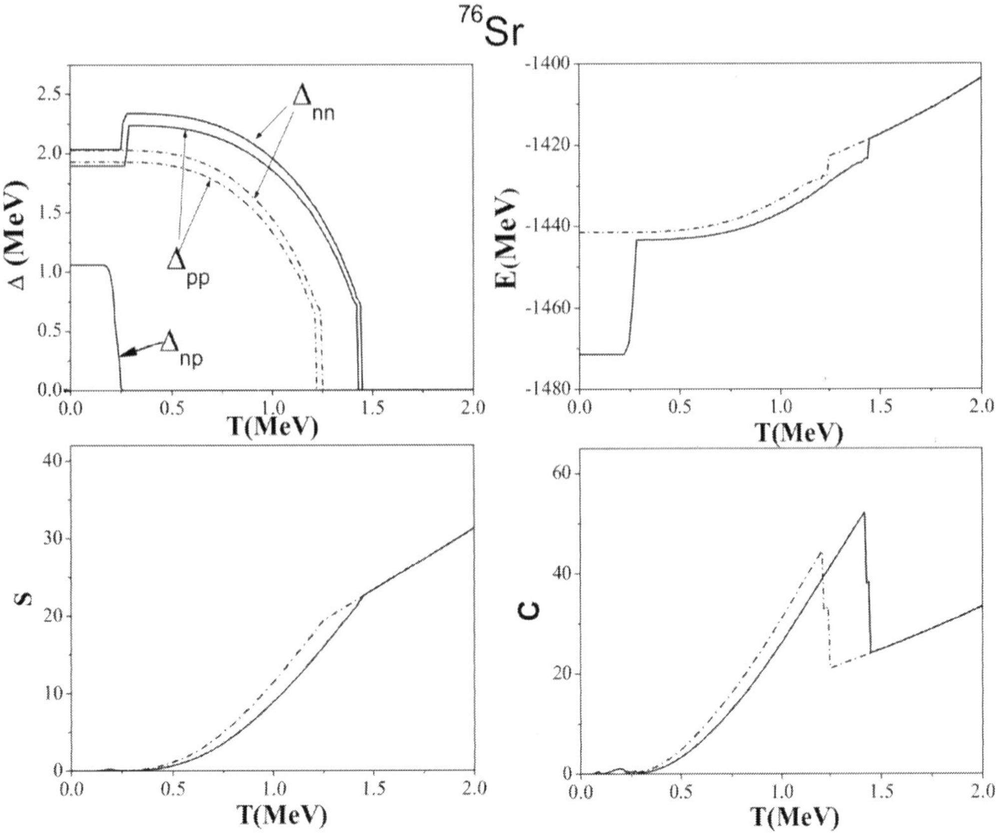

Figure 4. Same as Fig. 1 for the nucleus ^{76}Sr.

Journal of Physics: Conference Series **338** (2012) 012022 doi:10.1088/1742-6596/338/1/012022

Shell-model versus clustering effects in heavy nuclei

Doru S. Delion

Horia Hulubei National Institute for Physics and Nuclear Engineering, P.O. Box MG-6, Bucharest, Romania

E-mail: delion@theory.nipne.ro

Abstract. We analyze the interplay between the shell-model and α-clustering in heavy nuclei. In particular, we show that the standard shell-model approach underestimates B(E2) values and α-decay widths in α-like nuclei. By using surface Gaussian-like components in the single particle orbitals, mocking four body correlations, it becomes possible to enhance both electromagnetic and α transition probabilities up to the experimental values in the case of ^{212}Po.

1. Introduction

Nuclei are typical finite size mesoscopic systems with a specific interplay between volume and surface effects. One important property concerns the nuclear stability with respect to the α-particle and electromagnetic emission. In the simplest approach of the α-decay process the preformed α-particle was considered a cluster moving around the core since the time when Gamow proposed the penetrability theory [1]. Later on, the cluster was built microscopically from two proton and two neutron single particle configurations [2, 3]. Anyway, the decay widths were by several orders of magnitude smaller than the experimental values. In the late 1970's it was shown that the inclusion of many configurations increases the value of the absolute decay width by more than four orders of magnitude [4]. Yet, the calculated widths were two orders of magnitude too small. Only a theory where a combined shell- and cluster-model configurations were considered could reproduce the absolute decay width [5, 6]. This may justify the so-called extreme cluster models, where the α-particle moves around the core [7, 8, 9].

In this presentation we will show that a mixed shell-model and cluster representation is able to describe not only α decay witdths, but also electromagnetic transitions.

2. Shell-model description of decay processes

In the standard description of low-lying states the nuclear many-body Hamiltonian can be approximated by a mean-field plus a residual two-body part

$$H \approx H_{mf} + H_{res} \ . \tag{1}$$

The mean field can be estimated within the self-consistent Hartree-Fock-Bogoliubov scheme, but a good approximation is given by the Woods-Saxon ansatz [10]. The spin of the nucleon is strongly coupled to the orbital motion in the mean field. Therefore the mean field part generates a single particle shell structure of the following form

$$\psi^{(SM)}_{eljm}(\mathbf{r}, \mathbf{s}) \rightarrow a^{\dagger}_{eljm}|0\rangle \ , \tag{2}$$

Journal of Physics: Conference Series **338** (2012) 012022 doi:10.1088/1742-6596/338/1/012022

characterized by the eigenvalue (e), orbital angular momentum (l), total spin (**j=l+s**) and its projection (m). The single particle levels in the mean field potential are clustered in major shells characterized by the principal harmonic oscillator quantum number. Shell structure is a volume feature which is characteristic for mesoscopic finite size systems.

The most important residual interaction describing ground state is given by the monopole pairing force

$$H_{res} \approx -G \sum_{j} \left(\sum_{m=-j}^{j} a^{\dagger}_{jm} a^{\dagger}_{j-m} \right) \sum_{j'} \left(\sum_{m'=-j'}^{j'} a_{j'-m'} a_{j'm'} \right) , \tag{3}$$

where we denoted by G the strength of the pairing interaction and introduced the short-hand index notation $(e, l, j) \rightarrow j$. It turns out that the residual interaction is peacked on the nuclear surface, as can be shown in figure 1, where we plotted radial center of mass two-particle wave functions for protons (solid line) and neutrons (dashed line) in ^{220}Ra. This justify the use of some schematic interactions in describing spectroscopic properies, like the surface delta interaction. Surface effects are also characteristic for mesoscopic systems.

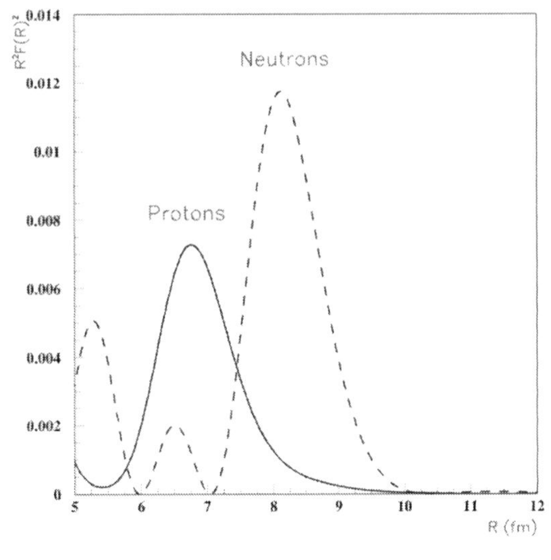

Figure 1. Center of mass two particle wave function for protons (solid line) and neutrons (dashed line) in ^{220}Ra.

Let us consider an α-decay process

$$P \rightarrow D + \alpha . \tag{4}$$

The amplitude to find an α-particle in a shell-model wave function of the parent nucleus is given by the following overlap

$$F(R_\alpha) = \langle \Psi_D \psi_\alpha | \Psi_P \rangle , \tag{5}$$

depending on the α-daugher distance R_α. Here, we denoted by $\Psi_{D/P}$ the many-body wave function of the daughter/parent nucleus and by ψ_α the intrinsic wave function of the α-particle. Its radial part is a product of two-proton and two-neutron Gaussians, while the spin component is a singlet state [2]. By using as a residual interaction the pairing force, the formation amplitude becomes also peacked on the nuclear surface, as can be seen from figure 2.

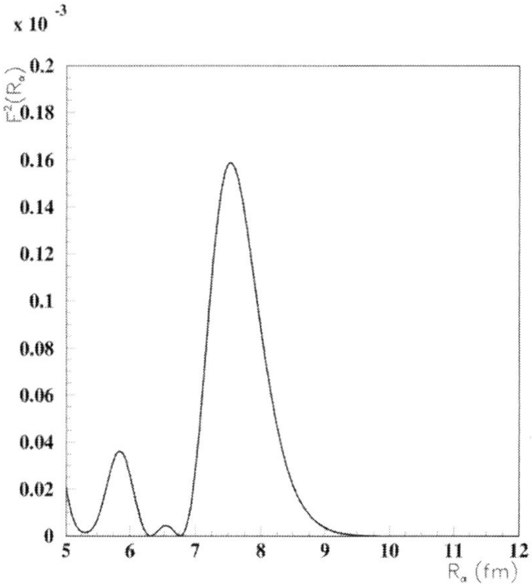

Figure 2. α-particle formation probability in ^{220}Ra versus the α-daughter radius.

For spherical emitters the decay width is defined in terms of the monopole scattering amplitude N [11] as follows

$$\Gamma = \hbar v|N|^2 = \hbar v \left| \frac{R_\alpha F(R_\alpha)}{G_0(R_\alpha)} \right|^2 , \tag{6}$$

where v is the velocity of the α-daughter system at infinity and $G_0(R_\alpha)$ is the monopole irregular Coulomb function.

Of course, the above ratio does not depend on the radius, because in principle both internal and external wave functions should satisfy the same Schrödinger equation. Actually the formation amplitude $F(R_\alpha)$ is estimated independently from the external wave function $G_0(R_\alpha)/R_\alpha$ and their ratio depends on the radius, but this dependence is rather week around the geometrical touching point, as can be seen from figure 3.

Traditionally the decay width is rewritten

$$\Gamma = 2P_0\gamma^2 , \tag{7}$$

in terms of the penetrability and reduced width squared

$$\begin{aligned} P_0 &= \frac{\kappa R_\alpha}{|G_0(R_\alpha)|^2} \\ \gamma^2 &= \frac{\hbar^2}{2\mu R_\alpha}|F(R_\alpha)|^2 , \end{aligned} \tag{8}$$

where κ is the momentum and μ the reduced α-daughter mass.

It turns out that the residual pairing interaction is not able to reproduce the absolute value of the decay witdh. Indeed, in figure 3 we give by solid line the logarithm of the ratio between the computed decay width according to Eq. (6) and the experimental value as a function of the radius, for a typical decay process ${}^{220}Ra \rightarrow {}^{216} Rn + \alpha$. One sees that the result underestimates the experiment by four orders of magnitude.

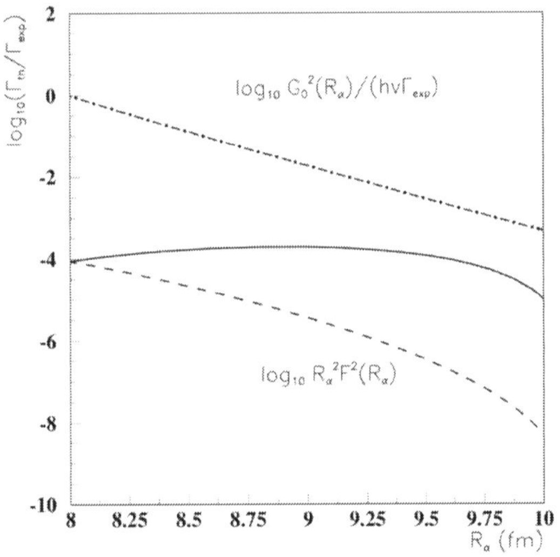

Figure 3. Logarithm of the ratio between the theoretical and experimental α-decay width for ${}^{220}Ra \rightarrow {}^{216} Rn+\alpha$ as a function of the α-daughter radius (solid line). By dashed and dot-dashed lines are given the two terms in Eq. (6).

3. Cluster description of decay processes

The binding energy per nucleon for an α-particle is much larger with respect to their neighbours. Due to a smaller nuclear density at surface, an α-cluster structure is energetically more favourable in this region. Therefore, clustering is a surface effect and it can be simulated by assuming that the α-particle moves in a pocket-like potential centered at some radius R_0 on the nuclear surface. By considering the Q-value as the first eigenvalue corresponding to a shifted harmonic oscillator wave function

$$|F(R_\alpha)|^2 = A^2 e^{-\beta(R_\alpha - R_0)^2} , \tag{9}$$

one easily obtains an universal law for the reduced width squared (8) [12]

$$\log_{10} \gamma^2(R_B) = -\frac{log_{10}e^2}{\hbar\omega} V_{frag}(R_B) + \log_{10} \frac{\hbar^2 A^2}{2e\mu R_B} , \tag{10}$$

depending on the fragmentation potential, defined as the difference between the Coulomb barrier and Q-value, i.e.

$$V_{frag}(R_B) = V_C(R_B) - Q , \tag{11}$$

where R_B is the radius of the Coulomb barrier. This law does not depend upon the pocket shift radius R_0 and it is satisfied with a good accuracy not only by all α-decays between ground states, but also by proton and heavy cluster reduced widths [12]. This law is a direct evidence that a cluster should exist on the nuclear surface with the amplitude A.

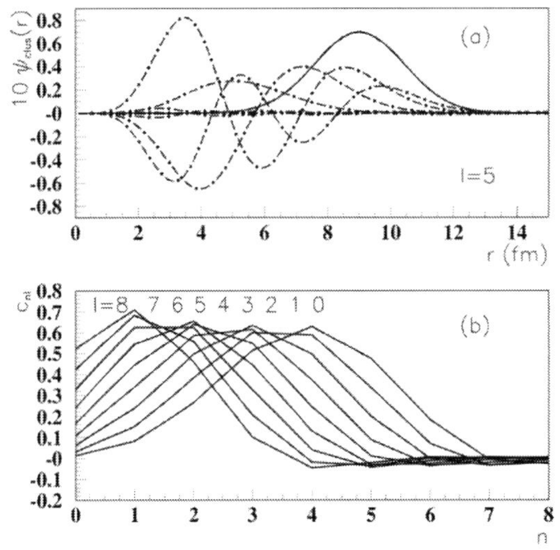

Figure 4. (a) Gaussian distribution centered on the nuclear surface (solid line) and various expansion terms (dot-dashed lines).
(b) Expanssion coefficients of a Gaussian centered on the nuclear surface given by Eq. (14).

Thus, in order to properly describe clustering properties at the single particle level one has to consider two components of the single particle wave function, i.e.

$$\psi_l(r) = \mathcal{N}_l^{(SM)}\psi_l^{(SM)}(r) + \mathcal{N}_l^{(clus)}\psi_l^{(clus)}(r) \ . \tag{12}$$

The shell model part has a standard spherical harmonic oscillator representation

$$\psi_l^{(SM)}(r) = \sum_n b_n \mathcal{R}_{nl}^{(\beta)}(r) \ , \tag{13}$$

where $\beta = M_N\omega/\hbar$ is the harmonic oscillator parameter and the principal quantum number has the values $N = 2n + l \leq 6$.

For the cluster part one uses a wave function of the shifted oscillator

$$\psi_l^{(clus)}(r) = e^{-\beta_c(r-r_0)^2/2} = \sum_n c_n \mathcal{R}_{nl}^{(\beta)}(r) \ , \tag{14}$$

where N is spread around a larger value $N \sim 8 - 10$, as can be seen from figure 4. Notice that in Ref. [6] the harmonic oscillator parameter of the cluster part has a smaller value, in order to reduce the dimension of the single particle basis (12). Here all coefficients are found by a diagonalization procedure of the Woods-Saxon mean field.

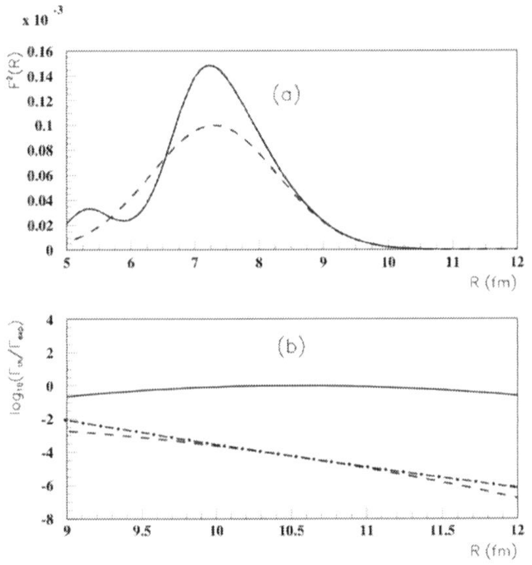

Figure 5. (a) α-particle formation probability for $^{212}Po \rightarrow^{208} Pb+\alpha$. Dashed line: shell model, Solid line: shell model + α-cluster.
(b) Logarithm of the ratio between the decay width and the experimental value as a function of the daughter-α-particle radius

The typical example of a heavy nucleus where the α-clustering effects are very important is ^{212}Po. Here, the structure of low-lying states can be explained in terms of two proton and two neutron orbitals above the double magic inert core ^{208}Pb [13]. Anyway, by using the standard components of the two protons and two neutrons the decay width is underestimated by two orders of magnitude. It is necessary to consider the second cluster component in Eq. (12), with $\mathcal{N}_l^{(clus)} = 0.3$, in order to describe the absolute value of the α-decay width, as can be seen in figure 5 (b) (solid line). In spite of the fact that in figure 5 (a) the difference between the formation probability within the shell model approach (dashed line) and mixed approach (solid line) is rather small, the enhancement of the decay width is about two orders of magnitude. Notice that the same value was obtained in Ref. [5], by using a diagonalisation procedure of the residual interaction.

This picture is also supported by analyzing electromagnetic transitions. The electric transition probability of the multipolarity λ is proportional to the reduced matrix element of the transition operator squared, i.e.

$$B(E\lambda : J \rightarrow J') = \frac{1}{2J+1}\left|\langle J'||\hat{T}_\lambda||J\rangle\right|^2 , \tag{15}$$

where $|J\rangle$ and $|J'\rangle$ denote initial and final states and the transition operator is proportional to spherical harmonics

$$\hat{T}_{\lambda\mu} = r^\lambda Y_{\lambda\mu} . \tag{16}$$

The radial part of the quadrupole transition matrix element is proportional to the principal

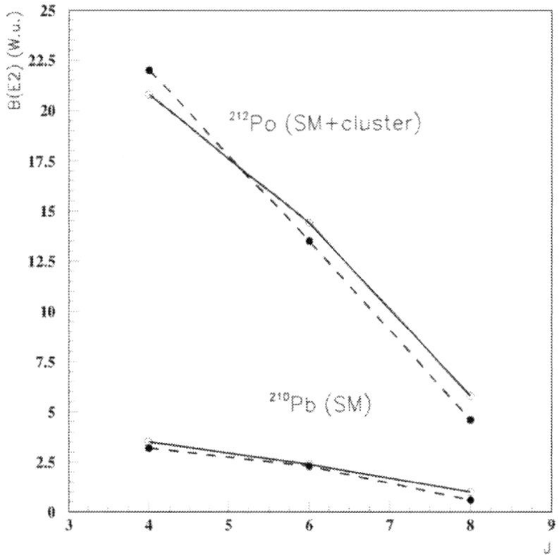

Figure 6. $B(E2: J \to J-2$ values (W.u.). Shell-model values in ^{210}Po and shell-model + cluster values in ^{212}Po (solid lines). By dashes are given experimental values.

quantum number

$$\langle n_1 l_1 | r^2 | n_2 l_2 \rangle \sim N = 2n + l \ . \tag{17}$$

As it is shown in figure 4, the clustering components are centered around the region $N \sim 8 - 10$ and therefore they should enhance transition probability.

It is known that B(E2) values are satisfactory described within the shell model for ^{210}Pb, but they are by one order of magnitude less than the experimental values in ^{212}Po. They are of the same order of magnitude as those in ^{210}Pb, given by the lower solid line in figure 6 [14]. In order to explain this discrepancy we used the mixed representation (12) with the same α-cluster amplitude. In figure 6 one indeed obtains a very good agreement of the computed values (upper solid line) with respect to experimental data (upper dashed line).

This is a very convincing evidence in favour of the α-clustering structure of this nucleus.

4. Conclusions

Electromagnetic and α-transitions in heavy nuclei cannot be described within the standard shell-model approach. The analysis of experimental α-decay reduced widths have shown that an α-cluster component centered on the nuclear surface is necessary in order to describe decay processes. In heavy nuclei this cluster component cannot be build from usual single particle orbitals because protons and neutron lie in different major shells. Thus, an additional gaussian centered on the nuclear surface is necessary to be included in the single particle basis. We described both α-decay width and B(E2) value in ^{212}Po by using a mixed single particle basis, containing shell-model and α-cluster components.

Acknowledgments
This work was performed in the frame of the CNCSIS IDEI 512 project of the Romanian Ministry of Education and Research.

References

[1] Gamow G 1928 *Z. Phys.* **51** 204
[2] Mang H J 1960 *Phys. Rev.* **119** 1069
[3] Săndulescu A 1962 *Nucl. Phys.* **37** 332
[4] Tonozuka I and Arima A 1979 *Nucl. Phys. A* **323** 45
[5] Varga K Lovas R G and Liotta R J 1992 *Phys. Rev. Lett.* **69** 37
[6] Delion D S Insolia A and Liotta R J 1996 *Phys. rev. C* **54** 292
[7] Săndulescu A Cusson R Y and Greiner W 1983 *Lett. Nuovo Cim.* **36** 321
[8] Buck B Merchant A C and Perez S M 1994 *Phys. Rev. Lett.* **72** 1326
[9] Ohkubo S 1995 *Phys. Rev. Lett.* **74** 2176
[10] Blomqvist J and Vahlborn S 1960 *Ark. Fys.* **16** 543
[11] D.S. Delion 2010 *Theory of particle and cluster emission* (Springer-Verlag, Berlin)
[12] Delion D S 2009 *Phys. Rev. C* **80** 024310
[13] Delion D S and Suhonen J 2000 *Phys. Rev. C* **61** 024304
[14] Astier A Petkov P Porquet M -G Delion D S and Schuck P 2010 *Eur. Phys. J. A* **47** 165

Integrable Richardson-Gaudin models in mesoscopic physics

Jorge Dukelsky
Instituto de Estructura de la Materia. C.S.I.C. Serrano 123. 28006 Madrid. Spain
E-mail: j.dukelsky@csic.es

Abstract. The exact solution of the SU(2) pairing Hamiltonian with non-degenerate single particle orbits was introduced by Richardson in the early sixties. The exact solution passed almost unnoticed till was recovered in the last decade in an effort to describe the disappearance of superconductivity in ultrasmall superconducting grains. Since then it has been extended to several families integrable models, called the Richardson-Gaudin (RG) models. In particular, the rational family of integrable RG models has been widely applied to mesoscopic systems like small grains, quantum dots and nuclear systems where finite size effects play an important role. We will first introduce these families of integrable models and then we will describe the first applications of the hyperbolic family to spinless cold fermionic atomes in two dimensional lattices and to heavy nuclei.

1. Introduction

The work of Bardeen, Cooper and Schrieffer (BCS) of 1957 [1] gave the first microscopic description of the superconducting phenomenon assuming a quantum pairing Hamiltonian and a variational wavefunction based on a coherent state of pairs. The following year, Bohr, Mottelson and Pines [2] noted that similar physics may underly the large gaps seen in the spectra of even-even atomic nuclei, emphasizing however that finite-size effects would be critical for a proper description of such systems. The program to include number conservation in the BCS theory within nuclear structure [3] started at roughly the same time at which Richardson [4] showed that for a pure pairing Hamiltonian it is possible to *exactly* solve the Schrödinger equation by following closely Cooper's original idea. Years later and from a different perspective, Gaudin introduced an integrable spin model having striking similarities with the Richardson exact solution [5]. In spite of the fact that exact solvability is linked to quantum integrability, he couldn't find the explicit relation between both models. Years later, we were able to find this relation through a generalization of the Gaudin integrals of motion giving rise to three classes of pairing-like models that we called the RG integrable models, all of which were integrable and all of which could be solved exactly for both fermion and boson systems [6]. During the last decade, the rational family of the RG models was extensively used to describe ultrasmall superconduction grains, heavy nuclei, quantum dots, ultracold atomic gases, etc [7]. More recently, we have found two physical realizations of the hyperbolic family, one for p-wave polarized atomic gases in two dimensional lattices [8, 9], and the other as a potentially useful realistic pairing Hamiltonian for heavy nuclei. In this contribution we will first introduce the RG integrable models and then we will briefly describe the exact solution for $p_x + ip_y$ superfluids as a first realization of the

Journal of Physics: Conference Series **338** (2012) 012023 doi:10.1088/1742-6596/338/1/012023

hyperbolic RG model. Finally, we will present preliminary results showing how the integrable hyperbolic Hamiltonian could reproduce Gogny Hartree-Fock-Bogoliubov (HFB) gaps in heavy nuclei.

2. Richardson's exact solution of the pairing model

We begin our discussion of Richardson's solution of the pairing model by assuming a system of N fermions moving in a set of L single-particle states l, each having a total degeneracy Ω_l, and with an additional internal quantum number m that labels the states within the l subspace. If the quantum number l represents angular momentum, the degeneracy of a single-particle level l is $\Omega_l = 2l+1$ and $-l \leq m \leq l$. In general, however, l could label different quantum numbers. The operators on which the pairing hamiltonian is based are

$$\widehat{n}_l = \sum_m a^\dagger_{lm} a_{lm} \ , \quad P^\dagger_l = \sum_m a^\dagger_{lm} a^\dagger_{l\overline{m}} = (P_l)^\dagger \ , \tag{1}$$

where $a^\dagger_{lm}$ (a_{lm}) creates (annihilates) a particle in the state (lm) and the state $(l\overline{m})$ is the corresponding time-reversed state or conjugate state in case of broken time reversal system. The number operator $\widehat{n}_l$, the pair creation operator $P^\dagger_l$ and the pair annihilation operator P_l close the commutation algebra

$$\left[\widehat{n}_l, A^\dagger_{l'}\right] = 2\delta_{ll'} \ , \quad \left[A_l, A^\dagger_{l'}\right] = 2\delta_{ll'} \left(\Omega_l - 2\widehat{n}_l\right) \ . \tag{2}$$

The reduced BCS model also known as the constant pairing model solved by Richardson is

$$H_P = \sum_l \varepsilon_l \widehat{n}_l + \frac{g}{2} \sum_{ll'} P^\dagger_l P_{l'} \ . \tag{3}$$

The approximation leading to the Richardson hamiltonian must be supplemented by a cutoff restricting the number of l states in the single-particle space. In condensed-matter problems this cutoff is naturally provided by the Debye frequency of the phonons. In nuclear physics, the choice of cutoff depends on the specific nucleus and on the set of active or valence orbits in which the pairing correlations develop.

A generic state of M correlated fermion pairs and ν unpaired particles can be written as

$$|n_1, n_1, \cdots, n_L, \nu\rangle = \frac{1}{\sqrt{\mathcal{N}}} \left(P^\dagger_1\right)^{n_1} \left(P^\dagger_2\right)^{n_2} \cdots \left(P^\dagger_L\right)^{n_L} |\nu\rangle \ , \tag{4}$$

where $\mathcal{N}$ is a normalization constant. The number of pairs n_l in level l is constrained by the Pauli principle to be $0 \leq 2n_l + \nu_l \leq \Omega_l$, where ν_l denotes the number of unpaired particles in that level. The unpaired state $|\nu\rangle = |\nu_1, \nu_2 \cdots \nu_L\rangle$, with $\nu = \sum_l \nu_l$, is defined such that

$$P_l |\nu\rangle = 0 \ , \quad \widehat{n}_l |\nu\rangle = \nu_l |\nu\rangle \ . \tag{5}$$

A state with ν unpaired particles is said to have seniority ν. The total number of collective (or Cooper) pairs is $M = \sum_l n_l$ and the total number of particles is $N = 2M + \nu$.

The dimension of the hamiltonian matrix in the Hilbert space of Eq. (4) grows exponentially with the number of pairs and it quickly exceeds the limits of large-scale diagonalization. For the pairing Hamiltonian (3) to be exactly solvable implies that the exponential complexity of the problem should be reduced to an algebraic problem. Indeed, Richardson showed that the exact unnormalized eigenstates of the hamiltonian of Eq. (3) can be written as

$$|\Psi\rangle = B^\dagger_1 B^\dagger_2 \cdots B^\dagger_M |\nu\rangle \ , \tag{6}$$

where the collective pair operators B_α have the form appropriate to the solution of the one-pair problem,

$$B_\alpha^\dagger = \sum_l \frac{1}{2\varepsilon_l - E_\alpha} P_l^\dagger \ . \tag{7}$$

In the one-pair problem, the quantities E_α that enter Eq. (7) are the eigenvalues of the pairing hamiltonian, *i.e.*, the *pair energies*. Richardson proposed to use the M pair energies E_α in the many-body wavefunction of Eq. (6) as parameters which are then chosen to fulfill the eigenvalue equation $H_P \left|\Psi\right\rangle = E \left|\Psi\right\rangle$.

Richardson showed that the ansatz (6) is an eigenstate of the Hamiltonian (3) if the pair energies E_α are a particular solution of the set of M nonlinear coupled equations

$$1 - 4g \sum_l \frac{d_l}{2\varepsilon_l - E_\alpha} + 4g \sum_{\beta(\neq\alpha)} \frac{1}{E_\alpha - E_\beta} = 0 \ , \tag{8}$$

where $d_l = \frac{\nu_l}{2} - \frac{\Omega_l}{4}$ is related to the effective pair degeneracy of single-particle level l.

The energy eigenvalue associated with a given solution for the pair energies is

$$E = \sum_l \varepsilon_l \nu_l + \sum_\alpha E_\alpha \ . \tag{9}$$

There are as many independent solutions of the equation (8) as the dimension of the Hilbert space. Therefore, the ansatz (6) provides the complete set of eigenstates of the Richardson Hamiltonian. Unlike the case of a single pair where the pair energy E corresponds to the energy eigenvalue and therefore it is real, for the many-body case the pair energies could be real or they could appear in pairs of complex conjugate values. Upon inspection of the pair wavefunction (7), we conclude that only pairs with complex pair energies represent truly correlated pairs.

3. Richardson-Gaudin integrable models

Classical integrability is a crucial concept for the study of dynamics of classical system. A classical system with M degrees of freedom is integrable if it possesses M independent integrals of motion that fulfill the Poisson brackets algebra. The important consequence of classical integrability is that the evolution of the system can be obtained by effective *integration* in the action-angle variables. A natural extension of the concept of integrability to quantum systems would require that a system with M quantum degrees of freedom should have M independent hermitian operators that commute among themselves, the integrals of motion. However, quantum integrability suffers from the serious drawback of the impossibility to proof the independency of two commuting hermitian operators. In fact, as shown by von Neumann in 1931 [10], two commuting hermitian operators could be expressed as a function of a third hermitian operator. Precisely, this difficulty makes quantum integrability a still debated concept and an active research field. In spite of this drawback, we will use the generalization of the definition of classical integrability to quantum systems, employing as an independent confirmation the study of the statics of level spacings which, as shown by Berry and Tabor [11], should be described by a Poisson distribution if the system is quantum integrable. The RG models are based on the SU(2) algebra with three generators S^z, S^+ and S^-. The three elements have the following commutation algebra:

$$\left[S_l^z, S_{l'}^\pm\right] = \pm\delta_{ll'} S_l^\pm \ , \quad \left[S_l^+, S_{l'}^-\right] = 2\delta_{ll'} S_l^z \ . \tag{10}$$

It is important to note that the SU(2) generators may have different representations in terms of fermionic creation and annihilations operators as well as the inherent angular momentum representation. Each representation gives rise to different physical problems. We are interested here to make contact with the pair representation associated with pair superfluidity and to the Richardson's exact solution. In doing so, we note that it could be established a relation with the three operators of the Richardson Hamiltonian (1) such that they fulfilled the commutation relations of SU(2).

$$S_l^z = 2\hat{n}_l - \frac{\Omega_l}{2}, \quad S_l^+ = \frac{1}{2}P_l^\dagger = (S_l)^\dagger \tag{11}$$

Taking into account that the SU(2) has one quantum degree of freedom, our task will be to find a set of L hermitian operators constructed in terms of the SU(2) generators such that they commute with the total z componenet $S^z = \sum_l S_l^z$. The reason for this restriction is that in the pair representation $S^z = 2\widehat{N} - \frac{1}{2}\sum_l \Omega_l$, and therefore the operators will preserve the total number of particles. Let us write the most general linear a quadratic operators R_l

$$R_l = S_l^z + 2g \sum_{l'(\neq l)} \left[\frac{X_{ll'}}{2} \left(S_l^+ S_{l'}^- + S_l^- S_{l'}^+ \right) Y_{ll'} S_l^z S_{l'}^z \right] . \tag{12}$$

Where the X and Y matrices are yet arbitrary. Using this freedom to impose integrability, it is straightforward to show that the conditions for these operators to commute among themselves $[R_l, R_l']$ are

$$Y_{ij}X_{jk} + Y_{ki}X_{jk} + X_{ki}X_{ij} = 0 . \tag{13}$$

These set of conditions are precisely the same found by Gaudin [5] in his integrable spin model known as the Gaudin magnet. There two generic solutions for the Gaudin conditions (13) giving rise to two families of integrable models

I. The Rational Model

$$X_{ij} = Y_{ij} = \frac{1}{\eta_i - \eta_j} \tag{14}$$

II. The Hyperbolic Model

$$X_{ij} = \frac{1}{\sinh(\eta_i - \eta_j)} , \quad Y_{ij} = \coth(\eta_i - \eta_j) \tag{15}$$

Each model defines a completely integrable family where the L operators R_l play the role of the integrals of motion. Any function of the integrals of motion define an integrable Hamiltonian. In particular one and two-body Hamiltonians arise as linear combinations of the integrals of motions $H = \sum_l \varepsilon_l R_l$. Both families are exactly solvable in the sense that one can formally write the form of the complete set of eigenstates common to the set of integrals of motion (see [6] for details). Moreover, the rational family has the same set of eigenstates represented by the Richardson ansatz (6) for the solution of the solution of the pairing model.

Within the rational family, if we choose the $\varepsilon_l = \eta_l$ the integrable Hamiltonian reduces to the Richardson Hamiltonian (3). In fact, this was precisely the way in which it was shown that the reduced BCS Hamiltonian (3) was integrable [12] even before the advent of the RG models and without having the knowledge of the Richardson exact solution.

One of the most important characteristics of the RG models is that they posses a large number of free parameters that could be adjusted to define a realistic pairing interaction adapted to the physical system of interest. Alternative, they could be chosen randomly still preserving quantum integrability. As mentioned before, Berry and Tabor showed in the semiclassical limit that the spectral distribution of quantum integrable Hamiltonians should be of the Poisson type. Though the level spacing distribution of all quantum integrable systems with more than one degree of

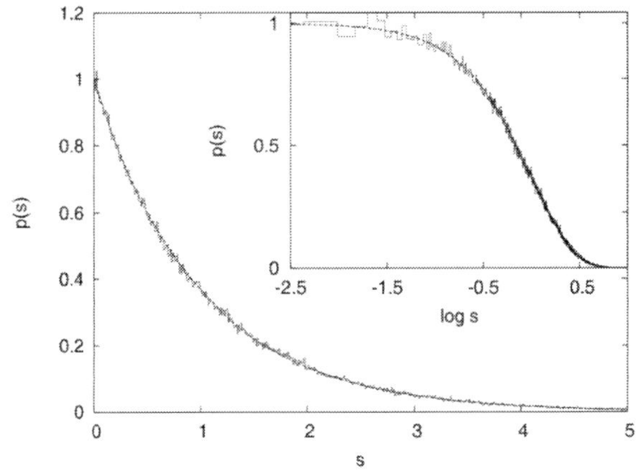

Figure 1. Nearest neighbor spacing distribution, $p(s)$, for 200 ensemble members each one with $L = 13$, $M = 6$ and randomly chosen parameters. The dashed curve correspond to the Poisson limit.

freedom approximately follow a Poisson distribution, the numerical tests are limited by size of the Hilbert space amenable to an exact diagonalization. Typically, just a few thousand of energy levels could be use to construct the level spacing histogram. In order to verify the correctness of the Berry and Tabor conjecture we have studied an ensemble of 200 Hamiltonians derived from the rational model selecting the free parameters randomly [13]. We used the case of $(L, M) = (13, 6)$ with a Hilbert space dimension $D = 1716$. In the Poisson limit, characteristic of a regular system, the nearest neighbor spacing distribution behaves as $p(s) = \exp(-s)$. Fig. 1 compares the $p(s)$ distribution of our ensemble with the expected Poisson limit. It can be seen that the histogram and the theoretical curve match perfectly. Actually, figure 1 constitutes the most precise numerical verification of the Berry and Tabor theoretical proof due to the fact that we were able to accumulate statistics by using an ensemble of random integrable Hamiltonians which wouldn't be possible for any other integrable model.

The rational model has been extensively exploited in the last decade in applications to nuclear structure, cold atomic gases, quantum dots, ultrasmall superconducting grains, quantum optical models, etc. We will not continue describing these applications which are summarize in two recent reviews [7, 14]. Instead we will present in the next section the first physical applications of the hyperbolic model.

4. The Hyperbolic Model

We start with the integrals of motion of the hyperbolic RG model [6], which can be written in a compact form [14] by making the replacements $\sinh(x) = \frac{\eta-1}{2\sqrt{\eta}}$ and $\coth(x) = \frac{\eta-1}{\eta+1}$ as

$$R_i = S_i^z + 2g \sum_{j \neq i} \left[\frac{\sqrt{\eta_i \eta_j}}{\eta_i - \eta_j} \left(S_i^+ S_j^- + S_i^- S_j^+ \right) + \frac{\eta_i + \eta_j}{\eta_i - \eta_j} S_i^z S_j^z \right], \tag{16}$$

Journal of Physics: Conference Series **338** (2012) 012023 doi:10.1088/1742-6596/338/1/012023

where S_i^z, $S_i^{\pm}$, are the three generators of the $SU(2)_i$ algebra of mode i, $i = 1, \cdots, L$, with spin representation s_i such that $\langle S_i^2 \rangle = s_i(s_i + 1)$. We assume that there are L copies of the $SU(2)$ algebra or equivalently L modes. Therefore, the L operators R_i contain L free parameters η_i plus the strength of the quadratic term g. The integrals of motion (17) commute with the z component of the total spin $S^z = \sum_{i=1}^{L} S_i^z$.

It is worthwhile to verify that the set of operators R_i commute among themselves, conforming a complete set of integrals of motion. Therefore, they have a complete set of common eigenstates which are parametrized by the ansatz

$$|\Psi\rangle = \prod_{\alpha=1}^{M} S_\alpha^+ |\nu\rangle\, , \; S_\alpha^+ = \sum_i \frac{\sqrt{\eta_i}}{\eta_i - E_\alpha} S_i^+, \tag{17}$$

where the E_α are the pair energies or pairons which are to be determined such that the ansatz (17) satisfies the eigenvalue equations $R_i |\Psi\rangle = r_i |\Psi\rangle$.

In the pairing representations each $SU(2)$ copy is associated with a single particle level i and M is the number of active pairs. The vacuum $|\nu\rangle$ is defined by a set of seniorities, $|\nu\rangle = |\nu_1, \nu_2, \ldots, \nu_l\rangle$, where the *seniority* ν_i is the number of unpaired particles in level i with single particle degeneracy Ω_i, such that $s_i = (\Omega_i - 2\nu_i)/4$.

Although any function of the integrals of motion generates an exactly solvable Hamiltonian, we will restrict ourselves in this presentation to the simple linear combination $H = \sum_i \eta_i R_i$ that after some algebraic manipulations reduces to

$$H = \sum_i \eta_i S_i^z - G \sum_{i,j} \sqrt{\eta_i \eta_j} S_i^+ S_j^- . \tag{18}$$

This separable Hamiltonian has the eigenvectors (17) and the eigenvalues $E = \sum_i < \nu | S_i^z | \nu > + \sum_\alpha E_\alpha$, where the pairons E_α are a solution of the set of non-linear Richardson equations

$$\sum_i \frac{s_i}{\eta_i - E_\alpha} - \sum_{\alpha'(\neq\alpha)} \frac{1}{E_{\alpha'} - E_\alpha} = \frac{Q}{E_\alpha}, \tag{19}$$

with $Q = \frac{1}{4\gamma} - \frac{L_c}{4} + \frac{M-1}{2}$, $L_c = 2\sum_i s_i$, and M is the number of pairons.

Each particular solution of Eq. (19) defines a unique eigenstate. For the remaining discussion we will assume that $\langle \nu | H_h | \nu \rangle = 0$, which amounts to a simple shift in the energy scale, without loss of generality.

5. The $p_x + ip_y$ pairing Hamiltonian

In recent years p-wave paired superfluids have attracted a lot of attention, in part due to their exotic properties [15]. Of particular interest is the chiral two-dimensional (2D) $p_x + ip_y$ superfluid of spinless fermions, that supports a topological phase with zero energy Majorana modes [16] and, unlike the s-wave superfluid, it has a quantum phase transition in the crossover from BCS to BEC whose properties are not yet well understood. Therefore, the derivation of an exactly solvable model could be essential for the understanding of this exotic superfluid.

In two spatial dimensions, one can define a representation of the $SU(2)$ algebra in terms of creation (annihilation) spinless fermions operators in momentum space, $c_k^\dagger$ (c_k). Each pair of states $(k, -k)$ is associated to a single-particle level η_k, where the index k now refers to the momentum in 2D (in order to avoid double counting we select $k_x > 0$ to label the levels). Furthermore, one can include a phase factor in the definition of $SU(2)$ generators:

$$S_k^z = \frac{1}{2}\left(c_k^\dagger c_k + c_{-k}^\dagger c_{-k} - 1\right), S_k^+ = \frac{k_x + ik_y}{|k|} c_k^\dagger c_{-k}^\dagger, S_k^- = \frac{k_x - ik_y}{|k|} c_{-k} c_k. \tag{20}$$

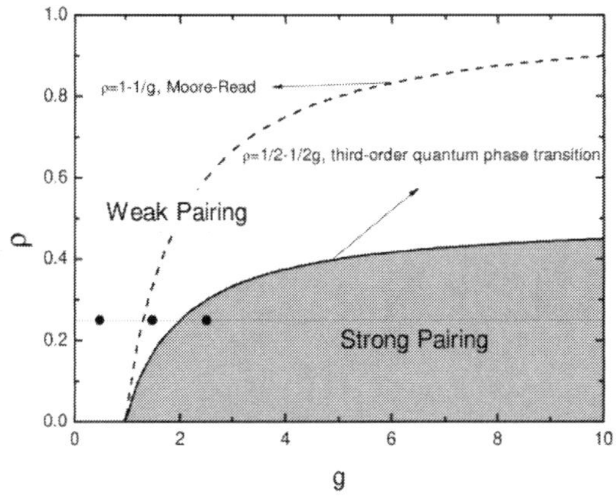

Figure 2. Phase diagram of the p_x+ip_y model in terms of the density ρ and the pairing strength g. The three circles at quarter filling indicate the configurations studied in the following figure.

By taking $\eta_k = k^2$, one obtains the exactly solvable $p_x + ip_y$ model first introduced Ibañez et. al [17]:

$$H_{p_x+ip_y} = \sum_{k,k_x>0} \frac{k^2}{2}\left(c_k^\dagger c_k + c_{-k}^\dagger c_{-k}\right) - G \sum_{k,k_x>0,k',k'_x>0} (k_x+ik_y)(k'_x-ik'_y)c_k^\dagger c_{-k}^\dagger c_{-k'}c_{k'}. \quad (21)$$

Coming back to the Richardson equations (19) that solves the Hamiltonian (21), we recognize two special cases: *case (i)* all pairons are real and negative if $\frac{1}{G} \leq L - 2M + 1$; we will see that the boundary coincides with the phase transition line. *case (ii)* all pairons converge to zero for $\frac{1}{G} = L - M + 1$; this situation determines the so called *Moore-Read* line [17, 18], with interesting properties associated with the fractional quantum Hall effect. Between these two regimes, a fraction of the pairons can converge to zero at integer values of G^{-1}. The phase diagram of the $p_x + ip_y$ Hamiltonian (21) depicted in Fig. 3 is characterized by the density $\rho = M/L$ and the scaled pairing strength $g = GL$. The transition between the strong pairing region (BEC) with all pairons real and negative and the weak pairing region (BCS) takes place when one pairon change sign implying that one of the bound molecules in the BEC gets unbounded. For this reason we characterized the transition as a confinement-deconfinement quantum phase transition.

In order to get a more quantitative picture of the pairon distribution in the three regions of the quantum phase diagram, we plot in Fig. 2 the pairon distributions for three representative values of the coupling strength, $g = 0.5, 1.5, 2.5$, at quarter filling for a disk of radius 18 corresponding to a total pair degeneracy $L = 504$. The positions of these points in the quantum phase diagram of Fig. 3 is indicated by the three filled circles. In the weak coupling BCS region part of the pairons stick to the lower part of the real positive axis, while the remaining pairons form an arc in the complex plane. Approaching the Moore-Read line it looks like the arc is going to close around the origin, but just at the Moore-Read line all pairons collapse to zero, and then a first real negative pairon emerges. In the intermediate weak pairing region a successive series of collapses ensues, at integer values of Q, each time producing one more real negative pairon and reducing the size of the arc around the origin. When the last pairon turns real and negative, the system enters the strong pairing phase. From then on the most negative pairon diverges

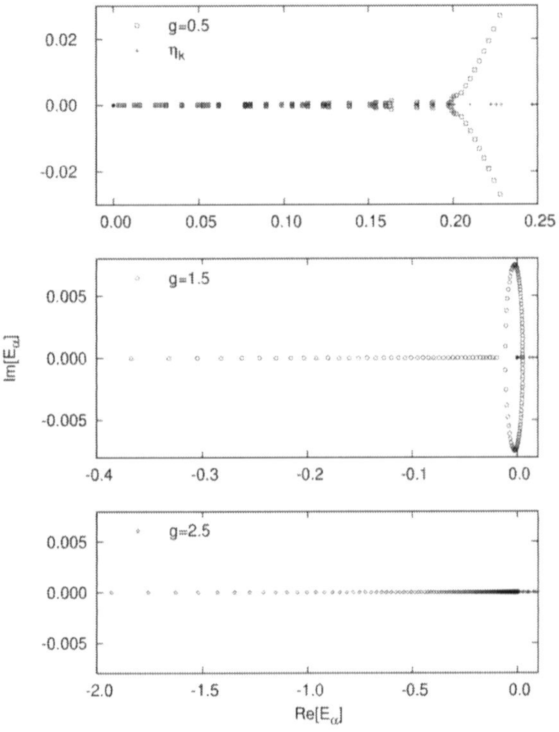

Figure 3. Pairon distribution for $L = 504$ at quarter filling, $\rho = \frac{M}{L} = 0.25$, for $g = GL = 0.5, g = 1.0$, and $g = 2.5$.

proportional to the interaction strength G, while the least negative pairon converges to a finite value.

In order characterize the quantum phase transition, we study the energy density derivatives as describe by the BCS theory which is exact in the thermodynamic limit. As can be seen in Fig. 3, the third derivative shows a discontinuity confirming that the phase transition is third order in the Ehrenfest classification.

6. The integrable nuclear pairing Hamiltonian

Let us come back to the separable pairing Hamiltonian (18) to note that if we interpret the parameters η_k as single particle energies corresponding to a nuclear mean field potential, the pairing interaction has the unphysical behavior of increasing the strength with energy. In order to reverse this unwanted effect we define $\eta_k = 2(\varepsilon_k - \alpha)$, where the free parameter α plays the role of an energy cutoff and ε_k is the single particle energy in the mean field level k. Making use of the pair representation of the $SU(2)$, $S_k^+ = c_k^\dagger c_{\bar{k}}^\dagger$, $S_k^z = \frac{1}{2}(c_k^\dagger c_k + c_{\bar{k}}^\dagger c_{\bar{k}} - 1)$, the exactly solvable pairing Hamiltonian (18) reduces to

$$H = \sum_{k>0} \varepsilon_k \left(c_k^\dagger c_k + c_{\bar{k}}^\dagger c_{\bar{k}} \right) - G \sum_{kk',>0} \sqrt{(\alpha - \varepsilon_k)(\alpha - \varepsilon_{k'})} c_k^\dagger c_{\bar{k}}^\dagger c_{\bar{k}'} c_{k'} \tag{22}$$

Our aim is to compare at the BCS level of approximation the results coming from the integrable Hamiltonian (22) with those from a Gogny HFB calculation. As a first step in

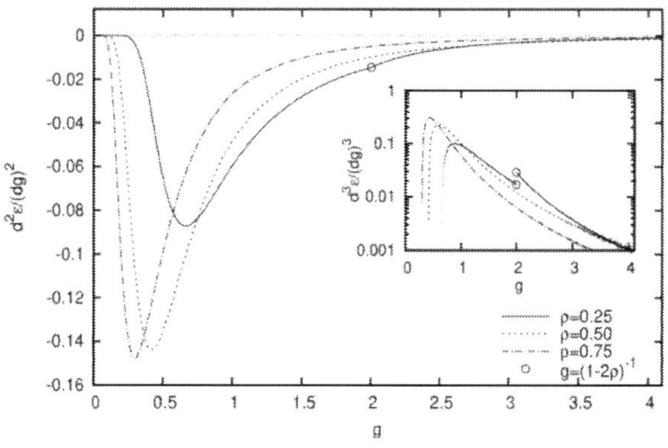

Figure 4. Higher order derivatives of the energy density as a function of g for various densities. The open circles mark the transition point at $g = (1 - 2\rho)^{-1}$.

ascertain the quality of the Hamiltonian (22) to reproduce the superfluid features of heavy nuclei we compare the pairing tensor $u_k v_k$ and the pairing gaps Δ_k with those of a self-consistent mean field Gogny calculation in the canonical basis. The pairing gaps and pairing tensor of the integrable pairing Hamiltonian in the BCS approximation are

$$\Delta_k^{Exact} = G\sqrt{\alpha - \varepsilon_k} \sum_{k'>0} \sqrt{\alpha - \varepsilon_{k'}} < c_{\overline{k'}} c_{k'} >= \Delta\sqrt{\alpha - \varepsilon_k} \tag{23}$$

$$u_k v_k = \frac{\Delta\sqrt{\alpha - \varepsilon_k}}{2\sqrt{(\varepsilon_k - \mu)^2 + (\alpha - \varepsilon_k)\Delta^2}}. \tag{24}$$

Note that the gaps Δ_k and the pairing tensor $u_k v_k$ depend on a single gap parameter Δ and have a square root dependence on the single particle energy. Hence, the model has a highly restricted form for both magnitudes that we will test against the Gogny gaps $\Delta_k^G = \sum_{k'} V_{k\bar{k},\bar{k'}k'} u_{k'}^G v_{k'}^G$ and pairing tensor $u_k^G v_k^G$, where $V_{k\bar{k},\bar{k'}k'}$ are the matrix elements of the Gogny force in the canonical basis and $(u^G v^G)$ is the HFB eigenvector. We take the single particle energies ε_k of the integrable Hamiltonian from the HF energies of the Gogny HFB calculations and we set up a fitting procedure for the two model parameters α and G. We performed the first application to ^{238}U obtaining the values $G = 1.99 \times 10^{-3} MeV$ and $\alpha = 25 MeV$ for the proton system. The number of resulting active orbits is $L = 148$ with $M = 46$ proton pairs. The corresponding dimension of the Hilbert space is $D = 4.83 \times 10^{38}$, well beyond the limits of large scale diagonalizations. However, the exact solution reduces to solve a problem of 46 non-linear coupled equations.

In figure 5 we plot the pairing tensor and the gaps for protons in ^{238}U. In spite of the significant dispersion of the Gogny gaps due to the details of the Gogny force in the canonical basis, it is clear that the integrable gaps follow correctly the global trend. It is interesting to note that a constant pairing interaction, extensively used in the past and also exactly solvable within the rational family of RG models, would give a non reliable constant gap (horizontal line).

These preliminary results suggest that the hyperbolic model could be extremely useful in nuclear structure calculations as a realistic exactly solvable benchmark to test approximations beyond HFB. On a more ambitious respect, it might be possible to fit the pairing strength G as

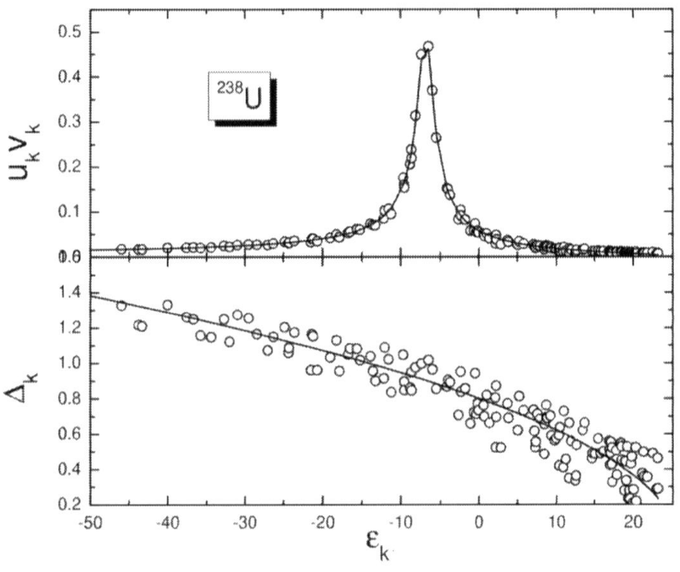

Figure 5. Pairing tensor and gaps for protons in ^{238}U. Open circles are Gogny HFB calculations in the canonical basis while the continuous lines correspond to the hyperbolic Hamiltonian

a function of N and Z to the whole table of nuclides and to set up a program of self-consistent Hartree-Fock plus exact pairing. Work along these lines is in progress.

Acknowledgments
This work has been supported by the Spanish Ministry for Science and Innovation Project No. FIS2009-07277.

7. References

[1] Bardeen J, Cooper L N and Schrieffer J R 1957 *Phys. Rev.* **108** 1175
[2] Bohr A, Mottelson BR and Pines D 1958 *Phys. Rev.* **110** 936
[3] Dietrich K, Mang H J and Pradal J H 1964 *Phys. Rev.* **135** B22
[4] Richardson R W 1963 *Phys. Lett.* **3** 277
[5] Gaudin M 1976 *J. Physique* **37** 1087
[6] Dukelsky J, Esebbag C and Schuck P 2001 *Phys. Rev. Lett.* **87** 066403
[7] Dukelsky J, Pittel S and Sierra G 2004 *Rev. Mod. Phys.* **76** 643
[8] Rombouts S M A, Dukelsky J and Ortiz G 2010 *Phys. Rev. B* **82** 224510
[9] Lerma H S, Rombouts S M A, Dukelsky J, and Ortiz G 2011 *Phys. Rev. B* **84** 100503
[10] von Neumann 1931 J *Ann. Math.* **32** 191
[11] Berry M V and Tabor M 1977, *Proc. Roy. Soc. Lon. A* 356
[12] Cambiaggio M C, Rivas A M F, and Saraceno M 1997 *Nucl. Phys. A* **424** 157
[13] Relaño A, Dukelsky J, Gómez J M G, and Retamosa J 2004 *Phys. Rev. E* **70** 026208
[14] Ortiz G, Somma R, Dukelsky J and Rombouts S M A 2005 *Nucl. Phys. B* **707** 421
[15] Gurarie V and Radzihovsky L 2007 *Ann. Phys.* **322** 2
[16] Read N and Green D 2000 *Phys. Rev. B* **61** 10267
[17] Ibañez M, Links J, Sierra G, and Zhao S-Y 2009 *Phys. Rev. B* **79** 180501
[18] Moore G and Read N 1991 *Nucl. Phys. B* **360** 362

Advanced many-body and statistical methods in mesoscopic systems IOP Publishing
Journal of Physics: Conference Series **338** (2012) 012024 doi:10.1088/1742-6596/338/1/012024

The Onset of Deformation in Atomic Nuclei and Atomic Clusters

Mihai Horoi

Department of Physics, Central Michigan University, Mount Pleasant, MI 48859, USA

E-mail: horoi@phy.cmich.edu

Abstract. It is well know that atomic nuclei and atomic clusters exhibit properties that can be related to intrinsic deformation. The physical signals of deformation are different, but one can show that for these two mesoscopic systems one can identify the sizes for which they undergo shape transitions using a simple criteria. We give examples of shape transitions in these system and we show how they can be identified experimentally looking to the binding energy per particle vs $n^{-1/3}$, where n is the number of particles. It is also know that most mesoscopic systems exhibit prolate deformations. It is interesting to understand why the emergence of prolate deformation is dominant. We present some quantum mechanisms that could explain the dominance of the prolate shapes in nuclei.

1. Introduction

Mesoscopic systems occupy an intermediate size range between individual particles on the one hand and very large systems on the other. This intermediate range is interesting because here properties can change dramatically with the number of particles in the system. For example, in quantum systems like atomic nuclei or clusters of metal atoms, there are strong shell effects in the relative stability of mesoscopic systems, corresponding to the complete filling of energy shells. The shape of a mesoscopic system is also sensitive to the number of particles. In a simple picture, one expects a system of mutually attractive particles to form a compact, quasi-spherical shape to maximize the attractive potential energy. But in the mesoscopic size range, strong deviations from a spherical shape are known. Perhaps the most spectacular case is that of silicon clusters, Si_n. These clusters have been extensively studied both experimentally [1, 2, 3, 4, 5] and theoretically [6, 7, 8, 9, 10, 11, 12, 13] in recent years. Data from ion mobility experiments [14] indicate a roughly spherical shape at $n = 10$ that evolves into a highly elongated shape by $n = 24$. The ten-atom cluster is particularly stable and is a common by-product in the fragmentation of larger clusters. Over the range $n = 25$ to 28 prolate and compact structures are seen to coexist in the experiments and beyond $n = 28$ only compact structures are observed. The isotope chain of Samarium nuclei is another example of a mesoscopic shape transition. The isotope at $A = 144$ corresponds to a closed shell of neutrons and has a near spherical shape. As neutrons are added, the nucleus becomes deformed, slowly at first, and then abruptly over the range $A = 148 - 152$. At $A = 152$ there is experimental evidence for an excited nucleus with a compact shape coexisting with the elongated ground state [15]. In this paper we show that a horizontal plateau on a curve of binding energy per particle (E_b/n) vs. $n^{-1/3}$ is a signature of the shape transition in these type of mesoscopic systems.

The majority of deformed nuclei have axially symmetric prolate deformation in their ground state (g.s.). The presence of rotational bands with typical energy intervals is the first signature of stable deformation. The states in a band are connected by strong quadrupole transitions obeying simple rigid rotor intensity rules. The same rules [16] determine the expectation values of multipole operators. This is essentially a projection of nearly constant within the band intrinsic (body-fixed) quantities onto the laboratory (space-fixed) coordinate frame. The overwhelmingly positive sign of the intrinsic quadrupole moment manifests the dominance of prolate deformation. The Nilsson diagram shows a pattern of split spherical single-particle orbitals that seems to be more or less symmetric with respect to the sign of deformation; the difference in the liquid drop energy for the two signs of deformation is pretty small. While the results of the numerous mean field calculations for specific nuclei in general correctly predict the presence and the sign of deformation, the underlying physical reason for the predominance of prolate shapes is unclear.

In this contribution we present a novel approach to this problem that was previously attacked from the viewpoint of the deformed mean field and corresponding quasiparticle motion. As the mean field itself is generated by the nucleon-nucleon interactions, it makes sense to take a step back and consider the many-body problem rather than the resulting s.p. motion. This would allow us to avoid questions of self-consistency and take into account not only pairing and spin-orbit forces but the full inter-particle interaction. The natural framework is provided by the shell model (SM) with effective interactions; modern versions of the SM give good agreement with the data. The exact diagonalization of the Hamiltonian matrix, with all conservation laws strictly respected, provides the eigenstates in the space-fixed frame; the spectrum and observables can be analyzed similarly to the experimental data. The problem requires exploration of the parameter space of the SM Hamiltonians in a sufficiently large orbital space.

2. Signature of shape transition in atomic nuclei and atomic clusters

In the introduction we indicated that one could use the behavior of E_b/n vs. $n^{-1/3}$ as a signal for shape transitions/shape coexistence in mesoscopic systems. It is easy to understand how a plot of E_b/n vs. $n^{-1/3}$ could contain indications of shape change. For systems dominated by forces that saturate, the binding energy can be written generically in the form

$$E_b = -\gamma S + \epsilon V, \tag{1}$$

where γ is the energy cost to create a unit area of surface, S is the total surface area, ϵ is the energy per unit volume and V is the total volume of the system. The number of particles in the system, n, is proportional to V, which in turn scales like R^3 in a quasi-spherical system. On the other hand, S scales like R^2, or $n^{2/3}$. Therefore, the binding energy per particle is expected to behave as

$$\frac{E_b}{n} = \epsilon_b(\infty) - cn^{-1/3} \tag{2}$$

for a compact, quasi-spherical system. Here $\epsilon_b(\infty)$ is the binding energy per particle for an infinite system and c is a constant, related to the surface tension of the system. For non-spherical particles, S does not scale like $n^{2/3}$, provided that the shape is changing. Therefore, a deviation from the linear behavior with slope $c \neq 0$ predicted in Eq. (2) should be expected to accompany a departure from spherical shapes.

In Ref. [17] we analyzed three examples for which the plateau on $\frac{E_b}{n}$ vs. $n^{-1/3}$ represents a generic signature of shape change. There, we used a combination of theoretical calculations and data taken from relevant experiments to plot the binding energy per particle vs $n^{-1/3}$ for Si_n, ${}^{A}Sm$, and Lennard-Jones (LJ_N), demonstrating a common signature of shape evolution in each case. We then discussed these results, showing how the energetic signature can be understood in terms of a prolate growth pattern. We also contrasted the unusual behavior of Si_n clusters

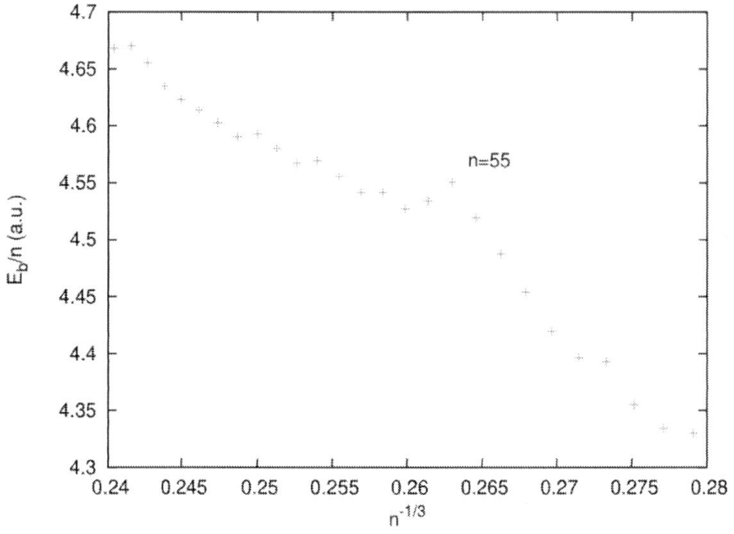

Figure 1. (Color online) Binding energies per atom vs $n^{-1/3}$ for the lowest energy Morse clusters ($\rho = 6$).

with the results for LJ_N. Here we examine two more examples, those of Morse clusters and Na clusters.

To further investigate the generality of the shape transition signature, we examine the case of heavy noble gas clusters whose binding energies can be described by a classical Morse potential,

$$V(\mathbf{R}) = \epsilon \sum_{i<j} \exp\left[\rho\left(1 - \frac{R_{ij}}{R_0}\right)\right]\left\{\exp\left[\rho\left(1 - \frac{R_{ij}}{R_0}\right)\right] - 2\right\}, \tag{3}$$

where $\mathbf{R}$ represents a collective coordinate of all internuclear distances R_{ij}, and R_0 is the equilibrium distance for which the pair potential reaches the minimum value ϵ (i.e., the well depth); here, we apply the Morse potential in reduced coordinates by imposing $R_0 = 1$ and $\epsilon = 1$. The parameter ρ describes the range of the potential, namely the potential well becomes wider as ρ decreases. Here we use $\rho = 6$, which is in the middle of the range considered for this parameter

As in the other cases, we do not attempt to describe the mechanism of formation of the two shape isomers. This process is complicated, and in general requires a more complex thermodynamical analysis [18]. Our point in treating Morse clusters is simply to show the generality of the shape transition signature described in the previous sections. The binding energies of the most stable Morse clusters are available for cluster sizes up to about a hundred [19]. The lowest two magic numbers for these clusters are 13 and 55, corresponding to completely filled geometrical shells of icosahedral symmetry. Plots of the binding energy per particle as a function of $n^{-1/3}$ for Morse clusters show clear inflections at n around 13 and 55, also exhibiting a visible plateau (see e.g. Fig. 1). Examining the corresponding structures shows that over the plateau region, atoms are being added sequentially to one side of the icosahedral $n = 55$ structure. By $n = 60$, a full cap is formed, making the structure clearly prolate, as illustrated in Fig. 2.

Another examples is that of Na clusters. Metallic clusters behave differently than the covalent bond clusters, such as Si_n clusters. However, we were able to use our "BigBang" algorithm [20]

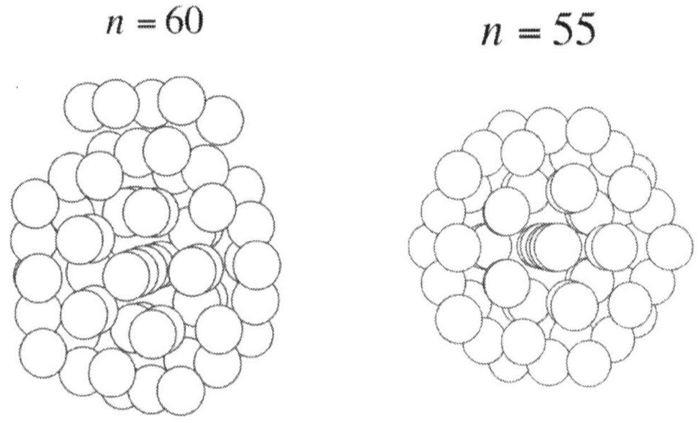

Figure 2. Shapes of the lowest energy Morse clusters ($\rho = 6$) with $n = 55$ (compact) and $n = 60$ (deformed).

to calculate the lowest energy structure of Na_n cluster with n up to 30. The same functional as for Si_n clusters [21, 22] was used. The binding energies per atom vs. $n^{-1/3}$ are presented in Fig. 3. The magic numbers 8 and 20 indicate the fermionic quantum behavior typical for an harmonic oscillator like mean field. The plot indicates once again that a plateau is developed after the magic number $n = 20$. Further indications that prolate structures appear at the end of the plateau, around $n = 25$, will be presented elsewhere, but the typical shape transition signature discussed above is also present in the case of Na_n clusters.

The different mesoscopic systems described above exhibit similar growth features despite very different binding forces: *(i)* a prolate growth regime after a particularly stable structure is formed; *(ii)* a plateau on the curve of $\epsilon_n(n)$ vs. $n^{-1/3}$ accompanying the prolate growth; and *(iii)* a competition/coexistence of different shapes near the end of a long plateau. Certainly, one should not read too much in the analogy with the surface energy of macroscopic liquid drops. The shape of mesoscopic systems are also influenced by the barrier energies. However, in all cases studied we found this generic signature of phase transition to be present.

It is interesting to consider the origin of the common signature of shape transition described here. The simplest model of prolate growth is to consider a cylindrical shape with constant cross-section radius, R. The volume of the system, $\pi R^2 L$, is proportional to the number of particles in the system. The surface area $S = (2\pi RL + 2\pi R^2)$ is also proportional to L (ignoring the endcaps), and thus to the number of particles. Therefore, according to Eq. (1), E_b/n will be independent of n, for ideal prolate growth.

Among all examples of shape evolution discussed, Si_n stands out as the most dramatic, as the prolate growth is far more pronounced [17]. Near $n = 25$, the aspect ratio of the clusters approaches 3:1. Furthermore, the end of the plateau region is marked by an abrupt transition back to compact shapes, whereas in ^ASm and LJ_N cases, in the linear region after the plateau, the clusters remain deformed, gradually decreasing in prolateness. It is instructive to contrast Si_n and LJ_N to understand the physical origin of this different behavior. In the case of LJ_N atoms added to the magic LJ_{147} core group together to form a cap, because of the attractive interaction between them. Once a cap is full, as in $N = 160$, the next atom can either go to a second layer on the cap, or to a cap edge to begin covering a new face of the core cluster. While

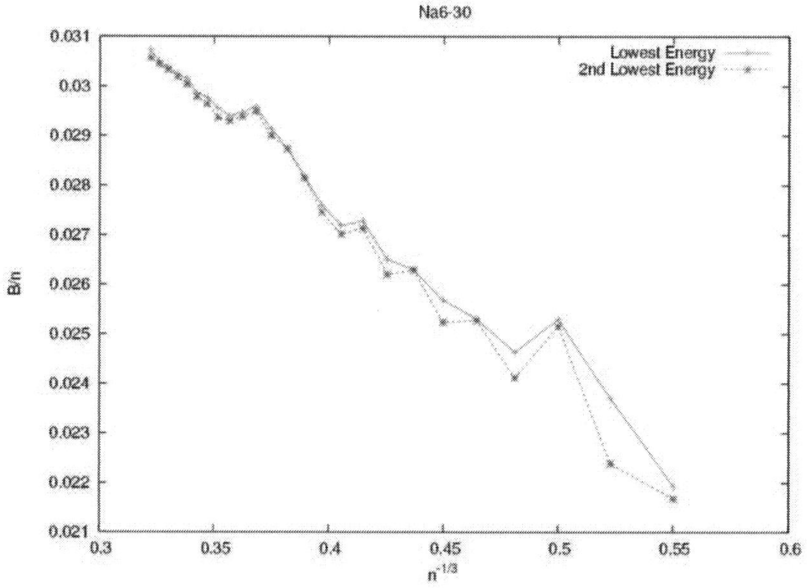

Figure 3. (Color online) Binding energies (B) per atom (in Hartree) vs $n^{-1/3}$ for the lowest and second lowest energy Na clusters.

the former would make the cluster still more prolate, the latter is energetically favored, as it maximizes the number of bonds to the ad-atom. As more atoms are added, more of the core is covered, and the cluster becomes less deformed. Finally, as the outer layer is completed, a new magic cluster is created with the previous magic cluster at its core.

In Si_n the situation is different because the magic Si_{10} structure that is the starting point for prolate growth does not form the core of larger compact structures. The ideal, four-fold covalent bonding of bulk Si cannot be realized in small clusters because of the number of dangling bonds that would be created on the cluster surface. Instead, the atoms rearrange to minimize the dangling bonds, leading to unusual bonding configurations in the small clusters. In larger compact structures, the interior atoms begin to adopt the tetrahedral arrangements very different from what is present in Si_{10}. As a result the growth in Si_n is qualitatively different than in LJ_N and Morse clusters. The first atoms beyond $n = 10$ form prolate caps, analogous to the situation in LJ_N, but as more atoms are added, their energy is lowered by creating stable subunits that stack up along a prolate axis, rather than by covering the Si_{10} core. The stacking process continues out to $n = 25$, with very little change in the binding energy per particle, until compact clusters with new core structures become lower in energy. Further growth proceeds via compact structures following the behavior expected in Eq. 2.

3. The predominance of prolate deformation in nuclei

The majority of deformed nuclei have axially symmetric prolate deformation in their ground state. The presence of rotational bands with typical energy intervals is the first signature of stable deformation. The states in a band are connected by strong quadrupole transitions obeying simple rigid rotor intensity rules. The same rules [16] determine the expectation values of multipole operators. This is essentially a projection of nearly constant within the band intrinsic (body-fixed) quantities onto the laboratory (space-fixed) coordinate frame. The overwhelmingly positive sign of the intrinsic quadrupole moment manifests the dominance of prolate deformation. The Nilsson diagram shows a pattern of split spherical single-particle orbitals that seems to be more or less symmetric with respect to the sign of deformation; the difference in the liquid drop energy for the two signs of deformation is pretty small. While the results of the numerous mean field calculations for specific nuclei in general correctly predict the presence and the sign of deformation, the underlying physical reason for the predominance of prolate shapes is unclear.

It was argued by Lemmer [23] that the kinetic energy should contain an additional contribution similar to the term $-D\vec{\ell}^2$ introduced by Nilsson [24] for interpolation between the harmonic oscillator and potential box. Being split to time-conjugate pairs $(m, -m)$ by deformation, the large-ℓ spherical orbitals determine the single-particle (s.p.) occupancies in the way that makes the prolate case energetically favorable. Related arguments were given by Castel and Goeke [25], who showed that the collective inertial parameter is larger for the prolate deformation, and later by Castel, Rowe and Zamick [26] with the help of self-consistency conditions [16]. There are also ideas based on the semiclassical analysis of the s.p. level density [27], simplest periodic orbits [28] and their bifurcations in a deformed cavity [29]. The predominance of prolate deformation analogously emerges in s.p. motion in metallic clusters [30] and for many biological objects, from molecular level to pollen grains [31]. The latest analysis by Hamamoto and Mottelson [32] starting with the statement that "the nature of the parameters responsible for the prolate dominance has not yet been adequately understood" proceeds along similar lines. Detailed calculations and comparison between the harmonic oscillator potential and that of a spheroidal cavity show a different character of the mixing of spherical orbitals for the prolate and oblate cases. Being a surface effect, the difference should decrease in large systems. The authors of Ref. [32] mention additional factors which were not sufficiently accounted for including the role of the spin-orbit potential, pairing effects [33], and the presence of two kinds of particles.

Below we suggest a new approach to this problem that was previously attacked from the viewpoint of the deformed mean field and corresponding quasiparticle motion. As the mean field itself is generated by the nucleon-nucleon interactions, it makes sense to take a step back and consider the many-body problem rather than the resulting s.p. motion. This would allow us to avoid questions of self-consistency and take into account not only pairing and spin-orbit forces but the full inter-particle interaction. The natural framework is provided by the shell model (SM) with effective interactions; modern versions of the SM give good agreement with the data. The exact diagonalization of the Hamiltonian matrix, with all conservation laws strictly respected, provides the eigenstates in the space-fixed frame; the spectrum and observables can be analyzed similarly to the experimental data. The problem requires exploration of the parameter space of the SM Hamiltonians in a sufficiently large orbital space.

To explore the parameter space we will work with random interactions. All two-body matrix elements allowed by angular momentum, parity and isospin conservation are taken as random uncorrelated quantities. Among many realizations of such an ensemble we can identify those with definite rotational characteristics and study the abundance of such cases and their dynamic origin. Random interactions in many-body systems are usually analyzed for studies of quantum chaos [34, 35] in relation to the random matrix theory. In Ref. [36] the ensemble was constructed with rotationally invariant interactions, and only the magnitudes of allowed matrix elements were taken randomly. The unexpected output was the discovery of the predominance, for an even number of particles, of the zero ground state spin, in spite of relatively low multiplicity of $J = 0$ states in Hilbert space. This result, valid not only in a single-j case but in the context of realistic SM space [37] as well, was fully understood in terms of geometry of the parameter space only for the exceptional case of the single $j = 7/2$ shell [38] and for the similar results in the interacting boson model [39]. In more realistic fermionic models the full theory is still absent; the most promising are the ideas of geometric chaoticity of random angular momentum coupling [40, 41].

	λ	$N(0,2)$	$\frac{N(Q<0)}{N(0,2)}$	$N(E4/E2)$	$N_{\rm rot}$	$\frac{N_{\rm prolate}}{N_{\rm rot}}$
(a)	0.05	1398	0.62	50	3	1.00
	0.5	3320	0.54	322	100	0.74
	1.0	3846	0.52	354	100	0.70
	1.5	4056	0.52	378	119	0.72
	2.0	4129	0.52	371	122	0.74
	3.0	4196	0.52	366	126	0.70
	4.0	4233	0.52	367	125	0.71
	10.0	4295	0.53	368	112	0.74
(b)	1.0	3156	0.55	289	39	0.77
	2.0	3153	0.53	264	34	0.79
	3.0	3140	0.52	266	34	0.82
	4.0	3156	0.53	240	35	0.89
(c)	1.0	4569	0.54	322	120	0.73
	2.0	4530	0.52	339	116	0.75
	3.0	4490	0.52	349	119	0.76
	4.0	4461	0.52	371	124	0.81
(d)	1.0	2170	0.43	176	55	0.07
	2.0	2185	0.37	193	70	0.06
	3.0	2212	0.35	188	73	0.05

Table 1. The probabilities of occurrence of collective prolate configurations (see text for details).

The details were discussed in the review articles [42, 43, 44].

The rotationally invariant random interaction models provide a wealth of information. Any given set of random interaction parameters gives rise to a specific version of the SM. It was noticed, for example, that some sequences of the lowest excited states, like $0^+ - 2^+ - 4^+ - ...$, appear with the enhanced probability. The energy ratios in such yrast sequences are spread around the values close to limiting cases of the collective models, such as vibrational and rotational bands [45]. In Ref. [42], the realizations with the lowest states $0^+ - 2^+$ were analyzed in terms of the Alaga ratio,

$$A = \frac{Q(2^+)^2}{B(E2; 2^+ \rightarrow 0^+)}, \tag{4}$$

of the squared expectation value of the quadrupole moment of the 2^+ state to the reduced transition probability from this state to the g.s. The distribution of the Alaga ratio reveals two pronounced peaks, close to zero and at the value corresponding to the rigid rotor [16]. This means that the stable mean fields generated by random interactions with high probability correspond either to near-spherical or to well deformed shape. The situation is similar in the case of the interacting boson model [39].

We define the space-fixed quadrupole moment of the axially symmetric rotor with spin J [16] as

$$Q(J) = Q_0 \frac{3K^2 - J(J+1)}{(J+1)(2J+3)} \Rightarrow -Q_0 \frac{J}{2J+3}, \tag{5}$$

where Q_0 is the intrinsic (body-fixed) quadrupole moment, and we assume $K = 0$ for the yrast band of an even-even nucleus [46]. Prolate intrinsic shapes, $Q_0 > 0$, correspond to squeezed shapes around the axis of collective rotation, $Q < 0$. With the standard [16] definition of the B(E2) probability for the yrast band of a well deformed rotor, the Alaga ratio (4) is equal to 4.10.

In the single-j model [42], where the splitting of a j-level proceeds fan-like with approximate symmetry of oblate and prolate sides, the sign of Q will be random. Therefore we expect the crucial role of mixing of spherical orbitals. It was often suggested [47] that the spherical SM can realistically describe deformation if one uses at least two s.p. orbits with $\Delta j = 2$. This approach successfully produces collective bands, such as in ^{48}Cr, and even in the double magic ^{56}Ni [48]. The collectivity was tested not only by the $J(J+1)$ behavior, but also by large and consistent in band quadrupole moments and $B(E2)$ strengths. As a generic example, we consider a model with four neutrons and four protons in the space of two spherical orbitals, $0f_{7/2}$ and $1p_{3/2}$, of the same parity. This corresponds to the oversimplified SM description of ^{48}Cr, the nucleus with well known collective properties. Without interactions, we would have half-filled $f_{7/2}$ orbitals with zero quadrupole moment. In this space the most general rotationally invariant two-body interaction is described by 30 matrix elements (see Table V of Ref. [49]). We choosed the random ensemble defined by the uniform distribution of uncorrelated matrix elements between $-V$ and $+V$, and the only parameter is the ratio $\lambda = V/\epsilon$, where ϵ is the (non-random) spacing between the two orbitals. With the statistics of 10000 realizations for each value of λ, the results are presented in Table I*a*.

The case of very small $\lambda = 0.05$ can be viewed as the single-j result. The column $N(0,2)$ shows the number of realizations with the sequence $0^+, 2^+$ of the g.s. and the first excited state. The probability of such sequences is strongly enhanced. Before applying additional requirements we can find the sign of the expectation value (5) of the quadrupole moment $Q(2^+)$. As expected, the fraction $N(Q<0)/N(0,2)$ of "prolate" cases, $Q<0$, is stable near 50% (the third column). Among these states there are cases with rotational properties. The constraint that the ratio $E(4^+)/E(2^+)$ be between 3 and 3.6 (the rigid rotor would give 3.33) leaves the number of states indicated in the column $N(E4/E2)$. In the column $N_{\rm rot}$ we count the states where, in addition to the energy criterion, the Alaga ratio (4) is between 3.90 and 4.30, an arbitrary but restrictive choice. The last column gives the fraction of states with negative sign of Q among all "rotational" states. The predominance of prolate configurations is practically independent on the relative strength of mixing between the orbitals if this strength is sufficient for the onset of deformation. The full histogram for the distribution of the Alaga ratio is shown in Fig. 1 for two values of the interaction strength; this figure shows also the statistics without any rotational selection cuts. The prolate deformed peak is quite narrow. With the only energy criterion $(0^+, 2^+)$ applied, we see also a large peak for spherical configurations with A close to zero, as well as intermediate background situations. In this large set, the prolate cases appear only in slightly more than 50% cases. We can also note that at very weak mixing, $\lambda = 0.05$, the numbers $N(0,2)$ and especially $N(E4/E2)$ fall sharply, and the number of rotational configurations satisfying our criteria is very small. This emphasizes the importance of orbital mixing.

Going from the isospin-symmetric case with $N = Z$ to asymmetric one, we consider next the system with six neutrons and four protons assuming the same set of two s.p. levels. The results are shown by Table I*b*. The fraction of states with $Q < 0$ among all sequences $(0^+, 2^+)$ is still close to 1/2. The total number of rotational cases is now significantly lower. At the same time, the fraction of prolate configurations is even higher than in the symmetric case. The results are essentially the same in the $N = Z$ case for the inverted level scheme when the spherical $p_{3/2}$ level is placed below the $f_{7/2}$ level, see Table I*c*. Again for the very weak mixing, $\lambda = 0.01$, we do not find any rotational states; in this case all eight nucleons occupy the lowest $p_{3/2}$ level creating the closed shell. With strong level mixing, 45% of all states belong to the $(0^+, 2^+)$ sequence. Clearly, the occurrence of prolate deformation is even higher than for the normal fp level sequence. This confirms the presence of the effect we mentioned in the introduction, namely the influence of the splitting of the level with higher ℓ when low-$|m|$ components steeply go down with the prolate deformation. The last case we consider here contains four protons and four neutrons in the space of two levels of opposite parity, $f_{5/2}$ and $g_{9/2}$, Table I*d*. Here the fraction of the sequences $0^+, 2^+$ is lower, and practically all rotational states have oblate deformation. It is clear that, due to parity conservation, direct mixing of split levels by the

Journal of Physics: Conference Series **338** (2012) 012024 doi:10.1088/1742-6596/338/1/012024

deformed mean field is impossible and only pairwise transfers of particles can play a role.

The main new dimension brought in by our approach is in the visualization of the entire parameter space. The mechanism of the prolate predominance is revealed as we identify the realizations of the ensemble responsible for the effect. Fig. 2. of Ref. [49] shows all 30 reduced matrix elements $\langle j_1 j_2 || V || j_3 j_4 \rangle (JT)$ allowed by angular momentum and isospin for the strong coupling limit, $\lambda = 4.0$, in case (a) of Table I. The average value of each matrix element over the whole ensemble is zero, with small fluctuations (solid line). The dashed line shows the average over prolate rotational samples. Table V of Ref. [49] displays the ordering of the matrix elements and their average numerical values. Although all matrix elements are enhanced compared to the level of fluctuations, one can conclude that certain matrix elements are crucial for the transition from spherical shape to predominantly prolate axial deformation. We see the exceptional role of amplitudes *9-12* describing the transfer of a single nucleon $f_{7/2} \leftrightarrow p_{3/2}$ in the interaction with another $f_{7/2}$ nucleon, regardless of its isospin. Such a process is forbidden for the orbitals of opposite parity. This agrees with the idea discussed in Ref. [32]. The monopole pairing given by amplitudes *5,15* and *29* is not effective in this transition. Contrary to that, we see a large negative amplitude of quadrupole pair transfer *16.* Large quadrupole-quadrupole forces in the particle-hole channel correspond to matrix elements *20-24* in the particle-particle channel. They induce the collectivization process and formation of the proper symmetry of the mean field after mixing p- and f-orbitals.

4. Conclusions

In this paper we analyzed three different mesoscopic systems, Silicon clusters, the Samarium isotope chain and Lennard-Jones clusters, that exhibit similar shape transition and coexistence phenomena. The forces involved in these three mesoscopic systems are very different: the Silicon clusters are bound by covalent bonds, Samarium nuclei by short range nuclear forces that create a quantum liquid drop, and noble gas clusters by classical interatomic potentials. Despite the diverse nature of the binding, each system exhibits the same signature of shape transition, namely a long plateau in a curve of e_b/n vs. $n^{-1/3}$ following a size corresponding to a stable, compact structure. Further, we find that in each system, energetic competition between deformed and compact shapes occurs near the end of long plateaus, resulting in the observed coexistence of the two shape isomers in the case of Si_n and $^A Sm$. We conclude that these are generic features for mesoscopic systems.

We also investigate the predominance of prolate deformations in nuclei within the shell model approach. To achieve that goal we performed the exploration of the parameter space that serves as an arena for effective nucleon-nucleon interactions building the stable deformed mean field. We show the decisive role of the mixing between the valence spherical orbitals of the same parity split as a function of deformation. This mixing, different for the two sides of the axial deformation diagram, makes the prolate deformation energetically favorable with high probability. This picture is supported by the statistical analysis of the random interaction ensemble and by singling out the responsible components of the effective interaction. The process is amplified by the presence of two kinds of nucleons. Although we have shown in detail in this contribution only cases of simple configurations, the results are quite generic for a small system. The effect is driven by the surface orbitals but it still remains to see if it indeed disappears in the macroscopic limit as predicted in Ref. [32]. The random interaction ensemble provides a new powerful tool for understanding the many-body mechanisms of collective phenomena.

Acknowledgments

Support from the NSF grants No. PHY-0758099 and PHY-1068217 is acknowledged.

References

[1] R. R. Hudgins I. Motoharu and M. F. Jarrold 1999 *J. Chem. Phys.* **111** p 7865.
[2] M. F. Jarrold 1991 *Science* **252** p 1085.
[3] M. F. Jarrold and V. A. Constant 1991 *Phys. Rev. Lett.* **67** p 2994.
[4] E. C. Honea *et al.* 1933 *Nature* **366** p 42.
[5] Th. Bachels and R. Schäfer 2000 *Chem. Phys. Lett.* **324** p 365.
[6] K. M. Ho A. A. Shvartsburgh B. Pan, Z. Y. Lu C. Z. Wang J. G. Wacker J. L. Fye and M. F. Jarrold 1998 *Nature* **392** p 582.
[7] E. Kaxiras and K. A. Jackson 1993 *Phys. Rev. Lett.* **71** p 727.
[8] B. Liu Z. Y. Lu B. Pan C. Z. Wang K. M. Ho, A. A. Shvartsburg and M. F. Jarrold 1998 *J. Chem. Phys.* **109** p 9401.
[9] I. Rata A. A. Shvartsburgh M. Horoi Th. Frauenheim K. W. M. Liu and K. A. Jackson 2000 *Phys. Rev. Lett.* **85** p 546.
[10] L. R. Marim M. R. Lemes and A. Dal Pino Jr. 2003 *Phys. Rev. A* **67** p 033203.
[11] S. Yoo J. Zhao J. Wang and X. C. Zeng 2004 *J. Am. Chem. Soc.* **126** p 13845.
[12] S. Yoo and X. C. Zeng 2005 *Angew. Chem. Intl. Ed.* **44** p 1491.
[13] A. Tekin and B. Hartke 2005 *J. Theor. Comp. Chem.* **4** p 1119.
[14] The transition was observed for both cations and anions, although for anions it takes place for n=27-29 (see e.g. Fig. 2 of Ref. [1]).
[15] N. V. Zamfir *et al.* 1999 *Phys. Rev. C* **60** p 054312.
[16] A. Bohr and B.R. Mottelson 1989 *Nuclear Structure*, vol. 2 (Benjamin, New York).
[17] M. Horoi and K.A. Jackson 2006 *Chem. Phys. Lett.* **427** p 147.
[18] J, P. K. Doye M. A. Miller and D. J. Wales 1999 *J. Chem. Phys.* **110** p 6896.
[19] D. J. Wales and J. P. K. Doye 1997 *J. Phys. Chem.* **A 101** p 5111.
[20] K.A. Jackson M. Horoi I. Chaudhuri Th. Frauenheim and A.A. Shvartsburg 2006 *Comp. Mat. Sci.* **35** p 232.
[21] K.A. Jackson M. Horoi I. Chaudhuri Th. Fraunheim and A. A. Shvartsburg 2004 *Phys. Rev. Lett.* **93** p 013401.
[22] J. P. Perdew K. Burke and M. Ernzerhof 1996 *Phys. Rev. Lett.* **77** p 3865.
[23] R.H. Lemmer 1960 *Phys. Rev.* **117** p 1551.
[24] S.G. Nilsson 1955 *Kgl. Dan. Vid. Selsk. Mat.-fys. Medd.* **29**, No. 16.
[25] B. Castel and K. Goeke 1976 *Phys. Rev. C* **13** p 1765.
[26] B. Castel D.J. Rowe and L. Zamick 1990 *Phys. Lett.* B **236** p 121.
[27] H. Frisk 1990 *Nucl. Phys.* **A511** p 309.
[28] M.A. Deleplanque *et al.* 2004 *Phys. Rev. C* **69** p 044309.
[29] K. Arita A. Sugita and K. Matsuyanagi 1998 *Czechoslovak J. Phys.* **48** p 821.
[30] I. Hamamoto B.R. Mottelson H. Xie and X.Z. Zhang 1991 *Zeitschr. Phys. D* **21** p 163.
[31] J. Martin M. Torrell A.A. Korobkov and J. Valles 2003 *Plant Biol.* No. 5 p 85.
[32] I. Hamamoto and B.R. Mottelson 2009 *Phys. Rev. C* **79** p 034317.
[33] N. Tajima Y.R. Shimizu and N. Suzuki 2002 *Prog. Theor. Phys. Suppl.* **146** p 628.
[34] T.A. Brody *et al.* 1981 *Rev. Mod. Phys.* **53** p 385.
[35] V. Zelevinsky B.A. Brown N. Frazier and M. Horoi 1996 *Phys. Rep.* **276** p 85.
[36] C.W. Johnson G.F. Bertsch and D.J. Dean 1998 *Phys. Rev. Lett.* **80** p 2749.
[37] M. Horoi B.A. Brown and V. Zelevinsky 2001 *Phys. Rev. Lett.* **87** p 062501.
[38] P. Chau Huu-Tai A. Frank N.A. Smirnova and P. Van Isacker 2002 *Phys. Rev. C* **66** p 061302.
[39] R. Bijker and A. Frank 2002 *Phys. Rev. C* **65** p 044316.
[40] D. Mulhall A. Volya and V. Zelevinsky 2000 *Phys. Rev. Lett.* **85** p 4016.
[41] T. Papenbrock and H.A. Weidenmüller 2005 *Nucl. Phys.* **A757** p 422.
[42] V. Zelevinsky and A. Volya 2004 *Phys. Rep.* **391** p 311.
[43] Y.M. Zhao A. Arima and N. Yoshinaga 2004 *Phys. Rep.* **400** p 1.
[44] H.A. Weidenmüller and G.E. Mitchell 2008 *arXiv 0807.1070*.
[45] C.W. Johnson H. Nam 2007 *Phys. Rev. C* **75** p 047305.
[46] In Ref. [49] the similar Eq. (2) contains a missing 2 typo that is fixed here. However, the reported calculations were performed with the correct formula.
[47] E. Caurier *et al.* 2005 *Rev. Mod. Phys.* **77** p 427.
[48] M. Horoi *et al.* 2006 *Phys. Rev. C* **73** p 061305(R).
[49] M. Horoi and V. Zelevinsky 2010 *Phys. Rev. C* **81** p 034306.

Quantum phase transitions in algebraic models of mesoscopic systems

F Iachello

Center for Theoretical Physics, Sloane Physics Laboratory, Yale University, New Haven, CT 06520-8120, USA

Abstract. The theory of quantum phase transitions (QPTs) and excited states quantum phase transitions (ESQPTs) is briefly reviewed. The phase structure and phase diagram of the interacting boson model of nuclei are discussed. Experimental evidence for QPTs in nuclei is presented.

1. Algebraic models of mesoscopic systems

Mesoscopic systems are many-body systems with a finite number of constituents, N. The constituents can be either bosonic or fermionic or mixtures of bosons and fermions. In the last 40 years, it has become apparent that a convenient way to treat these systems is by algebraic methods [1]. In algebraic models of many-body systems , the Hamiltonian, H, and other operators are expanded in elements $G_{\alpha\alpha'}$ of an algebra g,

$$H = E_0 + \sum_{\alpha\alpha'} \varepsilon_{\alpha\alpha'} G_{\alpha\alpha'} + \sum_{\alpha\alpha'\beta\beta'} u_{\alpha\alpha'\beta\beta'} G_{\alpha\alpha'} G_{\beta\beta'} + ... \tag{1}$$

States are characterized by the irreducible representations of g, labelled by the total number of particles, N. The representations are totally symmetric for bosons and totally antisymmetric for fermions.

For fermionic systems the algebra g is composed by the bilinear products of fermion creation and annihilation operators, satisfying $\{a_i, a_{i'}^\dagger\} = \delta_{ii'}; \{a_i, a_{i'}\} = \{a_i^\dagger, a_{i'}^\dagger\} = 0$,

$$G_{ii'} = a_i^\dagger a_{i'} \quad , \qquad i, i' = 1, ..., n \ ; \qquad G_{ii'} \in g \equiv U(n). \tag{2}$$

The Hamiltonian of any (number-conserving) fermionic system can be written in the form

$$H = E_0 + \sum_{ii'} \eta_{ii'} a_i^\dagger a_{i'} + \frac{1}{2} \sum_{ii'jj'} u_{ii'jj'} a_i^\dagger a_{i'}^\dagger a_j a_{j'} + ... \tag{3}$$

which, upon rearrangement of the fermion operators, can be cast in the form given above. For rotationally invariant systems, it is convenient to label the single-particle states by $i \equiv \ell_i, m_{\ell_i}, s_i, m_{s_i}$, and $n = \sum_i (2\ell_i + 1)(2s_i + 1)$, with $s_i =$ half-integer. Examples are the atomic shell model and the nuclear shell model.

Published under licence by IOP Publishing Ltd

For bosonic systems, the algebra g is composed of bilinear products of creation and annihilation operators for bosons, satisfying $[b_\alpha, b^\dagger_{\alpha'}] = \delta_{\alpha\alpha'}; [b_\alpha, b_{\alpha'}] = [b^\dagger_\alpha, b^\dagger_{\alpha'}] = 0$,

$$G_{\alpha\alpha'} = b^\dagger_\alpha b_{\alpha'} \ , \qquad \alpha, \alpha' = 1, ..., n \ ; \qquad G_{\alpha\alpha'} \in g \equiv U(n). \tag{4}$$

[Bosonic and fermionic systems have the same g. Their differ by its representations.] The Hamiltonian of any (number-conserving) bosonic system can be written in the form

$$H = E_0 + \sum_{\alpha\alpha'} \varepsilon_{\alpha\alpha'} b^\dagger_\alpha b_{\alpha'} + \frac{1}{2} \sum_{\alpha\alpha'\beta\beta'} v_{\alpha\alpha'\beta\beta'} b^\dagger_\alpha b^\dagger_{\alpha'} b_\beta b_{\beta'} + ..., \tag{5}$$

which, upon rearrangement of the boson operators, reduces to the Hamiltonian given above. [For non-number conserving bosonic systems (for example, ensembles of photons or phonons) bilinear products of creation and annihilation operators are added, $G'_{\alpha\alpha'} = b^\dagger_\alpha b^\dagger_{\alpha'}, G''_{\alpha\alpha'} = b_\alpha b_{\alpha'}$, and the algebra g is Sp(2n,$\mathbb{R}$).] Examples of number conserving bosonic systems are the Interacting Boson Model of nuclei [2], for which $n = 6$, and the Vibron Model of molecules [3], for which $n = 2, 3, 4$.

Many-body systems composed of a mixture of bosons and fermions can also be treated by algebraic methods by mapping onto an algebric superstructure, g^*, composed of elements

$$G_{\alpha\alpha'} = b^\dagger_\alpha b_{\alpha'} \ , \qquad G_{ii'} = a^\dagger_i a_{i'} \ , \qquad F^\dagger_{\alpha i} = b^\dagger_\alpha a_i \ , \qquad F_{i\alpha} = a^\dagger_i b_\alpha \ ; \qquad g^* \equiv U(n/m). \tag{6}$$

States are characterized by the totally supersymmetric representations of $U(n/m)$ labelled by the total number of bosons plus fermions $N = N_B + N_F$. Examples are the Interacting Boson-Fermion Model of nuclei [4] and the Electron-Vibron Model of molecules [5].

In this contribution only bosonic systems will be considered.

2. Quantum phase transitions

Quantum phase transitions (QPT) are qualitative changes in the structure of a physical system that occur as a function of one (or more) parameters, $\xi_1, \xi_2, ...$, the so-called *control parameters* that appear in the quantum Hamiltonian describing the system

$$H = \varepsilon \left(H_1 + \xi_1 H_2 + \xi_2 H_3 + ... \right). \tag{7}$$

Originally introduced within the context of nuclear physics [6], they have had many applications to various areas of physics [7]. In recent years, the concept of QPT has been extended to excited states quantum phase transitions (ESQPT), that is qualitative changes in the state of a system that occur as a function of excitation energy [8]. In this contribution, after a review of the theory of QPT and ESQPT, applications of these concepts to mesoscopic systems will be discussed. An important aspect of QPTs in mesoscopic systems is the study of finite size scaling and its large N limit. As a specific example, QPTs and ESQPTs in atomic nuclei ($N \sim 2 - 250$) will be discussed. The advantages of using atomic nuclei as examples are: (i) there are in this case specific models that can be studied in depth and (ii) the theory can be tested with experiments.

2.1. QPTs in algebraic models

Algebraic models are particularly well-suited to study QPTs, because they possess exact solutions, called *dynamic symmetries* (for purely bosonic or fermionic systems) or dynamical super-symmetries (for mixed Bose-Fermi systems) which describe properties of the system in explicit analytic form. These symmetries can be used to define the phases of the system. The possible phases of a system described by the algebra g, are obtained by breaking g into its subalgebras

$$\begin{aligned} g &\supset g_1 \supset g_1' \supset ... \\ g &\supset g_2 \supset g_2' \supset ... \\ &... \\ g &\supset g_\phi \supset g_\phi' \supset ... \end{aligned} \tag{8}$$

Dynamical symmetries are situations in which the Hamiltonian H does not contain all elements of g, but only invariant (Casimir) operators, C, of a subalgebra chain

$$H = aC(g) + a_1C(g_1) + a_1'C(g_1') + ... \tag{9}$$

In this case, the energy eigenvalues can be written in explicit analytic form in terms of quantum numbers $\lambda_1, \lambda_2, ...$ uniquely characterizing the states

$$E(\lambda_1, \lambda_2, ...) = a\left<C(g)\right> + a_1\left<C(g_1)\right> + a_1'\left<C(g_1')\right> + ... \tag{10}$$

Having defined the phases, one can study the phase diagram by considering the Hamiltonian

$$H = a_1C(g_1) + a_2C(g_2) + ... + a_\phi C(g_\phi) \tag{11}$$

written in terms of the Casimir operators of the subalgebras $g_i (i = 1, 2, ..., \phi)$ that define the phases. Taking out a scale factor, $\varepsilon = a_1$, and introducing the control parameters $\xi_1 = \frac{a_2}{a_1}, \xi_2 = \frac{a_3}{a_1}, ...$, the Hamiltonian can be written in the standard form $H = \varepsilon\left(H_1 + \xi_1H_2 + \xi_2H_3 + ...\right)$. Note that the number of control parameters is equal to the number of phases ϕ minus 1. [It is sufficient to take one Casimir operator for each g_i. The corresponding Hamiltonian is often referred to as the "essential" Hamiltonian.] For systems with only two phases, it is convenient to parametrize H as

$$H = \varepsilon\left[(1-\xi)H_1 + \xi H_2\right] \tag{12}$$

in such a way that the control parameter varies in the interval 0-1 instead of 0-∞.

QPTs have been studied in many bosonic systems. A class extensively investigated is the so-called s-b systems, defined in terms of a singlet boson $s^{(0)}$ and a $(2\ell+1)$-fold degenerate boson $b^{(\ell)}$. These models have had many applications in physics (see [9]). The algebraic structure of these models is $U(n)$ with $n = 2\ell + 2$, with elements $U(n) \doteqdot s^\dagger s, s^\dagger b_m, b_m^\dagger s, b_m^\dagger b_{m'}$, $(m, m' = -\ell, ..., +\ell)$. For applications to nuclear and molecular physics it is convenient to impose (3D) rotational invariance for problems with $2\ell+1$=odd and (2D) invariance for problems with $2\ell+1$=even. The decomposition of the algebra of $U(n)$ with rotational invariance has been carried out explicitly up to $n = 6$. QPTs in s-b models have been studied both classically and quantum mechanically [10].

The classical analysis is done by first introducing a geometry of algebraic models through the coset space $U(n)/U(n-1) \otimes U(1)$ and then introducing coherent (or intrinsic) states expressed in terms of complex variables α_m. For finite (mesoscopic) systems with a conserved number of particles it is convenient to use *number-projected* coherent states. The ground state of the system is written as a boson condensate

$$\left|N, \alpha\right> = \frac{1}{\sqrt{N!}}\left(s^\dagger + \sum_m \alpha_m b_m^\dagger\right)^N \left|0\right>. \tag{13}$$

The energy expectation value in this state is $\frac{\left<N,\alpha|H|N,\alpha\right>}{\left<N,\alpha|N,\alpha\right>} = E(\alpha)$. Coordinates and momenta are introduced as $q_m = [(-)^m\alpha_{-m} + \alpha_m^*]/\sqrt{2}$, $p_m = i[(-)^m\alpha_{-m}^* + \alpha_m]/\sqrt{2}$. A convenient set of coordinates in this $(n-1)$-dimensional space is the hyperspherical coordinates set (r, Θ) with $\Theta \equiv (\theta_1, \theta_2, ..., \theta_{n-2})$. The classical Hamiltonian associated with the quantal Hamiltonian H is

$$H_{cl} = T(p_r, p_\Theta, r, \Theta) + V(r, \Theta). \tag{14}$$

The classical potential function, $V(\Xi; r, \Theta)$ depends on coordinates (r, Θ) and on control parameters $\Xi \equiv (\xi_1, \xi_2, ..., \xi_{\phi-1})$. Minimizing V with respect to the coordinates, one obtains the equilibrium values (r_e, Θ_e), which are the *classical order parameters*, and the classical ground state energy, $V_e(\Xi; r_e, \Theta_e) \equiv E_0$. The study of V_e and its derivatives with respect to the control parameters determines the order of the phase transition. A classification often used is Erhenfest classification. A phase transition is called of 0th order if the ground state energy is discontinuous, 1st order if $\partial E_0/\partial \Xi$ is discontinuous, 2nd order if $\partial^2 E_0/\partial \Xi^2$ is discontinuous, etc.. If there is no discontinuity it is called a crossover. The behavior of the order parameters at (or around) the critical values of the control parameters also determines the order of the phase transitions and the associated critical exponents. A classification often used is Landau classification. A phase transition is called of 1st order if the order parameters (r_e, Θ_e) are discontinuous, 2nd order if $(\frac{\partial r_e}{\partial \Xi}, \frac{\partial \Theta_e}{\partial \Xi})$ are discontinuous, etc. If there is no discontinuity, it is called a crossover. Critical exponents describe the behavior around the critical point. For example, for 2nd order transitions

$$r_e \propto (\xi - \xi_c)^\mu \qquad \xi \geq \xi_c \ , \qquad r_e = 0 \qquad \xi < \xi_c. \tag{15}$$

For first order transitions the situation is more complex, since there are three points of importance, the spinodal, critical and antispinodal points [10]. Critical exponents have been evaluated for $U(n)$ models with at most quadratic Hamiltonians.

The quantal analysis is done by diagonalizing the Hamiltonian H as a function of the control parameters Ξ in one of the dynamical symmetry bases $g \supset g_1 \supset g_1' \supset ...$, thus evaluating numerically the ground state energy E_0 and its derivatives $\partial E_0/\partial \Xi, ...$ If the diagonalization is done for sufficiently large N, one can observe precursors of QPTs and determine the order of the phase transition. The *quantum order parameters* are the expectation values in the ground state of some suitably chosen operators. The most basic order parameter for s-b models is the expectation value of the number of b bosons $\omega = \frac{\langle N_b \rangle}{N}$. This is related to the classical order parameter by $\omega = \frac{r_e^2}{1+r_e^2}$. The behavior of the order parameter and other quantities as $N \to \infty$ is called finite size scaling. Finite size scaling in s-b models can be studied both numerically and analytically. For second order transitions all physical quantities scale as a power law [11]. For example, $\omega(\xi_c) = A_{n0} N^{-A_{n1}}$. [For first order transitions a complete theory of scaling is not yet available].

3. Excited state quantum phase transitions (ESQPT)

ESQPTs are qualitative changes in the state of the system as a function of excitation energy [8]. In QPTs one studies classically the potential V and quantum mechanically the ground state energy, E_0. In ESQPTs one needs to do dynamics, either semi-classically or quantum mechanically.

Semi-classical dynamics has been done completely for s-b models with pairing interactions with quantum Hamiltonian

$$H = \frac{(1-\xi)}{N} \hat{N}_b - \frac{\xi}{N^2} \left(s^\dagger \tilde{b} + b^\dagger \tilde{s} \right) \cdot \left(s^\dagger \tilde{b} + b^\dagger \tilde{s} \right) \tag{16}$$

and classical Hamiltonian

$$\begin{aligned} H_{cl} &= \frac{1-\xi}{2N} [p_r^2 + r^{-2} T_\Theta(v)] + \frac{\xi}{N^2} [r^2 p_r^2 + T_\Theta(v)] + V(r, \xi) \\ V(r, \xi) &= \frac{1-5\xi}{2} r^2 + \xi r^4. \end{aligned} \tag{17}$$

A study of V and its derivatives shows that s-b models with pairing interactions have a 2nd order QPT at $\xi_c = 1/5$, with critical exponent $\mu = 1/2$. In order to study ESQPTs, one considers the full Hamiltonain (kinetic plus potential) and studies the classical action $S(\xi; E)$ shown in figure 1.

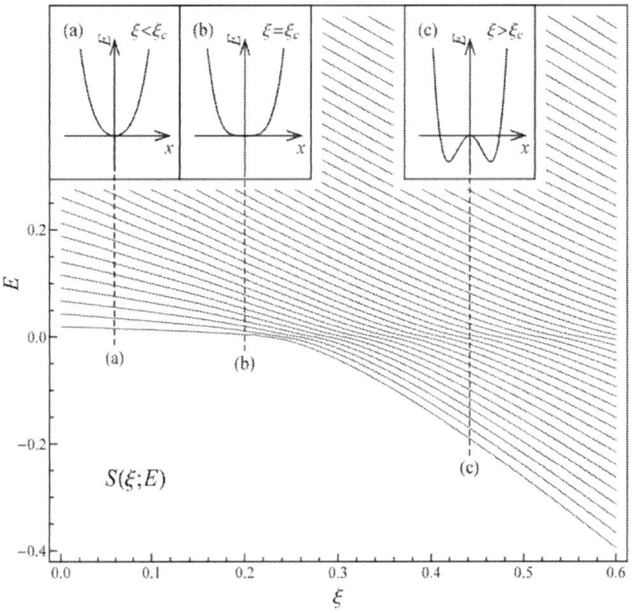

Figure 1. Contour plot of the classical action $S(\xi; E)$ for the classical Hamiltonian (17), through the different regimes determined by the shape of the potential energy function, which is shown for (a) $\xi < \xi_c$, (b) $\xi = \xi_c$, and (c) $\xi > \xi_c$. The individual contours are related semi-classically to the evolution of the level eigenvalues. Figure from Ref. [8].

At the top of the barrier in (c), the action has a logarithmic singularity, leading semi-classically to an infinite level density (ESQPT). The quantity $(\partial S/\partial E)_\xi$ diverges at $E = 0$.

The quantum Hamiltonian H can be diagonalized numerically for finite N. The evolution of the eigenvalues as a function of ξ is called *correlation diagram*, shown in figure 2. It can be seen that features (discontinuities when $N \to \infty$) appear by traversing the correlation diagram both in the horizontal direction, as a function of the control parameter ξ, QPT, and in the vertical direction as a function of excitation energy, ESQPT. An important property of excited states is that, at a critical value of a QPT, $\xi = \xi_c$, all excited states collapse to zero energy. For finite N, a gap develops. This gap scales also with a power law $\Delta = A_{\Delta 0} N^{-A_{\Delta 1}}$. The scaling exponent can be obtained either numerically or analytically.

4. Quantum phase transitions in nuclei: the Interacting Boson Model (IBM)

The interacting boson model (IBM) is a model of even-even nuclei in terms of correlated pairs of nucleons treated as bosons with $L = 0$ (s-boson) and $L = 2$ (d-boson) [2]. (S- and D-wave pairing in 3D).

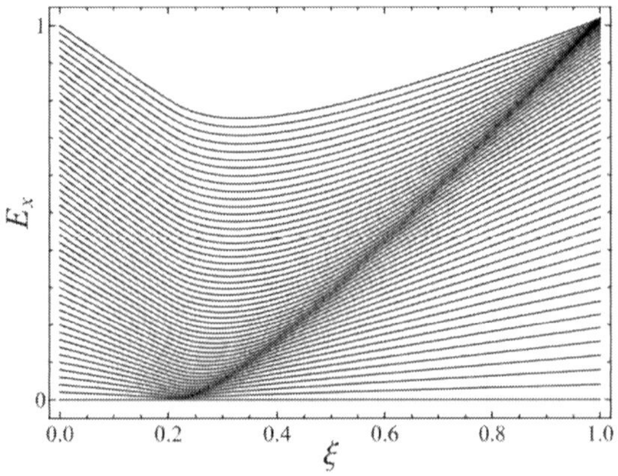

Figure 2. Correlation diagram for the two-level pairing model, illustrating the compression of level density associated with an ESQPT. Figure from Ref. [8].

4.1. Algebraic structure
The s and d bosons span a six-dimensional space with algebraic structure U(6) and dynamical symmetries

$$\begin{aligned} &U(6) \supset U(5) \supset SO(5) \supset SO(3) \supset SO(2) \quad (I)\,, \\ &U(6) \supset SU(3) \supset SO(3) \supset SO(2) \quad (II)\,, \\ &U(6) \supset SO(6) \supset SO(5) \supset SO(3) \supset SO(2) \quad (III)\,. \end{aligned} \tag{18}$$

The energy eigenvalues are

$$\begin{aligned} E(N, n_d, v, n_\Delta, L, M_L) &= E_0 + \alpha n_d(n_d+4) + \beta v(v+3) + \gamma L(L+1)\,, \\ E(N, \lambda, \mu, K, L, M_L) &= E_0 + \kappa\left(\lambda^2 + \mu^2 + \lambda\mu + 3\lambda + 3\mu\right) + \kappa' L(L+1)\,, \\ E(N, \sigma, \tau, \nu_\Delta, L, M_L) &= E_0 + A\sigma(\sigma+4) + B\tau(\tau+3) + CL(L+1)\,. \end{aligned} \tag{19}$$

4.2. Phase stucture and phase diagram
The essential Hamiltonian of IBM is

$$H = \varepsilon\left\{C_1[U(5)] + \xi_1 C_2[SU(3)] + \xi_2 C_2[SO(6)]\right\}, \tag{20}$$

where $C_p(g)$ is the Casimir operator of order p of g. The IBM has three structural phases, Eq.(18), and its phase diagram is two-dimensional. A convenient (physical) parametrization of H is

$$H = \left(\frac{\varepsilon_0}{N}\right)\left[(1-\xi)\,\hat{n}_d - \frac{\xi}{4N}\hat{Q}^\chi \cdot \hat{Q}^\chi\right]. \tag{21}$$

When $\xi = 0$ one has U(5) symmetry, when $\xi = 1, \chi = 0$ one has SO(6) symmetry and when $\xi = 1, \chi = \mp\sqrt{7}/2$, one has SU(3) symmetry.

The coset space of IBM is U(6)/U(5)⊗U(1), with five complex variables $\alpha_\mu (\mu = 0, \pm1, \pm2)$. The number projected coherent state is

$$|N, \alpha_\mu\rangle = \left(s^\dagger + \sum_\mu \alpha_\mu d_\mu^\dagger\right)^N |0\rangle\,. \tag{22}$$

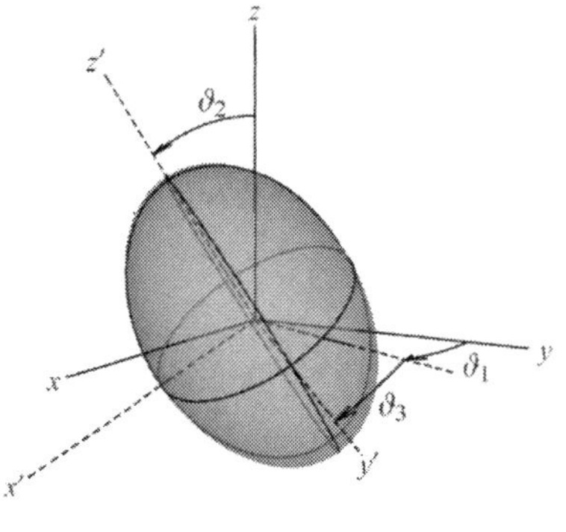

Figure 3. Liquid drop model of the nucleus with ellipsoidal shape. Intrinsic axis orientations are described by Euler angles ϑ_i, and axis lengths are determined by β and γ.

The variables α_μ can be transformed to Bohr variables (β, γ) and Euler angles $(\theta_1, \theta_2, \theta_3)$ by rotation. This is motivated by the liquid drop model of nuclei, shown in figure 3. The intrinsic state can be written in terms of Bohr variables as

$$|N;\beta,\gamma\rangle = \left(s^\dagger + \beta\left[\cos\gamma d_0^\dagger + \frac{\sin\gamma}{\sqrt{2}}\left(d_{+2}^\dagger + d_{-2}^\dagger\right)\right]\right)^N |0\rangle . \tag{23}$$

and the potential function is

$$V(\xi,\chi;\beta,\gamma) = \frac{(1-\xi)\,\beta^2}{1+\beta^2} - \frac{\xi\left[4\beta^2 - 4\sqrt{2/7}\chi\beta^3\cos 3\gamma + 2\chi^2\beta^4/7\right]}{(1+\beta^2)^2}. \tag{24}$$

The classical order parameters are the equilibrium deformations, β_e, γ_e. A study of V_e and its derivatives shows that there is in the phase diagram a line of 1st order transitions ending in a point of second order, as shown in figure 4 [12].

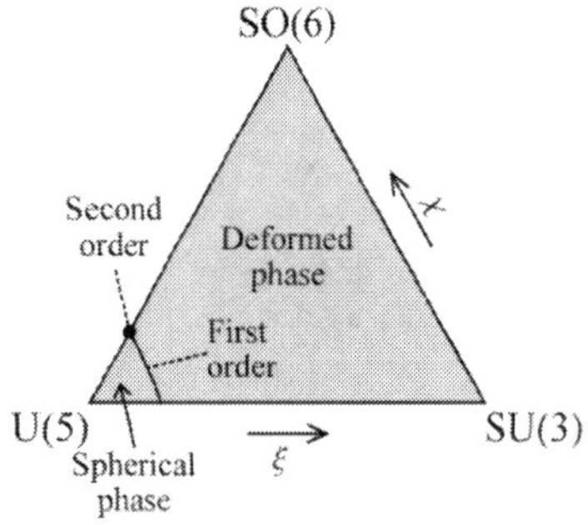

Figure 4. Phase diagram of the IBM. A curve of first-order QPT $[\xi_c(\chi) = (5 + \frac{1}{7}\chi^2)^{-1}]$ and point of second-order QPT separate spherical and deformed phases.

A quantal analysis of QPTs in IBM has been done, showing the smoothing out of critical properties. The order parameter ω is shown in figure 5 [13]. The ground state energy, E_0, and its derivatives are shown in figure 6 [14].

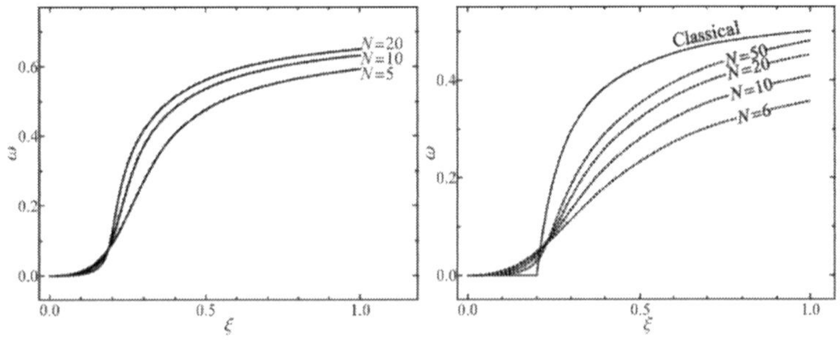

Figure 5. Quantum calculations of the order parameter ω, for the interacting boson model. (Right) The U(5)-SO(6) transition. (Left) The U(5)-SU(3) transition. Calculations are shown for various finite N.

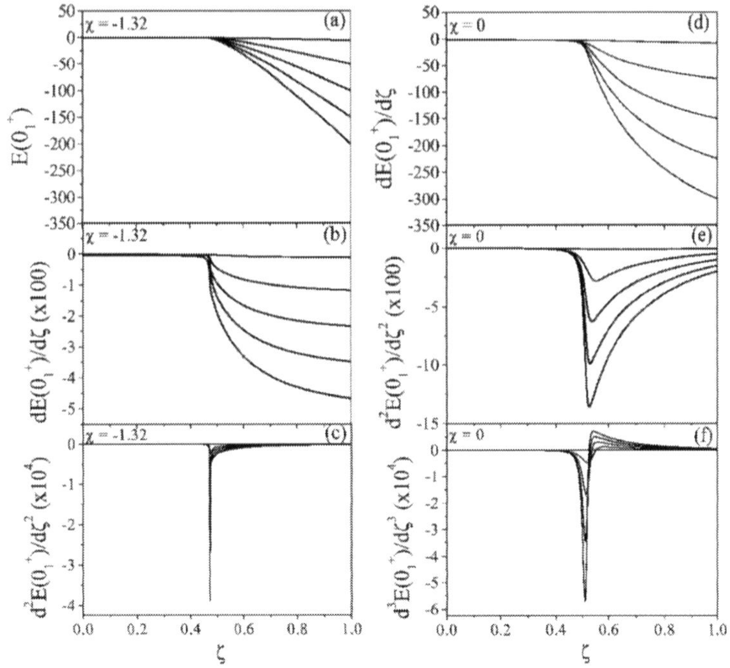

Figure 6. The ground state energy $E(0_1^+)$ and its derivatives for the IBM, across first-order U(5)-SU(3) (left) and second-order U(5)-SO(6) (right) QPTs, at finite N. The lines shown are for N=10,100, 200, 300, and 400. The control parameter used in these plots is $\zeta \equiv 4\xi/(1+3\xi)$. Thus, the critical point is at $\zeta_c \approx 0.47$ at left or $\zeta_c = 1/2$ at right. Figure from Ref. [14].

Scaling exponents for 2nd order transitions can be obtained either analytically or numerically [15]. Writing for any quantity $\Phi(\xi_c) = A_0 N^{-A_1}$ one obtains in the U(5)-SO(6) transition, for the ground state energy E_0, $A_1 = 4/3$, for the order parameter $\langle n_d \rangle$, $A_1 = -1/3$, for the gap Δ, $A_1 = 1/3$, and for the operator $T = B(E2)/5N$, $A_1 = -4/3$. These analytic values have been verified numerically as shown in figure 7. [For first order transitions a complete theory of scaling is not yet available].

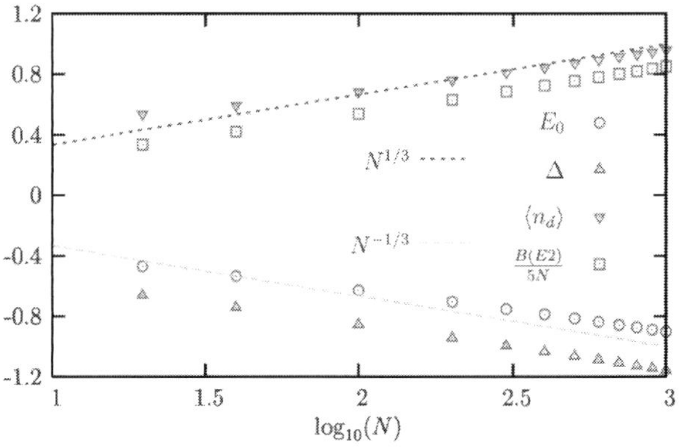

Figure 7. Log-log plot showing the finite size scaling of several calculated quantities (points) for the interacting boson model and the expected power law behavior (guide lines). Figure from Ref. [15].

4.3. Experimental evidence

In the quantal Hamiltonian of the IBM, $H = \varepsilon \hat{n}_d - \kappa \hat{Q}^\chi \cdot \hat{Q}^\chi$, the parameters ε and κ depend on proton and neutron number. It turns out that the control parameter depends (almost) linearly on proton and neutron number. Three quantities have been experimentally investigated: (a) The ground state energy, E_0. This can be measured easily. It is useful to plot the differences $S_{2n}(N) = E_0(N+1) - E_0(N) \propto \frac{\partial E_0}{\partial \xi}$. (b) The order parameter ω. This is difficult to measure, but its square, $B(E2; 2_1 \to 0_1) \propto \langle n_d \rangle^2$, can be measured easily. (c) The gap $\Delta = E(0_2) - E(0_1)$ easily measured. These three quantities are shown in figure 8. It is seen that critical behavior (precursors of QPT) is observed even for small values of N.

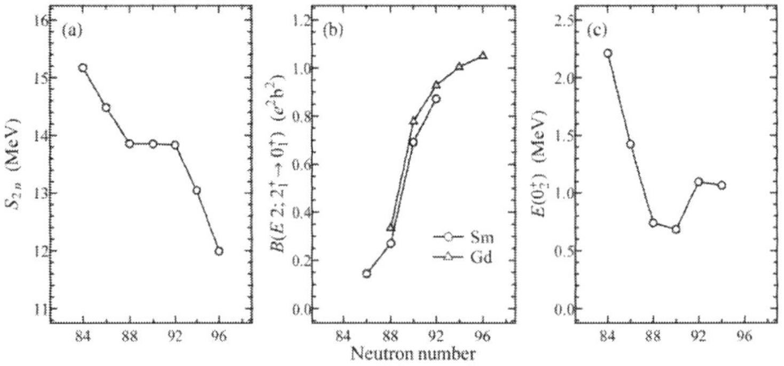

Figure 8. Experimental data for Sm isotopes (circles) and Gd isotopes (triangles): (a) two-neutron separation energy S_{2n}, (b) electromagnetic transition strength $B(E2; 2_1^+ \to 0_1^+)$, and (c) energy of the lowest $J = 0$ excitation $E(0_2^+)$.

5. Complex situations

In recent years, QPTs in systems more complex that a single bosonic system have been investigated, especially for coupled systems. Complete results are available for: (A) Quantum phase transitions in two-fluid bosonic systems [16], [17]. (B) Quantum phase transitions in two-level pairing models (bosonic and fermionic) [18]. The algebra g is in these cases, $g = U(n_1) \otimes U(n_2)$. The phase diagrams for these systems are more complicated, as there are more phases and more control parameters. Investigations have started for (C) Quantum phase transitions in mixed Bose-Fermi systems [19], [20], [21]. Algebraic methods are particularly useful in these cases. The characterization of the phases by means of dynamic symmetries, the branching of algebras, and the geometry of these algebraic structures are of such a complexity that require full exploitation of the mathematical techniques discussed above.

6. Conclusions

The study of QPTs and ESQPTs in s-b boson models is essentially completed. The only remaining areas for further work are: (i) the study of finite-size scaling and of (ii) level densities in 1st order transitions.

The class of models discussed here is similar to models in condensed matter physics, for example the quantum Ising model, with Hamiltonian, $H = J\left(-g\sum_i \sigma_i^x - \sum_{ij} \sigma_i^z \sigma_j^z\right)$. This model has the same structure of s-b models with algebra SU(2). The difference is that in condensed matter systems the spins are on a lattice, while in nuclear and molecular systems the bosons are in free 3D or 2D space.

Important aspects of algebraic models are: (i) a rigorous mathematical definition of phases; (ii) the possibility to do semi-classical dynamics; (iii) easy computation of critical values of the control parameters, critical exponents, etc.; and (iv) simple study of finite size scaling. Future work in QPT/ESQPT includes a generalization to multi-level mesoscopic systems and a generalization to infinite systems.

Acknowledgments

This work was supported in part by US Department of Energy Grant DE-FG02-91ER-40608.

References

[1] Iachello F 2006 *Lie algebras and applications* (*Lecture Notes in Physics* Vol 708) (Berlin: Springer)
[2] Iachello F and Arima A 1987 *The Interacting Boson Model* (Cambridge: Cambridge University Press)
[3] Iachello F and Levine R D 1995 *Algebraic Theory of Molecules* (Oxford: Oxford University Press)
[4] Iachello F and van Isacker P 1991 *The Interacting Boson Fermion Model* (Cambridge: Cambridge University Press)
[5] Frank A, Lemus R and Iachello F 1983 *J. Chem. Phys.* **91** 29
[6] Gilmore R 1979 *J. Math. Phys.* **20** 891
[7] Carr L D ed *Understanding Quantum Phase Transitions* (Boca Raton, FL: CRC Press)
[8] Caprio M A, Cejnar P and Iachello F 2008 *Ann. Phys.* **323** 1106
[9] Cejnar P and Iachello F 2007 *J. Phys. A: Math. Theor.* **40** 581
[10] Iachello F and Caprio M A 2010 in *Understanding Quantum Phase Transitions* (Boca Raton, FL: CRC Press) chapter 27
[11] Fisher M A and Barber M N 1972 *Phys. Rev. Lett.* **28** 1516
[12] Feng D S, Gilmore R and Deans S R 1981 *Phys. Rev. C* **23** 1254
[13] Iachello F and Zamfir N V 2004 *Phys. Rev. Lett.* **92** 21501
[14] Williams E T 2009 Ph D Thesis Yale University
[15] Dusuel S, Vidal J, Arias J M, Dukelsky J and Garcia-Ramos J E 2005 *Phys. Rev. C* **72** 011301 (R)
[16] Caprio M A and Iachello F 2004 *Phys. Rev. Lett.* **93** 242502
[17] Arias J M, Dukelsky J and Garcia-Ramos J E 2004 *Phys. Rev. Lett.* **93** 212501
[18] Caprio M A, Skrabacz J H and Iachello F 2011 *J. Phys. A: Math. Theor.* **44** 075303
[19] Jolie J, Heinze S, Van Isacker P and Casten R F 2004 *Phys. Rev. C* **70** 011305 (R)
[20] Alonso C E, Arias J M, Fortunato L and Vitturi A 2005 *Phys. Rev. C* **72** 061302
[21] Petrellis D, Leviatan A and Iachello F 2011 *Ann. Phys.* **326** 926

Advanced many-body and statistical methods in mesoscopic systems IOP Publishing
Journal of Physics: Conference Series **338** (2012) 012026 doi:10.1088/1742-6596/338/1/012026

Strong dissipation regime in nuclear disintegrations

Mihail Mirea
Horia Hulubei National Institute for Physics and Nuclear Engineering, P.O. Box Mg-6, Bucharest, Romania
E-mail: mirea@ifin.nipne.ro

Abstract. The coupling of collective degrees of freedom with the microscopic ones causes dissipation and a modification of the adiabatic potential. The term dissipation usually refers to exchange of energy (either linear or angular momentum) by all kind of damping from collective motion to intrinsic heat. A measure of the dissipated energy can be obtained solving the time dependent pairing equations. In this presentation, a generalization of the time dependent pairing equations is presented by including the Landau-Zener effect in the superfluid model. These new equations allows a mixing of seniority one configurations that allows us to obtain a ground state at the end of the process. An application concerning the ^{14}C emission is offered and its fine structure is explained. These new equations are furthermore used to evidence a dynamical pair breaking effect that could explain the fine odd-even effect in cold fission. Finally, the time dependent pairing equations are used to deduce a model for non-adiabatic cranking inertia.

1. Introduction

In a mean field approximation, the equations of motions take into account the two-body collisions to the extend to which the two-body collisions are incorporated in the mean field. In principle, the time-dependent HF approximation exactly treats the residual interactions only if the mean field is allowed to break all symmetries. Such an approach leads to huge computational difficulties and usually the mean field is constrained to have some symmetries. In such circumstances, two levels associated to the same symmetry cannot intersect and the level slippage is not allowed. That leads to the unpleasant feature that at the end of a nuclear process the systems cannot reach the ground state. In order to overcome this problem new approaches were developed leading to the extended mean field model [1] and to the stochastic HF equations, discussed in detail in Refs. [2, 3]. Alternatively, this problem was partially solved by taking into account the pairing interaction. The Time-Dependent Pairing Equations (TDPE) provide the possibility for the nucleon pair level slippage [4] and represent a tool to estimate the dissipation [5]. The dissipation is identified as the excitation derived from coupled channel equations, as in Refs. [6, 7, 8]. In Ref. [9] the dissipation was calculated for fission processes. For closed shell regions of fission fragments and for symmetric fission the results agree with experimental data. A deep connection with the Landau-Zener transitions is included in the time dependent pairing equations: pairs undergo Landau-Zener transitions on virtual levels with coupling strengths given by the magnitude of the gap. A problem appears in solving these equations in the case of seniority one nuclear systems: the unpaired nucleon located on the blocked level will remain on the same level during the deformation of the system. Due to this phenomenon, the system cannot end in its ground state after a disintegration. In this presentation, the first part addresses a generalization of the time dependent pairing equations by including the Landau-Zener effect in the superfluid model [10, 11]. These new equations allows a mixing of seniority one configurations that allows us to obtain a ground state at the end of the process. An application concerning the ^{14}C emission is offered and its fine structure is explained. Furthermore, the same formalism is used to evidence a dynamical pair breaking effect that could explain the fine odd-even effect in cold fission [12]. Finally, the time dependent pairing equations are used to deduce a model for non-adiabatic cranking inertia [13].

2. The Landau-Zener effect

Single particle levels are functions of the deformation parameters that define the mean field. Levels characterized by the same quantum numbers associated with the same symmetries of the system cannot intersect and exhibit avoided level crossing regions. The transition probability of a nucleon from one adiabatic level to another is strongly enhanced in an avoided crossing region. This promotion mechanism is known as the Landau-Zener effect [14]. Using the time-dependent Schrödinger equation, the following system of coupled channel equations is obtained [15]:

$$\dot{c}_j = \frac{1}{i\hbar}\sum_{j=1}^{n} c_j(t)h_{ij}(t)\exp\left(-\frac{1}{\hbar}\int_0^t(\epsilon_j(\tau)-\epsilon_i(\tau))d\tau\right) \tag{1}$$

where c_j is the amplitude of the diabatic state j in the total wave function, ϵ_j is its single particle energy and h_{ij} is the interaction energy in an avoided levels crossing region between two state i and j. The probability to find the the unpaired nucleon on the level i is $p_i =\mid c_i\mid^2$. These equations have already been used to explain the resonant fine structure in the fission cross section [16, 17, 18], the resonant-like structure of the inelastic cross section in heavy ion collisions [19, 20], the fine structure in cluster emission [21, 22] and in α decay [23].

3. Generalization of the Landau-Zener effect in superfluid systems

An effect analogous to the Landau-Zener one can be obtained by generalizing the TDPE for the case of seniority one nuclear systems. The problem will be explored in the simplest possible way: a monopole pairing force, and a sufficiently weak pairing such that the nucleons are not redistributed to change significantly the mean field potential. Using quasiparticle creation and annihilation operators $\alpha_{\bar{k}}^+$, and $\alpha_{\bar{k}}$

$$\begin{array}{ll} \alpha_{k(j)} = u_{k(j)}a_k - v_{k(j)}a_{\bar{k}}^+; & \alpha_{\bar{k}(j)} = u_{k(j)}a_{\bar{k}} + v_{k(j)}a_k^+; \\ \alpha_{k(j)}^+ = u_{k(j)}a_k^+ - v_{k(j)}^*a_{\bar{k}}; & \alpha_{\bar{k}(j)}^+ = u_{k(j)}a_{\bar{k}}^+ + v_{k(j)}^*a_k; \end{array} \tag{2}$$

it is possible to construct some interactions that help us to promote the nucleon from one diabatic level to another. Note that these operators are associated to each blocked level j. Here a_k^+ and a_k denote operators for creating and destroying a particle in the state k, respectively. The state characterized by a bar signifies the time-reversed partner of a pair. The parameters v_k and u_k are the occupation and vacancy amplitudes, respectively. Because only the relative phase between the parameters u_k and v_k matters, in the following, u_k is considered to be a real quantity and v_k a complex one. The two situations plotted in Fig. 1 can be modeled. In the plot 1(a), the single particle follows the diabatic level ϵ_j while in 1 (b) it remains on the adiabatic one e_i. The interaction able to promote the unpaired nucleon from one adiabatic level to another must be given by some products of operators of the type (2).

In order to obtain the equations of motion, we shall start from the variational principle taking the following energy functional

$$\delta L = \delta <\varphi\mid H - i\hbar\frac{\partial}{\partial t} + H' \mid\varphi>, \tag{3}$$

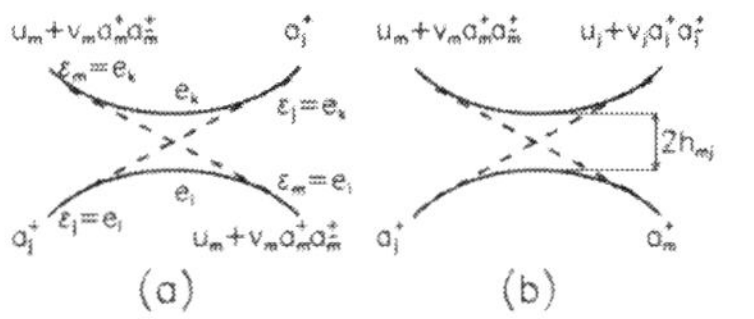

Figure 1. Ideal avoided level crossing region and possible transition states for an unpaired nucleon in the superfluid model. (a) The single particle follows the diabatic state. (b) The single particle remains on the same adiabatic level. Figure from Ref. [11].

and assuming the many-body state formally expanded as a superposition of n time dependent BCS seniority one diabatic wave functions

$$\mid \varphi(t) >= \sum_{m}^{n} c_m(t) a_m^+ \prod_{l \neq m} \left(u_{l(m)}(t) + v_{l(m)}(t) a_l^+ a_{\bar{l}}^+ \right) \mid 0 > . \quad (4)$$

Sometimes, this energy functional is called Lagrangian [4]. The functional contains several terms. The first one is the many body Hamiltonian with pairing residual interactions

$$H(t) = \sum_{k>0} \epsilon_k[q(t)](a_k^+ a_k + a_{\bar{k}}^+ a_{\bar{k}}) - G \sum_{k,i>0} a_k^+ a_{\bar{k}}^+ a_i a_{\bar{i}}. \quad (5)$$

The residual interactions between diabatic levels characterized by the same quantum numbers that are responsible for the Landau-Zener effect are assumed on the form:

$$\begin{aligned} H'(t) &= \textstyle\sum_{i,j\neq i}^{n} h_{ij}(t) \alpha_{i(j)}^+ \alpha_{j(i)} \prod_{k \neq i,j} \alpha_{k(j)} a_k^+ a_k \alpha_{k(i)}^+ \\ &= \textstyle\sum_{i,j\neq j}^{n} h_{ij}(t) (u_{i(j)} a_i^+ - v_{i(j)}^* a_{\bar{i}}) (u_{j(i)} a_j + v_{j(i)} a_{\bar{j}}^+) \prod_{k \neq i,j} \alpha_{k(j)} a_k^+ a_k \alpha_{k(i)}^+, \end{aligned} \quad (6)$$

The sum runs over diabatic levels i and j.

After some calculations, as detailed in Ref. [11], a new set of pairing equations which account for configuration mixing results:

$$i\hbar \dot{\rho}_{l(m)} = \kappa_{l(m)} \Delta_m^* - \kappa_{l(m)}^* \Delta_m, \quad (7)$$

$$i\hbar \dot{\kappa}_{l(m)} = \left(2\rho_{l(m)} - 1 \right) \Delta_m + 2\kappa_{l(m)} \epsilon_l - 2G \rho_{l(m)} \kappa_{l(m)}, \quad (8)$$

$$i\hbar \dot{p}_m = \sum_{j \neq m}^{n} h_{mj} \left(S_{mj} - S_{jm} \right), \quad (9)$$

$$\begin{aligned} i\hbar \dot{S}_{jm} = S_{jm} \Big\{ & -\tfrac{1}{G} \left(\mid \Delta_m \mid^2 - \mid \Delta_j \mid^2 \right) + \left(\epsilon_m(t) - \epsilon_j(t) \right) \\ & + G \big(\textstyle\sum_{k \neq m} \rho_{k(m)}^2 - \sum_{k \neq j} \rho_{k(j)}^2 \big) \\ & - \tfrac{1}{2} \textstyle\sum_{k \neq m} \left(\Delta_m \kappa_{k(m)}^* + \Delta_m^* \kappa_{k(m)} \right) \left(\frac{\rho_{k(m)}^2}{|\kappa_{k(m)}|^2} - 1 \right) \\ & + \tfrac{1}{2} \textstyle\sum_{k \neq j} \left(\Delta_j \kappa_{k(j)}^* + \Delta_j^* \kappa_{k(j)} \right) \left(\frac{\rho_{k(j)}^2}{|\kappa_{k(j)}|^2} - 1 \right) \Big\} \\ & + \textstyle\sum_{l \neq m,j}^{n} \left[h_{ml}(t) S_{jl} - h_{jl}(t) S_{lm} \right] + h_{mj}(t) (p_j - p_m). \end{aligned} \quad (10)$$

Here, the following notations are used:

$$\begin{array}{rl} \Delta_m = G\sum_{k\neq m}\kappa_k; & \Delta_m^* = G\sum_{k\neq m}\kappa_k^*; \\ \kappa_k = u_k v_k; & \rho_k =\mid v_k \mid^2 \\ p_m =\mid c_m \mid^2; & S_{jm} = c_j^* c_m. \end{array} \tag{11}$$

where ρ_k are the single particle densities, κ_k are the pairing moment components, and p_m denotes the probability to have an unpaired nucleon on the level m. ρ_k and p_m are real quantities while κ_k and S_{jm} are complex ones. In analogy with the pairing moment components κ_k, S_{jm} can be called as unpairing moment components, having the property $\mid S_{jm} \mid^2 = p_j p_m$. Whenever the upper limit n is specified for a sum, it is implicitly assumed that the operation is realized on the n possible diabatic states of the unpaired nucleon. In this paper, the sum over pairs generally runs over the index k. When the single-particle sum over k is realized only for one partner of each reversed pair the result is multiplied with a factor 2. The index k runs over a workspace that allows the pairing force to operate only within a finite number of active levels around the Fermi energy. If the blocked levels are eliminated, the system (7)-(8) reduces to:

$$\begin{array}{c} i\hbar\dot{\rho}_l = \kappa_l\Delta^* - \kappa_l^*\Delta, \\ i\hbar\dot{\kappa}_l = (2\rho_l - 1)\Delta - 2\kappa_l\left[\epsilon_l(t) - \lambda(t) - 2G\rho_l\kappa_l\right], \end{array} \tag{12}$$

that is, the well known TDPE [4, 5]. The label m is removed from Δ_m because this quantity is now a sum over the remaining pairwise occupied levels. On the other hand if the pairing is neglected, the third equation of the system (9) can be written

$$i\hbar\left(\dot{c}_m c_m^* + \dot{c}_m^* c_m\right) = \sum_{j\neq m}^{n} h_{mj}\left(c_j c_m^* + c_j^* c_m\right). \tag{13}$$

Introducing explicitly the time dependence of the amplitudes c_m

$$c_m(t) = c_{0m}(t)\exp\left(-\frac{i}{\hbar}\int_0^t \epsilon_m(\tau)d\tau\right), \tag{14}$$

the next relation is obtained:

$$\begin{array}{c} i\hbar(\dot{c}_{0m}c_{0m}^* + \dot{c}_{0m}^* c_{0m}) = \sum_{j\neq m}^{n} h_{jm}\left[c_{0j}c_{0m}^*\exp\left(-\frac{i}{\hbar}\int_0^t(\epsilon_j - \epsilon_m)d\tau\right)\right. \\ \left. - \, c_{0j}^* c_{0m}\exp\left(\frac{i}{\hbar}\int_0^t(\epsilon_j - \epsilon_m)d\tau\right)\right]. \end{array} \tag{15}$$

The last relation gives also an equivalent form of the Landau-Zener equations for single particle systems. So, the Landau-Zener equation for single particle systems (without residual interactions) and the TDPE for quasiparticles are two particular cases of the coupled channel equations (7)-(10). So, this system represents a generalization of the TDPE in the case of seniority one nuclear systems.

4. Single particle model

To solve the TDPE (similar to TDHFB equations), only the variations of the single-particle energies ϵ_k are needed. The simplest way to obtain the evolutions of single-particle energies is to consider a time-dependent single particle potential in which the nucleons move independently. The basis [24, 25] for the two-center oscillators can be used for various ranges of models which are more or less phenomenological ones [26, 27]. On the other hand, they are different ways to obtain the single-particle energies for a two-center Woods-Saxon potential. Other recipes are given in Ref. [28] where the potentials are expanded in terms of harmonic oscillators functions. A microscopic potential must be constructed to be consistent within our nuclear shape parametrization. The simplest way it to use a semi-phenomenological Woods-Saxon potential. In order to take into account nuclear deformations going over to separate shapes and obtain two separated fragments, a two-center shell model with a Woods-Saxon potential was used:

$$V_0(\rho,z) = -\frac{V_c}{1+\exp\left[\frac{\Delta(\rho,z)}{a}\right]} \tag{16}$$

where $\Delta(\rho,z)$ represents the distance between a point (ρ,z) and the nuclear surface. This distance is measured only along the normal direction on the surface and it is negative if the point (ρ,z) is located in the interior of the nucleus. V_c is the depth of the potential while a is the diffuseness parameter. In our work, the depth is $V_c = V_{0c}[1 \pm \kappa(N_0 - Z_0)/N_0 + Z_0)]$ with plus sign for protons and minus sign for neutrons, V_{0c}= 51 MeV, a=0.67 fm, κ=0.67. Here A_0, N_0 and Z_0 represent the mass number, the neutron number and the charge number of the parent, respectively. This parametrization, referred as the Blomqvist-Walhlborn one, is adopted because it provides the same radius constant r_0 for the mean field and the pairing field. That ensures a consistency of the shapes of the two fields at hyperdeformations, i.e., two tangent ellipsoids. The Hamiltonian is obtained by adding the spin-orbit and the Coulomb terms to the Woods-Saxon potential. The eigenvalues are obtained by diagonalization of the Hamiltonian in the semi-symmetric harmonic two center basis. The two center shell model represents a valuable instrument to investigate the role of individual orbitals for the treatment of a wide variety of nuclear processes. like cold fission [29, 30, 31, 32], formation of superheavy elements [33, 34] or superasymmetric disintegration processes, pertaining to cluster- [35, 36] and alpha-decays [23].

5. Cluster radioactivity

A fine structure in the ^{14}C radioactivity of the ^{223}Ra was observed in 1989 [37]. In the first experiment, the results indicate that 15±3 % of ^{14}C decays are transitions on the ground state of the daughter, while 81±6% are transitions on the first excited state. In Ref. [38], using the M3Y potential, it was evidenced that the preformation probability must be more favorable for the excited state than for the ground state with a factor of 180. Such a value cannot be accounted from theoretical models [39] without taking into account for dynamical ingredients. This is the main reason why the fine structure phenomenon was selected to validate our equations.

In order to solve the time dependent pairing equations, the single-particle level schemes for neutrons and protons must be computed along the minimal action trajectory. It is known that ^{223}Ra has the spin $\frac{3}{2}$ emerging from $1i_{11/2}$. Adiabatically, the unpaired neutron reaches the $2g_{9/2}$ level of the daughter ^{209}Pb. As also evidenced in Refs. [15, 22], the fine structure in the ^{14}C radioactivity can be understood by an enhanced transition probability of the unpaired neutron from the adiabatic level $\Omega = 3/2$ emerging from $1i_{11/2}$ to the adiabatic level with the same spin projection Ω that emerges from $1j_{15/2}$, in terms of the Landau-Zener effect. The involved levels were selected and the time dependent pairing equations were solved as indicated in Ref. [11]. The branching ratio r between the partial half-life for transitions to the ground state of the daughter and the partial half-live to the first excited state was calculated. The experimental values of r range between 5.4 and 5.9. Our theoretical value is r=5.48, which is in an excellent agreement with experimental data.

6. New dynamical pair breaking effect

The dynamical pair breaking effect emerges from a new set of coupled channel equations deduced for the time-dependent probability to find the system in a seniority-one state or in a seniority-two one. In such an approximation, if the interactions produced in the avoided crossing regions or those due to the Coriolis coupling are not taken into consideration, other perturbations between

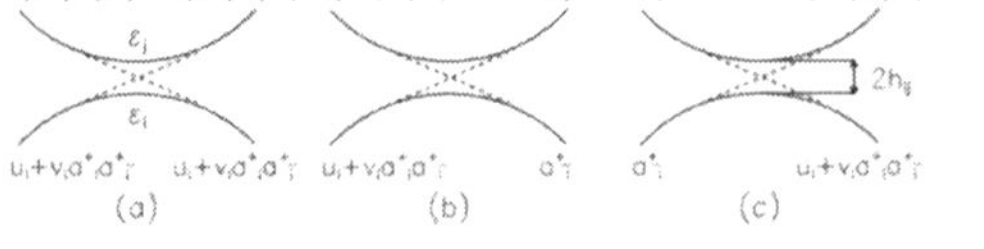

Figure 2. Ideal avoided crossing regions between two adiabatic levels ϵ_i and ϵ_j. Three possible transitions between configurations in an avoided crossing region in the superfluid model are displayed: (a) The configuration remains unchanged after the passage through the avoided crossing region. (b) A pair is broken. (c) A pair is created. Figure from Ref. [12].

two different states are not possible. In the following, the calculations are restricted only for perturbations produced in the avoided crossing regions. This approximation does not affect the essential features of the model but leads to a considerable simplification of the mathematical apparatus.

Using quasiparticle annihilation and creation operators, it is possible to construct some interactions able to break a Cooper pair when the system traverses a avoided crossing region. The three situations plotted in Fig. 2 can be modeled within products of such creation and annihilation operators acting on Bogoliubov wave functions. In the plot 2(a), the Cooper pairs remain on the adiabatic levels ϵ_i and ϵ_j after the passage through the avoided crossing region, in Fig. 2 (b) the pair destruction is illustrated, while in Fig. 2 (c) two fermions generate a pair after the passage through an avoided crossing region. Formally, to describe these three situations, an interaction in the avoided crossing can be postulated as follows:

$$H'(t) = \sum_{i,j\neq i}^{n} h_{ij}[q(t)] \left[\alpha_{i(0)}\alpha_{\bar{j}(0)} \prod_{k\neq i,j} \alpha_{k(0)} a_k^+ a_k \alpha_{k(ij)}^+ + \alpha_{i(0)}^+ \alpha_{\bar{j}(0)}^+ \prod_{k\neq i,j} \alpha_{k(ij)} a_k^+ a_k \alpha_{k(0)}^+ \right], \quad (17)$$

where h_{ij} is the interaction between the levels. Acting on a suited Bogoliubov wave function, the product over k transforms the seniority-two configuration in the seniority-zero one in the case of the first term in the left hand of Eq. (17), and vice-versa in the case of the second term. If the product $\alpha_{i(0)}\alpha_{\bar{j}(0)}$ acts on a seniority-zero function, then it annihilates a pair and creates two unpaired fermions in states i and $\bar{j}$. If the product $\alpha_{i(0)}^+\alpha_{\bar{j}(0)}^+$ acts on a seniority-two wave function, then it creates a pair distributed on both orbitals i and j. In order to obtain the equations of motion, we shall start from the variational principle taking the energy functional (3) and by assuming the many-body state formally expanded as a superposition of time dependent BCS seniority-zero and seniority-two adiabatic wave functions

$$\begin{aligned} |\varphi(t)> &= \left[c_0(t) \prod_k \left(u_{k(0)}(t) + v_{k(0)}(t) a_k^+ a_{\bar{k}}^+ \right) \right. \\ &\left. + \sum_{j,l\neq j} c_{jl}(t) a_j^+ a_{\bar{l}}^+ \times \prod_{k\neq j,l} \left(u_{k(jl)}(t) + v_{k(jl)}(t) a_k^+ a_{\bar{k}}^+ \right) \right] |0>, \end{aligned} \quad (18)$$

where c_0 and c_{jl} are amplitudes of the two kind of configurations. To minimize, the energy functional is derived with respect the independent variables $v_{k(0)}$, $v_{k(jl)}$, c_0, c_{jl}, together with their complex conjugates, and the resulting equations are set to zero. Eventually, eight coupled-channel equations are obtained:

$$i\hbar\dot{\rho}_{k(0)} = \kappa_{k(0)}\Delta_0^* - \kappa_{k(0)}^*\Delta_0;$$

$$
\begin{aligned}
i\hbar\dot{\rho}_{k(\gamma)} &= \kappa_{k(\gamma)}\Delta_\gamma^* - \kappa_{k(\gamma)}^*\Delta_\gamma; \\
i\hbar\dot{\kappa}_{k(0)} &= \left(2\rho_{k(0)} - 1\right)\Delta_0 + 2\kappa_{k(0)}\epsilon_k - 2G\rho_{k(0)}\kappa_{k(0)}; \\
i\hbar\dot{\kappa}_{k(\gamma)} &= \left(2\rho_{k(\gamma)} - 1\right)\Delta_\gamma + 2\kappa_{k(\gamma)}\epsilon_k - 2G\rho_{k(\gamma)}\kappa_{k(\gamma)}; \\
i\hbar\dot{P}_0 &= \sum_\gamma h_\gamma(S_{0\gamma}^* - S_{0\gamma}); \\
i\hbar\dot{P}_\gamma &= h_\gamma(S_{0\gamma} - S_{0\gamma}^*); \\
i\hbar\dot{S}_{0\gamma} &= S_{0\gamma}(E_0 - E_\gamma) + S_{0\gamma}\left(\sum_{k\neq\gamma} T_{k(\gamma)} - \sum_k T_{k(0)}\right) + \sum_\beta h_\beta S_{\beta\gamma} + h_\gamma(P_\gamma - P_0); \\
i\hbar\dot{S}_{\beta\gamma} &= S_{\beta\gamma}(E_\beta - E_\gamma) + S_{\beta\gamma}\left(\sum_{k\neq\beta} T_{k(\beta)} - \sum_{k\neq\gamma} T_{k(\gamma)}\right) + h_\beta S_{0\gamma} - h_\gamma S_{0\beta}^*;
\end{aligned}
\tag{19}
$$

where the partial derivatives with respect the time are denoted by a dot. If (γ) denotes the indexes (j,l) and (β) the indexes (m,n), the sums are restricted by the conditions $j \neq l$, $m \neq n$, $m \neq j$, and $n \neq l$. E_γ are exactly the expected values of the Hamiltonian (5) for the seniority-zero or seniority-two configurations:

$$
\begin{aligned}
E_0 &= 2\sum_k \rho_{k(0)}\epsilon_k - \frac{|\Delta_0|^2}{G} - G\sum_k \rho_{k(0)}^2; \\
E_{jl} &= 2\sum_{k\neq j,l} \rho_{k(jl)}\epsilon_k - \frac{|\Delta_{jl}|^2}{G} - G\sum_{k\neq j,l} \rho_{k(jl)}^2 + \epsilon_j + \epsilon_l;
\end{aligned}
\tag{20}
$$

and $T_{k(\gamma)}$ are energy terms associated to single-particle states:

$$
T_{k(\gamma)} = 2\rho_{k(\gamma)}(\epsilon_k - \lambda_\gamma) - 2G\rho_{k(\gamma)}^2 + \frac{\kappa_{k(\gamma)}\Delta_\gamma^* + \kappa_{k(\gamma)}^*\Delta_\gamma}{2}\left(\frac{\rho_{k(\gamma)}^2}{|\kappa_{k(\gamma)}|^2} - 1\right).
$$

The following notations are used in Eqs. (19):

$$
\begin{aligned}
\Delta_0 = G\sum_k \kappa_{k(0)} \quad &; \quad \Delta_{jl} = G\sum_{k\neq j,l} \kappa_{k(jl)}; \\
\kappa_{k(\gamma)} = u_{k(\gamma)}v_{k(\gamma)} \quad &; \quad \rho_{k(\gamma)} = |v_{k(\gamma)}|^2; \\
P_\gamma = |c_\gamma|^2 \quad &; \quad S_{\gamma\beta} = c_\gamma c_\beta^*;
\end{aligned}
\tag{21}
$$

where $\rho_{k(\gamma)}$ are single-particle densities and $\kappa_{k(\gamma)}$ are pairing moment components. P_γ denote the probabilities to find the system in the configurations γ. $S_{\gamma\beta}$ are moment components between two configurations γ and β and have the property $|S_{\gamma\beta}|^2 = P_\gamma P_\beta$. Δ_γ is the gap parameter. The values of $\rho_{k(\gamma)}$ and P_γ are reals. The particle number conservation conditions $2\sum_k \rho_{k(0)} = 2N$, $2\sum_{k\neq\gamma}\rho_{k(\gamma)} = 2N-2$ and $P_0 + \sum_\gamma P_\gamma = 1$ are fulfilled by Eqs. (19).

7. Application to fission processes

A direct application of the system (19) is related to the pair breaking and the odd-even structure in fission fragment yields. In fission, it is considered that the paired configuration is preserved until an interaction breaks some pairs in combination with the existence of a sufficient high excitation energy. The odd-even structure in fission is explained usually within statistical arguments, as for example in Refs. [40]. Alternatively, the probability to break a pair can be determined dynamically by taking into account only the interaction available in the avoided crossing regions within the present model. To solve the pair breaking equations (19), the variations of single-particle energies ϵ_k together with perturbations h_{ij} must be supplied. These quantities are obtained within the two center Woods-Saxon model along the minimal action fission path [41]. We used a method initiated in Ref. [42] which minimizes numerically the action integral. Alternatively, it is also possible to solve the associated Euler-Lagrange equations of motion [43, 44]. The results [12] are summarized in Fig. 3 for different constant values of the inter-nuclear velocity $\dot{R}$ ranging from 10^4 to 10^6 m/s. The results exhibit a clear decrease of

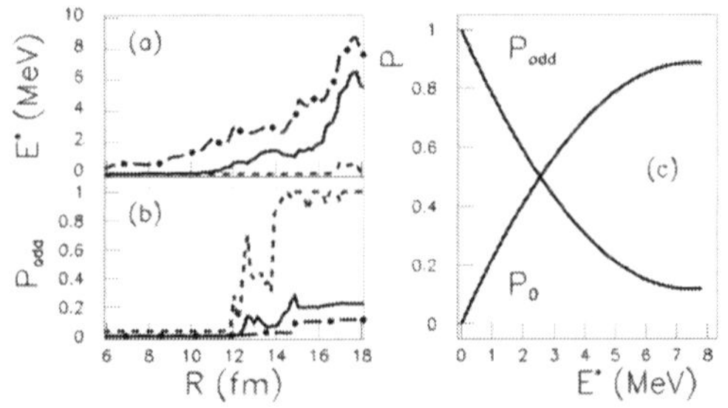

Figure 3. (a) The average excitation energy E^* as function of the elongation R. The inter-nuclear velocities $\dot{R}$ are 10^4, 10^5 and 10^6 m/s for the dashed, full and dot-dashed lines, respectively. (b) The probability P_{odd} of a seniority-two state with respect to the elongation R. The same line types and inter-nuclear velocities as in panel (a) are used. (c) The probabilities to obtain a seniority-zero state P_0 and a seniority-two state P_{odd} as function of the excitation energy E^* of the fragments at the elongation R=20 fm. Figure from Ref. [12].

P_{odd} as function of E^*. It is interesting to note that at zero excitation energy, the probability to find the system in a seniority-two state is practically one. In cold fission, at very low excitation energies of the fragments, the odd-even yields are always larger than the even-even ones [45, 46]. The even-even fragmentation dominates at larger excitation energies of the fragments, above 4-6 MeV. It is a very strange phenomenon because in cold processes the system doesn't possess enough energy to break a pair and because the penetrability is hindered for odd-systems due to the specialization energies associated to the two unpaired nucleons. if one assumes that the odd-even effect in the fission fragments distribution is strongly correlated to the seniority-two state probability, this phenomenon can be alternatively explained by solving the coupled-channel system of time-dependent pair breaking equations as evidenced above. In this work, only the Ω=1/2 subspace of the proton level diagram is treated, but the same formalism can be applied to other subspaces.

8. Effective mass

A semi-adiabatic cranking approximation is derived [13] for even-even systems to investigate the general features of the cranking mass parameters when the nucleus is internally excited. It is assumed that the system deforms slowly in time. Therefore, the matrix elements of the time-derivative of the wave functions are small. Because these matrix elements are responsible for quasi-particle excitations, our approximation allows to consider the system in a seniority-zero state during fission. In other words, the contributions to the inertia originating from seniority-two configurations are considered negligible. At each deformation, instantaneous values of the single-particle densities and of the pairing moment components are deduced. The semi-adiabatic effective mass is evaluated using these values. In the above approximation, the semi-adiabatic cranking formula can be obtained for the elements of the effective mass tensor $B_{\nu\mu}$:

$$B_{\nu\mu} = 2\hbar^2 \sum_{m,n\neq m} \frac{(E_{mn} - E_0) \mid \frac{\kappa_m \sqrt{\rho_m}|\kappa_n|}{|\kappa_m|\sqrt{\rho_n}} - \frac{\kappa_n \sqrt{\rho_n}|\kappa_m|}{|\kappa_n|\sqrt{\rho_m}} \mid^2 \langle m \mid \frac{\partial H}{\partial q_\nu} \mid n\rangle\langle n \mid \frac{\partial H}{\partial q_\mu} \mid m\rangle}{(E_{mn} - \sum_{k\neq m,n} T_k - E_0 + \sum_k T_k)^2(\epsilon_m - \epsilon_n)^2}$$
$$+2\hbar^2 \sum_m \frac{(E_{mm} - E_0)(\frac{\kappa_m}{\rho_m}\frac{\partial \rho_m}{\partial q_\nu} - \frac{\kappa_m}{\kappa_m^*}\frac{\partial \kappa_m^*}{\partial q_\nu})(\frac{\kappa_m^*}{\rho_m}\frac{\partial \rho_m}{\partial q_\mu} - \frac{\kappa_m^*}{\kappa_m}\frac{\partial \kappa_m}{\partial q_\mu})}{(E_{mm} - \sum_{k\neq m} T_k + T_m - E_0 + \sum_k T_k)^2} \quad (22)$$

where the values of ρ_k and κ_k are solutions of the time dependent pairing equations and the index ν is associated to the collective coordinate q_ν. E_0, E_{jl} and E_{jj} are exactly the expected values of the Hamiltonian (5) for seniority zero and seniority-two configurations, respectively.

The mass parameters can be obtained from formula (22) only if the derivatives $\partial\rho_m/\partial q_\nu$ and $\partial\kappa_m/\partial q_\nu$ are supplied. If the system follows a trajectory in the configuration space and the motion is characterized by some given collective velocities, then the derivatives depend also on $\dot{q}_\nu$. Therefore, the second term in Eq. (22) depends on the history of the system and can be calculated only if the variations of ρ_k and κ_k are known. The inertia along the trajectory is

$$B = \sum_{\nu,\mu} B_{1\nu\mu} \frac{\partial q_\nu}{\partial R}\frac{\partial q_\mu}{\partial R} + 2\hbar^2 \sum_m \frac{(E_{mm} - E_0) \mid \frac{\kappa_m}{\rho_m}\frac{\partial \rho_m}{\partial R} - \frac{\kappa_m}{\kappa_m^*}\frac{\partial \kappa_m^*}{\partial R} \mid^2}{(E_{mm} - \sum_{k\neq m} T_k + T_m - E_0 + \sum_k T_k)^2} \quad (23)$$

where one coordinate R is taken as the independent variable. The other coordinates q_ν are taken to be function of R [41]. ¿From definition, the collective kinetic energy is $E_c = B\dot{R}^2/2$ and the last term in Eq. (23) becomes:

$$E_{c0} = \hbar^2 \sum_m \frac{(E_{mm} - E_0) \mid \frac{\kappa_m}{\rho_m}\dot{\rho}_m - \frac{\kappa_m}{\kappa_m^*}\dot{\kappa}_m^* \mid^2}{(E_{mm} - \sum_{k\neq m} T_k + T_m - E_0 + \sum_k T_k)^2} \quad (24)$$

This term depends explicitly only on the derivatives with respect to time $\dot{\kappa}_m$ and $\dot{\rho}_m$. Their expressions are given by the coupled channel system of equations (19). That means, the term (24) is practically independent on a specific collective velocity and depends only on the values of ϵ_m, κ_m and ρ_m, that is on the structure of the system and its intrinsic excitation and represents the minimal collective kinetic energy shared among all degrees of freedom. In other words, if the nucleus is internally excited and the derivatives of the probabilities $\dot{\kappa}$, $\dot{\rho}$ are different from zero, the system possesses a minimal collective kinetic energy.

In Ref. [13], it was shown that the effective mass decreases when the excitation energy increases. It can be also mentioned that using BCS approximation, the formula (22) gives the classical result for the cranking inertia [41]. As remarked in Ref. [47], the results given by this non-adiabatic cranking are similar to that given by the Gaussian overlap approximation [48].

9. Conclusions

In this contribution, three approaches were used to generalize the the time dependent pairing equations and to derivate an excitation dependent inertia. The new formalisms are valid for any kind of mean field approximations that include a monopole pairing field. These new equations offer information about the spectroscopic amplitudes and the dissipated energies in different final channels. Using these equations, qualitative and quantitative features of the fine structure phenomenon in cluster decay were reproduced. Moreover, the odd-even effect in cold fission processes was explained. The main trends concerning the dependence of the odd-even effect in fragments yields versus the fragments excitation energy were reproduced. Finally, the values of the inertia obtained from the time dependent pairing equations exhibit a strong dependence on the intrinsic excitation energy. In general, the semi-adiabatic inertia have a similar shell structure as the adiabatic cranking model. It is worth to mention, that other succesfull models describe collective excitations within the stationary coupled channel approach [49].

Acknowledgments
Work supported by CNCS UEFISCDI, project number PN-II-ID-PCE-2011-3-0068

References

[1] Urban M, and Schuck P 2006 *Phys. Rev.* A **73** 013621
[2] Abe Y, Ayik S, Reinhard P-G and Suraud L 1996 *Phys. Rep.* **275** 49
[3] Lacroix D, Ayik S and Chomaz P 2004 *Prog. Part. Nucl. Phys.* 52 497
[4] Blocki J and Flocard H 1976 *Nucl. Phys.* A **273** 45
[5] Koonin S E and Nix J R 1976 *Phys. Rev.* C **13** 209
[6] Ledergerber T, Paltiel Z, Fraenkel Z and Pauli H C 1977 *Nucl. Phys.* A **275** 280
[7] Ledergerber T, Patiel Z, Pauli H C, Schutte G, Yariv Y and Fraenkel Z 1975 *Phys. Lett.* B **56** 417
[8] Schutte G and Wilets L 1975 *Nucl. Phys.* A **252** 21
[9] Mirea M, Tassan-Got L, Stephan C and Bacri C O *Nucl. Phys.* A **735** 21
[10] Mirea M 2003 *Mod. Phys. Lett.* A **18** 1809
[11] Mirea M 2008 *Phys. Rev* C **78** 044618
[12] Mirea M 2009 *Phys. Lett* B **680** 316
[13] Mirea M and Bobulescu R C 2010 *J. Phys.* G **37** 055108
[14] Hill D L and Wheeler J A 1953 *Phys. Rev.* **89** 1102
[15] Mirea M 1998 *Phys. Rev.* C **57** 2484
[16] Mirea M, Tassan-Got L, Stephan C, Bacri C O, Stoica P and Bobulescu R C 2005 *J. Phys.* G **31** 1165
[17] Mirea M, Tassan-Got L, Stephan C, Bacri C O and Bobulescu R C 2006 *Europhys. Lett.* **73** 705
[18] Mirea M, Tassan-Got L, Stephan C, Bacri C O and Bobulescu R C 2007 *Phys. Rev.* C **76** 064608
[19] Cha M H, Park J Y and Scheid W 1987 *Phys. Rev.* C **36** 2341
[20] Park J Y, Greiner W and Scheid W 1980 *Phys. Rev.* C **21** 958
[21] Mirea M and Clapier F 1997 *Europhys. Lett.* **40** 509
[22] Mirea M 1999 *Eur. Phys. J.* A **4** 335
[23] Mirea M 2001 *Phys. Rev.* C **63** 034603
[24] Mirea M 1996 *Phys. Rev.* C **54** 302
[25] Mirea M 2006 *Nucl. Phys.* A **780** 13
[26] Geng L-S, Meng J, and Hiroshi T 2007 *Chin. Phys. Lett.* **24** 1865
[27] Samarin V V 2010 *Phys. At. Nucl.* **73** 1416
[28] Diaz-Torres A and Scheid W 2005 *Nucl. Phys.* A **757** 373
[29] Mirea M, Delion D S and Sandulescu A 2010 *Phys. Rev.* C **81** 044317
[30] Mirea M 2011 *Phys. Rev.* C **83** 054608
[31] Companis I, Mirea M, and Isbasescu A 2011 *Rom. J. Phys.* **56** 63
[32] Mirea M, and L. Tassan-Got 2011 *Centr. Europ. J. Phys.* **9** 116
[33] Mirea M, Delion D S and Sandulescu A 2009 *EPL* **85** 12001
[34] Stoica P 2011 *Rom. Rep. Phys.* **63** 76
[35] Mirea M, Sandulescu A and Delion D S 2011 *Proc. Rom. Acad.* Series A **12** 203
[36] Mirea M, Sandulescu A and Delion D S 2011 *Nucl. Phys. A* **870-871** 23
[37] Brillard L, El Ayi A G, Hourani E, Hussonnois M, Le Du J F, Rosier L H, and Stab L 1989 *C.R. Acad. Sci. Paris* **39** 1105
[38] Gupta R K, Horoi M, Sandulescu A, Greiner M, and Scheid W 1993 *J. Phys. G* **19** 2063
[39] Lovas R G, Liotta R J, Insolia A, Varga K, and Delion D S 1998 *Phys. Rep.* **294** 265
[40] Avrigeanu V, Florescu A, Sandulescu A, and Greiner W 1995 *Phys. Rev.* C **52** R1755
[41] Brack M, Damgaard J, Jensen A S, Pauli H C, Strutinsky V M, and Wong C Y 1972 *Rev. Mod. Phys.* **44** 320
[42] Poenaru D N, Mirea M, Greiner W, Cata I, and Mazilu D 1990 *Modern Physics Letters* A **5** 2101
[43] Mirea M, Poenaru D N, and Greiner W 1992 *Nuovo Cimento A* **105** 571
[44] Mirea M, Poenaru D N, and Greiner W 1994 *Z. Phys. A* **349** 39
[45] Schwab W, Clerc H-G, Mutterer M, Theobald J P, and Faust H 1994 *Nucl. Phys.* A **577** 674
[46] Hambsch F-J, Knitter H-H, and Budtz-Jorgensen C 1993 *Nucl. Phys.* A **554** 209
[47] Mirea M 2011 *Rom. Rep. Phys.* **63** 676
[48] Gozdz A, Pomorski K, Brack M, and Werner E 1985 *Nucl. Phys.* A **442** 26
[49] Peltonen S, Delion D S, and Suhonen J 2008 *Phys. Rev. C* **78** 034608

Journal of Physics: Conference Series **338** (2012) 012027 doi:10.1088/1742-6596/338/1/012027

Competition of different coupling schemes in atomic nuclei

C. Qi, R. J. Liotta, R. Wyss
KTH, Alba Nova University Center, SE-10691 Stockholm, Sweden
E-mail: chongq@kth.se

Abstract. Shell model calculations reveal that the ground and low-lying yrast states of the $N = Z$ nuclei $^{92}_{46}$Pd and ^{96}Cd are mainly built upon isoscalar spin-aligned neutron-proton pairs each carrying the maximum angular momentum $J = 9$ allowed by the shell $0g_{9/2}$ which is dominant in this nuclear region. This mode of excitation is unique in nuclei and indicates that the spin-aligned pair has to be considered as an essential building block in nuclear structure calculations. In this contribution we will discuss this neutron-proton pair coupling scheme in detail. In particular, we will explore the competition between the normal monopole pair coupling and the spin-aligned coupling schemes. Such a coupling may be useful in elucidating the structure properties of $N = Z$ and neighboring nuclei.

The low-lying structures of many nuclei can be understood as the outcome of the competition between the pairing (or seniority) coupling scheme and the aligned coupling of individual particles in a non-spherical average potential. The seniority coupling dominates the low-lying states of semi-magic nuclei, where the driving force behind is the strong pairing interaction between like particles. Meanwhile, many open-shell nuclei have quadrupole moments that are much larger than could be attributed to a single particle, which implies the sharing of angular momentum between many particles. A remarkable feature of nuclear structure physics is that essential ingredients of the single-particle model could be retained by assuming that the nucleons move in an average nuclear field deviates from spherical symmetry, which removes the spurious degrees of freedom corresponding to the collective spectrum [1]. The transition between there two couplings in open-shell nuclei is expected to be rather fast [2, 3, 4].

Near the $N = Z$ line, nuclear collectivity may be further enhanced by the interactions arising between neutrons and protons occupying orbitals with the same quantum numbers, which can form np pairs with angular momenta $J = 0$ to $2j$ and isospin quantum numbers $T = 0$ (isoscalar) and $T = 1$ (isovector). The isovector np channel manifests itself in a fashion similar to like-nucleon correlations. There has been longstanding and intensive interest in exploring the signature of isoscalar np pairing (in particular np pairs with $J = 1$ and $S = 0$) in the structure of self-conjugate nuclei [5, 6, 7, 8, 9, 10]. Possible fingerprints may be inferred from the delayed alignments in the ground-state rotational bands of deformed $N = Z$ nuclei, but it is still controversial. In this contribution we would like to discuss the so-called spin-aligned neutron-proton pair coupling from a nuclear shell-model perspective. In some spherical $N = Z$ nuclei it may overcome the normal pairing as the dominant coupling scheme of the ground state structure. We will show that the aligned np pairs can generate striking regular evolution patterns in the energy spectra and transition probabilities along the yrast states of $N = Z$

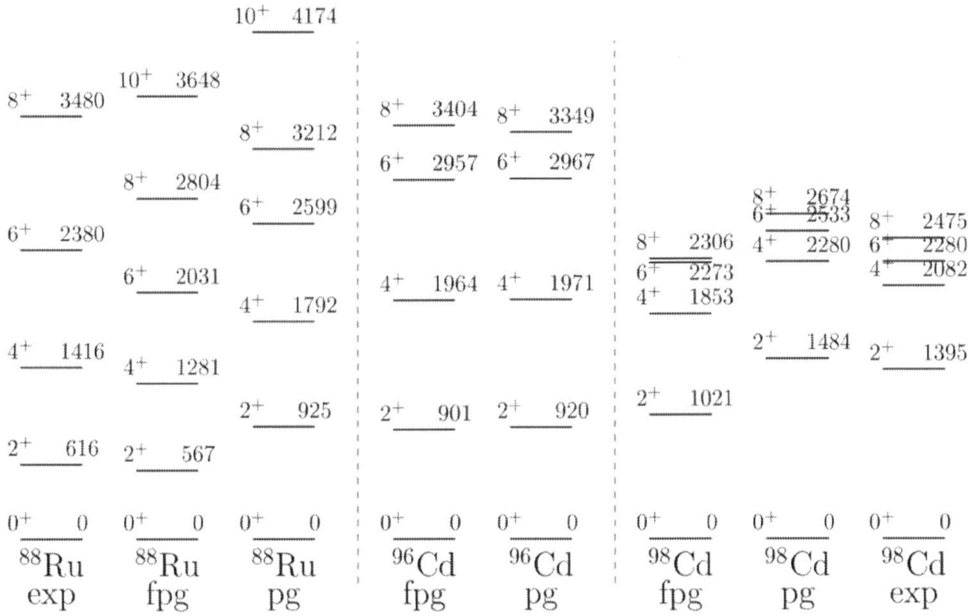

Figure 1. Experimental [19, 20] and calculated yrast level schemes of $N = Z$ nuclei ^{88}Ru and ^{96}Cd and ^{98}Cd with two proton holes inside the core ^{100}Sn.

nuclei. These may be deemed as a new kind of collective mode with isoscalar character that is unique in the atomic nucleus. In a recent work the low-lying yrast states in $^{92}_{46}$Pd were observed and it was inferred that for the first time a transition from the isovector pair coupling mode to such spin-aligned np coupling scheme may have occurred [11, 12]. For related earlier works, see Refs. [13, 14, 15, 16, 17].

I would like to remind you that, as one would expect, near the closed shell nucleus ^{100}Sn, i.e., in ^{96}Pd (^{98}Cd) with four (two) proton holes outside of ^{100}Sn, the positions of the energy levels correspond to a $(g_{9/2})^2_\lambda$ pairing spectrum, as can be seen from Fig. 1 of Ref. [12] and Fig. 1. When the number of neutron holes increases, the levels tend to be equally separated, which is a characteristic of vibrational spectra. This are specially the cases for ^{92}Pd and ^{96}Cd which show equally-spaced level schemes up to $I = 12$ and $I = 6$, respectively. Already the systematics of experimental data suggests gradual decrements in both the quadrupole deformation β_2 and $E(4^+_1)/E(2^+_1)$ ratio in $N = Z$ nuclei when approaching the ^{100}Sn shell closure [18, 19]. Many open shell $N = Z$ nuclei, like ^{88}Ru shown in Fig. 1, exhibit rotational-like ground state bands. The nuclei ^{76}Sr and ^{80}Zr are the most deformed $N = Z$ ones in this region.

Table 1. The $0g_{9/2}$ interaction matrix elements for different model spaces [21, 22, 24].

Space	$J=0$	$J=2$	$J=4$	$J=6$	$J=8$	$J=1$	$J=3$	$J=5$	$J=7$	$J=9$
$0g_{9/2}$	0	1.458	2.283	2.549	2.688	1.220	1.592	1.882	1.930	0.626
pg	-2.089	-0.919	-0.094	0.172	0.311	-1.038	-0.785	-0.495	-0.447	-1.751
fpg	-1.691	-0.959	-0.087	0.152	0.269	-1.138	-0.599	-0.383	-0.561	-2.207

To understand the yrast structures of ^{92}Pd, ^{96}Cd and neighboring nuclei, we perform nuclear shell model calculations within the $1p_{3/2}0f_{5/2}1p_{1/2}0g_{9/2}$ (fpg), $0g_{9/2}1p_{1/2}$ (pg) and $0g_{9/2}$ model spaces using the Hamiltonian given in Ref. [21] and a variety of other interactions which are quoted therein. In particular, we perform calculations using as single-particle states the orbits pg and $0g_{9/2}$, with the interactions of Refs. [22, 23, 24], in order to explore the importance of configuration mixing in determining the structure of the spectrum.

In Table 1 we compare the strengths of the interaction matrix elements corresponding to different shell model spaces [21, 22, 24]. In the case of $0g_{9/2}$ space, only the relative strengths are given for simplicity. The absolute strength of the monopole centroid has no influence on the coupling of the wave functions and excitation energies. Also it should be mentioned that the relative strength of the $T = 0$ and $T = 1$ monopole interactions does not play any role on the structure of nuclear states. That is, the wave functions remain unchanged by adding a constant to the $T = 0$ or 1 part of the interaction. This modification only affects the relative energies of the states with different isospin quantum numbers.

The matrix element for the fpg model space, for which we took from Ref. [21], are defined in the particle-particle channel by assuming ^{56}Ni as the core. The mass dependence of the interaction is assumed to be $(A/58)^{-0.3} = 0.87$ for $A = 92$ where A is the mass number of the nucleus to be calculated. For $A = 92$ we take the single-hole energies (relative to that of $0g_{9/2}$) of $\varepsilon(0g_{9/2}) = 0$ MeV, $\varepsilon(1p_{1/2}) = 0.144$ MeV, $\varepsilon(1p_{3/2}) = 1.443$ MeV and $\varepsilon(0f_{5/2}) = 4.407$ MeV.

The calculated results for Ru and Cd isotopes are plotted in Fig. 1. Calculations are done with the shell-model code described in Ref. [25]. Calculations for the spectra of Pd isotopes could be found in Refs. [11, 12]. Present shell model calculations are able to include a large number of shells with the help of modern computers. In Fig. 2 we calculated the average number of holes in each orbitals of the fpg model space in which four orbitals are involved. It is thus seen that it is the $0g_{9/2}$ shell that dominates the occupancy. Also comparison with pg and $0g_{9/2}$ calculations tend to suggest that many properties of nuclei in this region can be explained by calculations restricted the the single $0g_{9/2}$ shell only. This does not mean that the other shells (the $0f1p$ orbitals and even other higher lying shells) has no contribution to the wave function. But normally one may safely expect the effect from these shells can be taken into account through the renormalization of the effective interaction and effective operators (c.f., Table 1). The contributions from other shells may be deemed as the background whose effect on the nuclear structure is hard to identify. This is because our knowledge of complex objects like atomic nuclei are obtained through systematic studies of neighboring states. As a result, only relevant degrees of freedom that distinguish one state from the others can be identified. Simple models are only for us to understand the physics and the residual degrees of freedom. The real situation can be much more complex. Even the shell-model single-particle orbitals are very complicated from a microscopic point of view and can be very different from the real single-particle wave functions of the nucleus [26, 27]. Thus the shell occupancy plotted in Fig. 2 is a model dependent quantity. Also full fp shell model calculation has been feasible for the nucleus ^{48}Cr since almost 20 years ago. But it is realized that the $0f_{7/2}$ and $1p_{3/2}$ shells would be enough to explain the bulk properties of this deformed nucleus [28].

The regularly-spaced energy spectra in ^{92}Pd and ^{96}Cd might naively be interpreted as a collective vibrational motion. At present it is not clear how nuclear vibration can be described properly from a microscopic shell model perspective. On the other hand, it is suggested that the apparent collectivity in the spectrum of ^{92}Pd may have a different origin than that resulting from a vibration or rotational motion usually attached to nuclear collective phenomena [11, 12]. Since there is no evidence for shell interferences in the spectra of Pd isotopes, we will analyze other sources which control the structure of those spectra, namely the different coupled pair modes. This can be done rather easily in this case where the shell $0g_{9/2}$ is very dominant, by

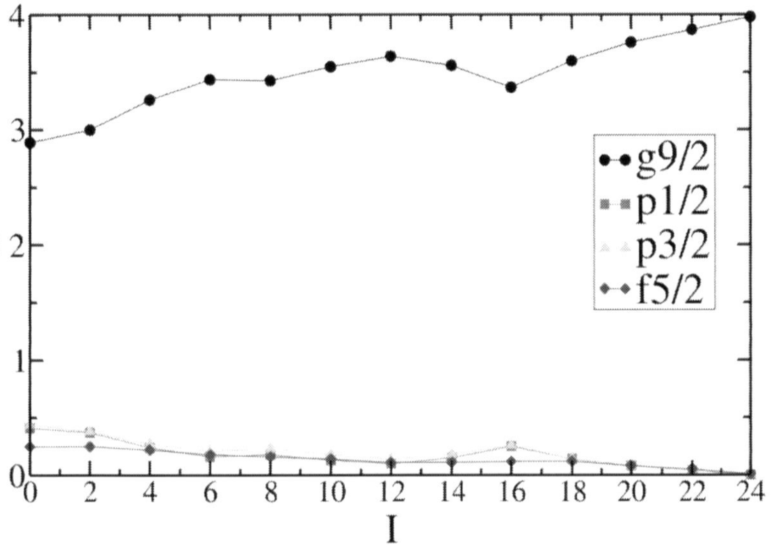

Figure 2. Average number of proton (or neutron) holes in different orbitals for shell-model calculations on nucleus ^{92}Pd in the fpg shell [21].

considering separately the various components of the interaction. We found that when the $T = 0$ np interaction is switched off, the corresponding spectrum is of seniority type. This underlines the importance of the np interaction to induce collective spectra.

To probe the pair content in the many-fermion wave function Ψ one may evaluate the two-particle transfer amplitude $\langle\Psi_N||(a_i^\dagger a_j^\dagger)_{J^\pi}||\Psi_{N-2}\rangle$ or the average number of pairs $\langle\Psi_N||((a_i^\dagger a_j^\dagger)_{J^\pi}\times(a_i a_j)_{J^\pi})_0||\Psi_N\rangle$ or project the shell-model wave functions onto a pair coupled basis with the help of the two-particle coefficients of fractional parentage. In Refs. [11, 12, 29] we applied the two-particle coefficients of fractional parentage technique. Two sets of the orthonormal bases are constructed starting from the monopole $J = 0$ and $J = 9$ pairs. In Refs. [24, 30, 31] all possible combinations are considered within an non-orthogonal basis by applying the so-called multistep shell model. After projecting the yrast wave functions into a product of isoscalar $J^\pi = 9^+$ pairs we found the most striking feature of this case: *all* low-lying yrast states are built mainly from $J^\pi = 9^+$ spin-aligned np pairs as $|((\nu\pi)_{J=9})_I^4\rangle$. In contrast to the isovector pairing excitations, in the aligned isocalar pair state there is no pair breaking when the nucleus is excited from one state to higher lying states in the spectrum. Instead, angular momentum in ^{92}Pd is determined by a new form of coupling [11]. The $((\nu\pi)_9)_I^N$ mode can be deemed as a *collective* pair coupling scheme which can be written as a coherent superposition of all isovector neutron and proton pairs occupying the $g_{9/2}$ shell in the form of $((\nu_J^2)^{N/2}\otimes(\pi_J^2)^{N/2})_I$. The dynamics of the system is determined by the active four $(\nu\pi)_9$ hole pairs aligned differently according to the total angular momentum of the system.

In a restricted single j shell, the calculations of the average pair number can be much simplified with the help of two-particle coefficients of fractional parentage as

$$C_J^I = \frac{n(n-1)}{2}\sum_{\alpha\beta} X_{\alpha\beta} M_{\alpha\beta}^I(J) X_\beta, \tag{1}$$

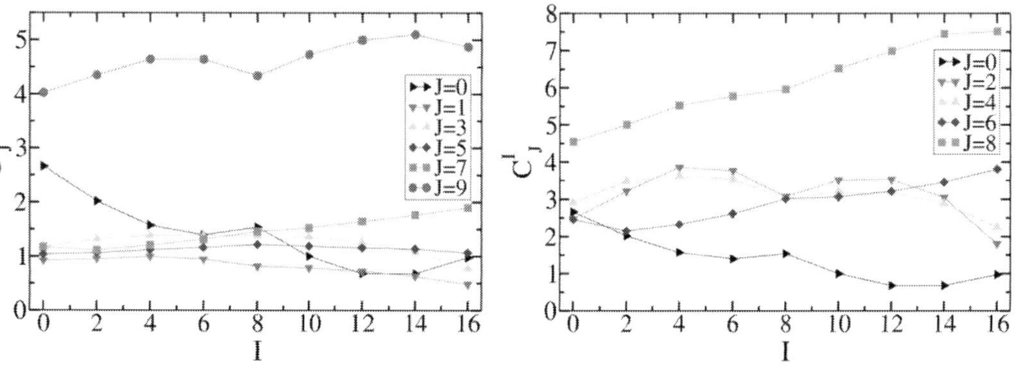

Figure 3. Left: Average number of isoscalar $(0g_{9/2}^2)_J$ interacting pairs, C_J^I, as a function of total angular momentum I for the wave functions of the yrast states of ^{92}Pd; Right: Same as the left panel but for the numbers of isovector $(0g_{9/2}^2)_J$ pairs.

and

$$M_{\alpha\beta}^I(J) = \sum_{\alpha_{n-2}I_{n-2}} [j^{n-2}(\alpha_{n-2}I_{n-2})j^2(J)I|\}j^n\alpha I] \times [j^{n-2}(a_{n-2}I_{n-2})j^2(J)I|\}j^n\beta I], \quad (2)$$

where n is the number of particles (holes) of the system and X are the expansion amplitudes of the wave function. The Greek letters denote the basis states and the summation runs over all possible states.

For systems with four pairs the total number of interacting pairs is

$$N = n(n-1)/2 = 28. \quad (3)$$

For a low-lying yrast state in ^{92}Pd which has total isospin $T = 0$, the total number of isoscalar pairs is

$$[n/2(n/2+1) - T(T+1)]/2 = 10. \quad (4)$$

If isospin symmetry is assumed, the numbers of isovector neutron-neutron, proton-proton and np pairs are the same and the total number of pairs is

$$[3n/2(n/2-1) + T(T+1)]/2 = 18. \quad (5)$$

The results for the isoscalar pairs thus obtained in the pg space are shown in the left panel of Fig. 3. One sees that *all* low-lying yrast states are built mainly from $J^\pi = 9^+$ spin-aligned np pairs. In contrast to the isovector pairing excitations, in the aligned isoscalar pair state there is no pair breaking when the nucleus is excited from one state to higher lying states in the spectrum. Instead, the angular momentum of excited yrast states in ^{92}Pd is generated by the rearrangement of the angular momentum vectors of the aligned np pairs while the isoscalar pair character itself is preserved. For the ground state the four pairs are maximally aligned in opposite directions as allowed by the Pauli principle. When exciting the nucleus from one yrast state to another, the dynamics of the system is determined by the active four $(\nu\pi)_9$ hole pairs aligned differently according to the total angular momentum of the system. It should be mentioned that the microscopic origin of the equidistant pattern generated by the aligned np

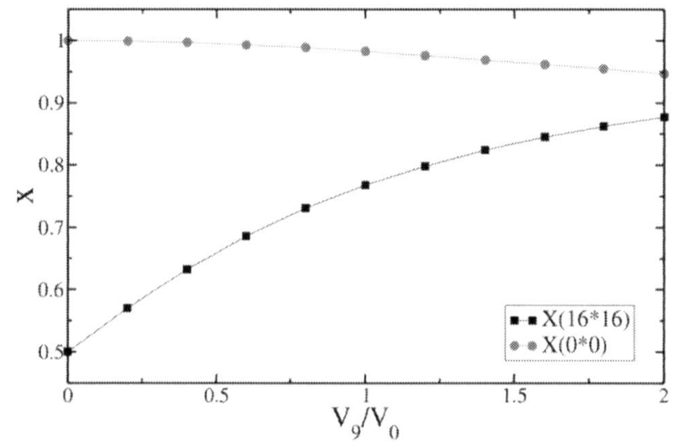

Figure 4. Wave function amplitudes for the components $|^{96}\mathrm{Cd(gs)}\otimes^{96}\mathrm{Cd(gs)}\rangle$ and $|^{96}\mathrm{Cd}(16^+)\otimes^{96}\mathrm{Cd}(16^+)\rangle$ in the ground state wave function of ^{92}Pd as a function of the ratio between the strengths of V_9 and V_0. The calculation is done in the $0g_{9/2}$ shell with all other matrix elements are set to zero.

pairs remain unclear at present. From a simple semiclassical point of view, one may argue that the corresponding energies vary approximately linearly with I for small total angular momenta, inducing similar energy spacings between consecutive levels.

To gain deeper physical insight into the structure of ^{92}Pd we also calculated the numbers of isovector pairs which are plotted in the right panel of Fig. 3. It is seen that already in the ground state we have much more pairs with $J > 0$ than the normal $J = 0$ pair. In particular, the dominating component is the $J = 8$ pair which is maximally aligned in the isovector channel. This is due to the large overlap between states generated by the spin-aligned np pairs and those generated by the isovector pairs. This phenomenon is not seen in systems with two np pairs where the contributions from the isovector aligned pair is practically zero for low-lying yrast states [24]. It should be emphasized that the dominating component in the wave functions of low-lying yrast states in ^{92}Pd is still the spin-aligned np pair coupling. But it may be interesting to clarify the role played by the isovector aligned pair in $N = Z$ systems with more than two pairs. Work on this direction is underway.

The four $J = 9$ np pairs in ^{92}Pd can couple in various ways. With the help of two-particle cfp one may express the wave function in terms of $((((\nu\pi)_9 \otimes (\nu\pi)_9)_{I'} \otimes (\nu\pi)_9)_{I''} \otimes (\nu\pi)_9)_I$. It is thus found that, among the various aligned np pair configurations, the dominating components can be well represented by a single configuration $((((\nu\pi)_9 \otimes (\nu\pi)_9)_{I'=16} \otimes (\nu\pi)_9)_{I''=9} \otimes (\nu\pi)_9)_I$. In the $0g_{9/2}$ shell, this configuration is calculated to occupy around 66% of the ground state wave function of ^{92}Pd, i.e., with amplitude $X(0_1^+) = 0.81$ [12]. Also it should be mentioned that the $J = 9$ term is not the generator for the full aligned np coupling. It only generates the stretch configuration [13] shown above. The maximal $I = 24$ state corresponds to a pure stretch configuration. On the other hand, by rewriting the wave function of ^{92}Pd as a product of two group with two np pairs (i.e., ^{96}Cd each) within an non-orthornormal basis it is thus found the leading component is coupling $|^{96}\mathrm{Cd(gs)}\otimes^{96}\mathrm{Cd(gs)}$, as can also be seen from Fig. 4. This is consistent with the weak coupling calculations for light nuclei [16].

Usually the vibrational spectrum is characterized by collective E2 transitions between states differing by one phonon, where the corresponding transition strengths increase linearly as a function of the spin of the initial state. One of the most important criteria to decide whether a

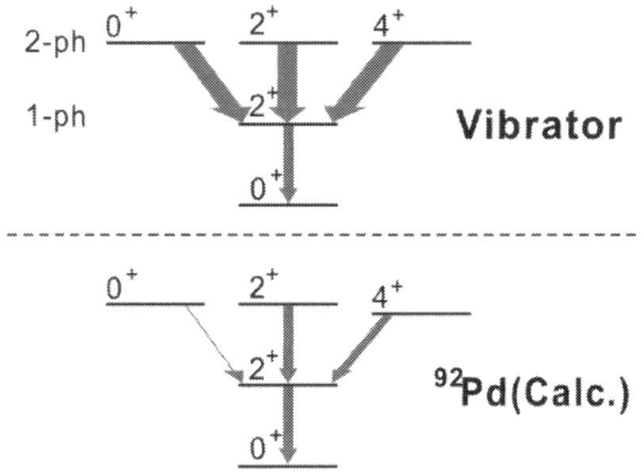

Figure 5. Upper: The two-phonon triplet states predicted by the vibrational model. Lower: The calculated triplet of states in ^{92}Pd with the relative B(E2) values (proportional to the width of the arrow) connecting them to the 2_1^+ state. The calculation is done in the fpg shell [21].

spectrum corresponds to the one of a vibrator is that there should be a nearly-degenerate two-phonon triplet with collective E2 transitions to the one-phonon state 2_1^+. In Fig. 5 we plotted the low-lying level scheme of ^{92}Pd predicted by fpg shell model calculations with the interaction of Ref. [21]. It is seen that there is indeed a triplet of states in the calculated spectrum which are nearly degenerate and lie at twice the energy of the state 2_1^+. The vibrational model predicts that the E2 transitions from the two-phonon states to the state 2_1^+ should all have the same strength and its value should be twice that of the $2_1^+ \to 0_1^+$ strength. This prediction differs very much from the calculated values for the aligned coupling scheme shown in the Figure. In particular, the strength of the $0_2^+ \to 2_1^+$ transition is one order of magnitude smaller than the predicted value for a vibrator. This may be due to the fact that the calculated second 0^+ state contain large contributions from shells (in particular the $1p_{1/2}$ shell) other than the $0g_{9/2}$ shell. That is, this state does not correspond to the spin-aligned np coupling scheme proposed in this work. Indeed, this 0_2^+ state cannot be reproduced in calculations within the $0g_{9/2}$ shell.

The three yrast states in Fig. 5 are mainly composed of the spin-aligned np pairs [12]. For example, the 2_1^+ state is dominated by the configuration of $^{96}\mathrm{Cd(gs)} \otimes ^{96}\mathrm{Cd}(2_1^+)$ with $x = 0.973$. The projection of this state on the stretch configuration $^{96}\mathrm{Cd}(16^+) \otimes ^{96}\mathrm{Cd}(16^+)$ is $x = 0.862$. It should be mentioned that the second 2^+ state in ^{92}Pd is also dominated by the spin-aligned np pair coupling. In this state the largest two components are $^{96}\mathrm{Cd}(2_1^+) \otimes ^{96}\mathrm{Cd}(2_1^+)$ with $x = 0.904$ and $^{96}\mathrm{Cd}(14_1^+) \otimes ^{96}\mathrm{Cd}(16^+) = 0.766$. This is in contrast with the coupling of $J = 0$ pairs where only one state can be generated from the monopole pairs irrespective of the number of pairs involved. An interesting question would be how many states can we construct from the aligned np pairs. A detailed analysis will be done elsewhere.

The proposed spin-aligned np coupling may also shed some new light on the structure of odd-odd $N = Z$ nuclei. Here we take the nucleus ^{94}Ag as an example. It has three np hole pairs in the ^{100}Sn core. The calculated spectra of ^{94}Ag in different model spaces are plotted in Fig. 6. The ground state is calculated to be $J^\pi = 0^+$ and $T = 1$ and is dominated by the coupling of two aligned np pair and a normal isovector $J = 0$ pair. The lowest $T = 0$ state is calculated to be a 7^+ state with excitation energy $E* \sim 0.7$ MeV. This state is dominated by the aligned np pair coupling of $(((\nu\pi)_9 \otimes (\nu\pi)_9)_{I'=16} \otimes (\nu\pi)_9)_I$ with $|X| \approx 0.97$ in restricted calculations performed within the $0g_{9/2}$ shell. This component decreases to $|X| = 0.83$ for the first 9^+ state, which is otherwise dominated by the coupling $(((\nu\pi)_9 \otimes (\nu\pi)_9)_{I'=0} \otimes (\nu\pi)_9)_I$. Again the wave

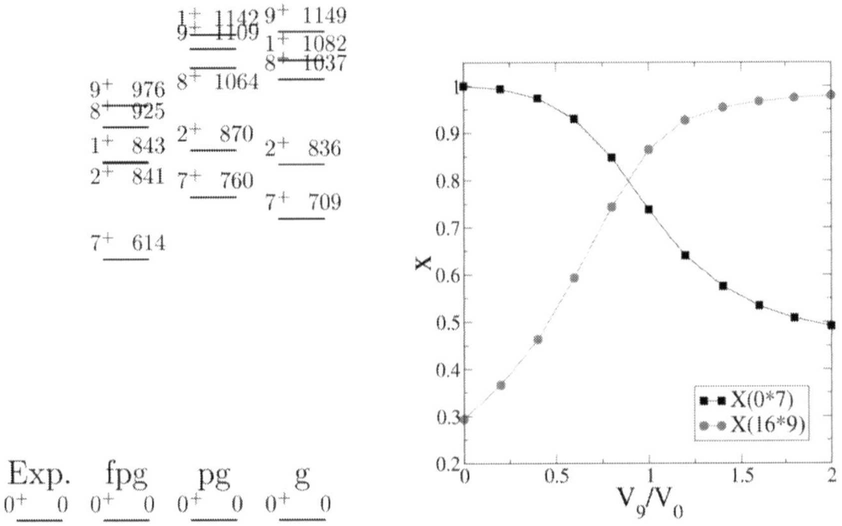

Figure 6. Left: Experimental and calculated low-lying states in the odd-odd nucleus ^{94}Ag. Right: Wave function amplitudes for the components $|^{96}\mathrm{Cd(gs)}\otimes(\nu\pi)_7\rangle$ and $|^{96}\mathrm{Cd}(16^+)\otimes(\nu\pi)_9\rangle$ in the 7_1^+ state of ^{94}Ag as a function of the ratio between the strengths of V_9 and V_0. The calculation is done in the $0g_{9/2}$ shell with all other matrix elements are put as zero for simplicity.

functions of these states are mixtures of many components if we express them in the normal proton-proton and neutron-neutron coupling scheme.

In the right panel of Fig. 6 we evaluated the wave function of the 7_1^+ state in ^{94}Ag in the $0g_{9/2}$ shell with a schematic Hamiltonian containing two interaction terms with $J = 0$ and $J = 9$. It is thus found that the wave function shows a pure configuration of $|^{96}\mathrm{Cd(gs)}\otimes(\nu\pi)_7\rangle$ for the calculation with a $J = 0$ term only, while the stretch configuration $|^{96}\mathrm{Cd}(16^+)\otimes(\nu\pi)_9\rangle$ gradually takes over by enhancing the strength of the aligned pair interaction. Similar calculations were also done for the 9_1^+ state. For $V_9 = 0$ the wave function is characterized by the pure configuration $|^{96}\mathrm{Cd(gs)}\otimes(\nu\pi)_9\rangle$ which is calculated to be the dominated component over a large range of V_9/V_0. The contribution from the stretch configuration $|^{96}\mathrm{Cd}(16^+)\otimes(\nu\pi)_9\rangle$ is also enhanced by enlarging the interaction V_9.

To gain deeper insight into the properties of the aligned np pair structure and the regions of the nuclear chart where this exotic mode can be observed, we also performed calculations for the N, $Z = 28$ mass region. It is noted that, when the calculations are restricted to the $0f_{7/2}$ shell only, both the $N = Z$ nuclei ^{44}Ti and ^{48}Cr exhibit an equally-spaced pattern similar to the one obtained with the shell $0g_{9/2}$, as can be seen in Fig. 7. However, the aligned $T = 0$ np coupling is not manifested in the spectra of the fp shell since the proximity of the shell $1p_{3/2}$ and the strong quadrupole interaction thus developed leads to the mixing of shell model configurations that break the spin-aligned np pair structure. As a result of this a combination of $T = 1$ pairing and quadrupole interaction forms a more favorable description and a deformed mean field is generated. To explore further the role played by the aligned np pair in this shell, we calculated the yrast spectra of ^{48}Cr with different strengths for the $J = 7$ interaction. It is thus found that a transition from rotational-like spectrum to vibrational-like spectrum occurs by enhancing the strength of the $J = 7$ interaction element. On the contrary, our calculations show that in the mass 90-100 region and the $0g_{9/2}$ shell the quadrupole interaction with the $1d_{5/2}$ level is not strong enough to scatter nucleons appreciably across the energy gap associated with the magic numbers $N, Z = 50$. Work underway is to explore in a more quantitative way the influence of deformation on the aligned np pair coupling.

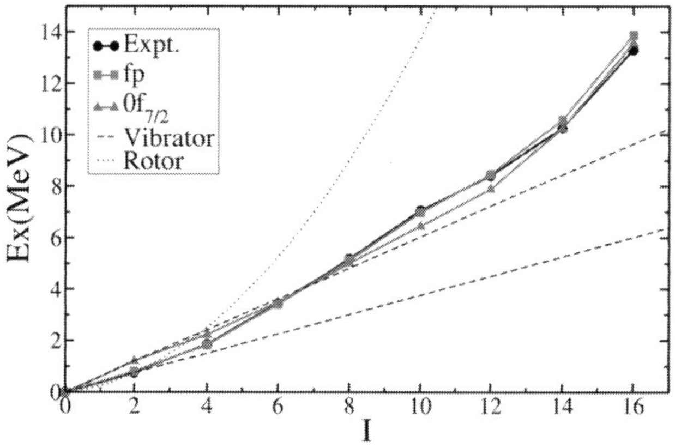

Figure 7. Experimental [20] and calculated level schemes of the nucleus ^{48}Cr. Calculations in the fp and $0f_{7/2}$ shells are done with the interactions from Ref. [32] and Ref. [33], respectively. The dashed and dotted lines correspond to the predictions of the the geometric collective model normalized to the 2_1^+ states.

Another ongoing work is to generalize the aligned np pair coupling to systems with more than one shell. This generalization may not be straightforward within the shell model framework due to the fact that the corresponding aligned np pairs can carry different angular momenta. One possibility to overcome this drawback is through introducing as building blocks four-body quartets which are composed of aligned np pairs [34].

Acknowledgments
This work has been supported by the Swedish Research Council (VR) under grant Nos. 623-2009-7340 and 621-2010-4723. CQ thanks Prof. Honma for sending him the interaction matrix elements of the fpg model space.

References
[1] Bohr A 1976 *Rev. Mod. Phys.* **48** 365.
[2] Bohr A and Mottelson B R 1980 *Phys. Scr.* **22** 468.
[3] Rowe D J, Bahri C, Wijesundera 1998 *Phys. Rev. Lett.* **80** 4394.
[4] Rosensteel G and Rowe D J 2007 *Nucl. Phys. A* **797** 94.
[5] Goodman A L 1979 *Adv. Nucl. Phys.* **11** 263.
[6] Satuła W and Wyss R 1997 *Phys. Lett. B* **393** 1.
[7] Sheikh J A and Wyss R 2000 *Phys. Rev. C* **62** 051302(R).
[8] Dukelsky J, Gueorguiev V G, Van Isacker P, Dimitrova S, Errea B, Lerma S 2006 *Phys. Rev. Lett.* **96** 072503.
[9] Lerma S, Errea B, Dukelsky J, and Satuła W 2007 *Phys. Rev. Lett.* **99** 032501.
[10] Sandulescu N, Errea B, Dukelsky J 2009 *Phys. Rev. C* **80** 044335.
[11] Cederwall B et al. 2011 *Nature* **469** 68.
[12] Qi C, Blomqvist J, Bäck T, Cederwall B, Johnson A, Liotta R J, and Wyss R 2011 *Phys. Rev. C* **84** 021301(R).
[13] Danos M and Gillet V 1967 *Phys. Rev.* **161** 1034.
[14] Danos M and Gillet V 1966 *Phys. Rev. Lett.* **17** 703.
[15] Bhatt K H and McGrory J B 1971 *Phys. Rev. C* **3** 2293.

[16] Wong S K M and Zuker A P 1971 *Phys. Lett.* **36B** 437.
[17] Chen H T, Feng D H and Wu C L 1992 *Phys. Rev. Lett.* **69** 418.
[18] Garnsworthy A B et al. 2008 *Phys. Lett. B* **660** 326.
[19] Marginean N et al. 2001 *Phys. Rev. C* **63** 031303(R).
[20] NuDat 2.5. http://www.nndc.bnl.gov/nudat2/.
[21] Honma M, Otsuka T, Mizusaki T and Hjorth-Jensen M 2009 *Phys. Rev. C* **80** 064323.
[22] Johnstone I P and Skouras L D 2001 *Eur. Phys. J. A* **11** 125.
[23] J. P. Schiffer and W. W. True, *Rev. Mod. Phys.* **48**, 191 (1976).
[24] Qi C 2010 *Phys. Rev. C* **81** 034318.
[25] Qi C and Xu F R 2008 *Chin. Phys. C* **32** (S2) 112.
[26] Talmi I 2011 *AIP Conf. Proc.* **1355** 121.
[27] Talmi I 1993 *Simple Models of Complex Nuclei* (Harwood Academic Publishers, Chur, Switzerland).
[28] Juodagalvis A 2000 arXiv:nucl-th/0004050.
[29] Qi C 2011 *Phys. Rev. C* **83** 014307.
[30] Qi C, Wang X B, Xu Z X, Liotta R J, Wyss R, and Xu F R 2010 *Phys. Rev. C* **82** 014304.
[31] Xu Z X, Qi C, Blomqvist J, Liotta R J, Wyss R 2011 arXiv:1108.0269.
[32] Poves A, Zuker A P 1981 *Phys. Rep.* **70** 235
[33] Escuderos A, Zamick L, Bayman B F 2005 arXiv:nucl-th/0506050
[34] Shah J A and Danos M 1969 *Phys. Rev.* **183** 899.

Statistical nuclear properties (level densities, spin distributions)

T. von Egidy
Physik Department, Technische Universität München, D-85748 Garching, Germany
E-mail: egidy@ph.tum.de

D. Bucurescu
Horia Hulubei National Institute of Physics and Nuclear Engineering, P.O. Box MG-6, 077125 Bucharest - Măgurele, Romania
E-mail: bucurescu@tandem.nipne.ro

Abstract. A general overview is given on the phenomenological methods used to describe the level densities in nuclei. Two well-known two-parameter formulas of level densities, the Back-Shifted Fermi Gas (BSFG) model and the Constant Temperature (CT) model, were used. A common ingredient of both is the spin distribution function, which contains in Ericsons's parametrization the spin-cutoff parameter σ. A realistic description of the parameters of both spin distribution function and the two level density models has been obtained by fitting the experimental data of 310 nuclei between ^{18}F and ^{251}Cf, consisting of the complete level schemes at low excitation energies and the s-wave neutron resonance spacings at the neutron binding energy. We determine a simple formula for the spin-cutoff parameter as a function of mass number and excitation energy. Also, an even-odd spin staggering in the spin distribution of the even-even nuclei was observed, and described with a simple formula. Using this newly defined spin distribution function, an empirical set of parameters of the BSFG and CT models was determined by fitting both the low-energy levels and the neutron resonance spacings. For these parameters, simple formulas were proposed that involve only quantities available from the mass tables, and allow reasonable estimations of the level density parameters for nuclei far from stability. Both the BSFG and CT models describe equally well the level densities at energies up to at least the neutron binding energy. Finally, we discuss recent experimental evidence that the CT model is the more correct description of the nuclei in the low-excitation energy (pairing) regime.

1. Introduction

The nuclear level densities represent the basic statistical information on nuclei both at low and higher excitation energies. The formulas for the level density are usually separated in a part with the total level density $\rho(E)$ (which increases exponentially with the excitation energy E) and a function $f(J)$ for the spin distribution,

$$\rho(E, J) = f(J)\rho(E). \tag{1}$$

Here a possible dependence on the level parity is neglected.

Two formulas are frequently used for the description of the total level density, the back-shifted Fermi gas formula (BSFG) [1],

$$\rho_{BSFG}(E) = \frac{e^{2\sqrt{a(E-E_1)}}}{12\sqrt{2}\sigma a^{1/4}(E-E_1)^{5/4}} \tag{2}$$

with the free parameters a and E_1, and the constant temperature formula (CT) [1],

$$\rho_{CT}(E) = \frac{1}{T}e^{(E-E_0)/T} \tag{3}$$

with the free parameters T and E_0.

These formulas were found equivalent in describing the experimental data at excitation energies generally below 10 MeV [2, 3, 4]. The parameters of the level density formulas can be experimentally obtained by fitting known energy levels of complete level schemes at low excitation energies together with neutron resonances at the neutron binding energy. This approach was described in refs. [2, 3, 4]. In the present article we review the recent work along this line. The free parameters of both the spin distribution function and the two level density models were determined by a fit to updated experimental level scheme data. We use a database of complete low-energy level schemes for 310 nuclei between ^{19}F and ^{251}Cf. For most of these nuclei the neutron resonance densities were also known. A list of these nuclei is given in [4]. For the sets of empirical values for both the spin-cutoff parameter of the spin distribution function and the parameters of the BSFG and CT models, we propose simple formulas that can be easily calculated using only quantities from the mass tables. These formulas can be used to obtain reasonable estimations for the level densities of nuclei farther from stability. Finally, recent experimental evidence is reviewed that at low excitation energy the Constant Temperature model is the more correct description of nuclei.

2. Spin distribution function

The spin distribution function $f(J)$ in eq. (1) is described by the formula proposed by Ericson [6]

$$f(J,\sigma) = e^{-J^2/2\sigma^2} - e^{-(J+1)^2/2\sigma^2} \approx \frac{2J+1}{2\sigma^2}e^{-J(J+1/2)/2\sigma^2} \tag{4}$$

with a single free parameter σ. The spin-cutoff parameter σ is generally related to an effective moment of inertia. This parameter may depend on the nuclear mass A, the level density parameter a or the nuclear temperature T. Also, it is predicted to increase with the excitation energy E. Different formulas were proposed for these dependencies [1, 5, 6, 7, 8, 9, 10, 11, 12, 13, 14, 15, 16]. Because of these ambiguities and in addition due to the lack of systematic experimental information we performed studies of the spin distribution function, using the actual knowledge of the discrete levels at low excitation energies.

2.1. Staggering of spin distribution function in even-even nuclei

In a first study [17], we concentrated on the general evolution of the spin-cutoff parameter σ at low excitation energies, with the mass number and possibly other quantities, without considering a dependence on the excitation energies. From our database with 310 nuclei, we selected complete level schemes in a given energy (usually up to 1 – 3 MeV excitation) and spin range. This procedure yielded 8116 levels (with known spin value) in 1556 spin groups (levels with the same spin in a nucleus).

An important outcome of this study has been the observation of a spin staggering of the spin distribution function for the even-even nuclei. This is illustrated in Fig. 1 for the nucleus ^{112}Cd where it can be seen that the spin distribution cannot be accurately described by the smooth function (4). This oscillation in spin has been observed for all even-even nuclei, but is absent

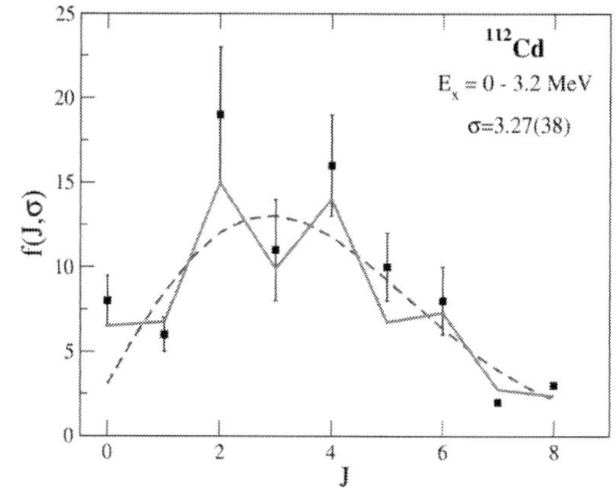

Figure 1. Experimental and calculated spin distribution for ^{112}Cd (symbols). Dashed line: fit with Ericson's formula (4). Full line: fit with the newly proposed formula (5).

in the odd-mass and the odd-odd nuclei. While the odd- and even-spin value deviations from the average description given by formula (4) are rather symmetrical, the deviation of the spin 0 value was found systematically larger. To describe this even-odd spin staggering, we proposed the following formula for the spin distribution of the even-even nuclei:

$$f_{ee}(J,\sigma) = f(J,\sigma)\cdot(1+x), \tag{5}$$

where $f(J,\sigma)$ is given by eq. (4), and

$$x = \begin{cases} +0.227(14), & \text{for even spin values,} \\ -0.227(14), & \text{for odd spin values,} \\ +1.02(9), & \text{for zero spin levels.} \end{cases} \tag{6}$$

The staggering parameter x was not found to depend on the mass number. The corresponding fit procedure of the experimental data (the number of levels of each spin in a nucleus) is described in ref. [17]. We investigated the dependence of σ on several quantities, like mass, level density parameter a, temperature T, moment of inertia, and deformation β, and found that only the mass dependence was important. The data could be well described by the simple formula $\sigma^2 = 2.61A^{0.28}$, which thus represents an average description of "low excitation energy" region.

The even-odd spin staggering of the spin distribution function in even-even nuclei was mentioned some time ago [2] but not systematically investigated and never discussed in connection with predictions of theoretical models. It appears, however, that, at least some theoretical models intrinsically contain this spin oscillation. This is shown in Fig. 2 for two models. In the left-side graph, experimental spin distribution functions for ^{168}Er in two excitation energy ranges (0 – 1.5 MeV and 1.5 – 2.3 MeV) are compared with predictions of the IBA (Interacting Boson Approximation) model [18]. We have used the *spdf* version of this model, employing bosons of spin 0, 1, 2, and 3 [19]. The Hamiltonian parameters of the model were determined by fitting both positive- and negative-parity low-energy levels in this nucleus. In the lower energy range the model predicts rather well the experimental situation (as described by eq. (5)). At higher excitation energies the model predicts a lower number of states, as expected. The right-side graph shows predictions of Shell Model Monte Carlo (SMMC) calculations for ^{56}Fe, at several excitation energies, up to 11.5 MeV [15]. These calculations show that the spin staggering is strongest at low energy (and consistent with our formulas (5) and (6)), and gradually diminishes, being practically absent above 10 MeV excitation. Similar calculations for the neighboring nuclei ^{55}Fe and ^{60}Co do not show any spin oscillation (Fig. 5 in ref. [17]). Thus, realistic SMMC calculations correctly predict the spin staggering, very probably an effect of the pairing interactions [15], but also the fact that it decreases with excitation energy and disappears at a certain energy. Due to the lack of experimental data at higher excitation energies we could not investigate this behaviour.

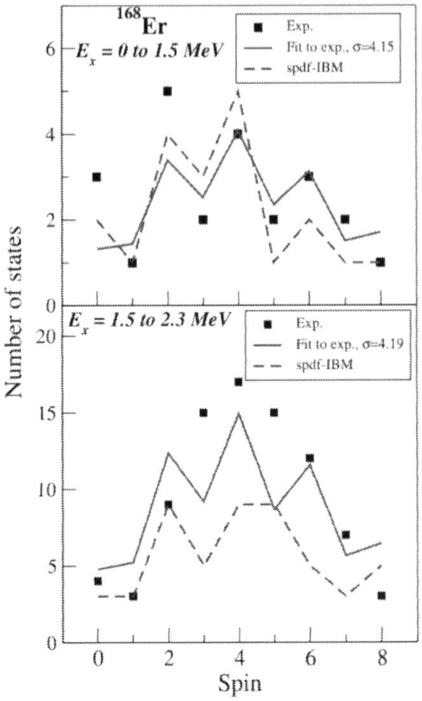

Figure 2. Spin distribution in ^{168}Er for two energy intervals. Symbols: experiment; full line: fits with eqs. (5)-(6); dashed line: predictions of the *spdf*-IBA model.

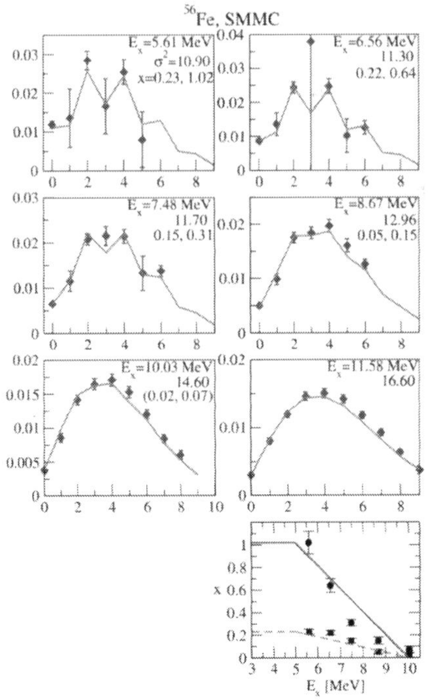

Figure 3. Shell model Monte Carlo predictions for ^{56}Fe, for different excitation energies. The solid lines are fits with eq. (5), yielding the displayed pairs of x values (eq. (5)). The lowest graph shows the evolution of x with the excitation energy (the straight lines are drawn to guide the eye). From [15].

2.2. Energy dependence of the spin-cutoff parameter

The method mentioned in the preceding section, of counting levels in energy bins, is not adequate to deduce the dependence of σ on the excitation energy because there are not enough levels with known spin. In a subsequent paper [20] we propose the "moment method" which exploits both the energy and spin dependence of a given nuclear level scheme, as shown in Fig. 4.

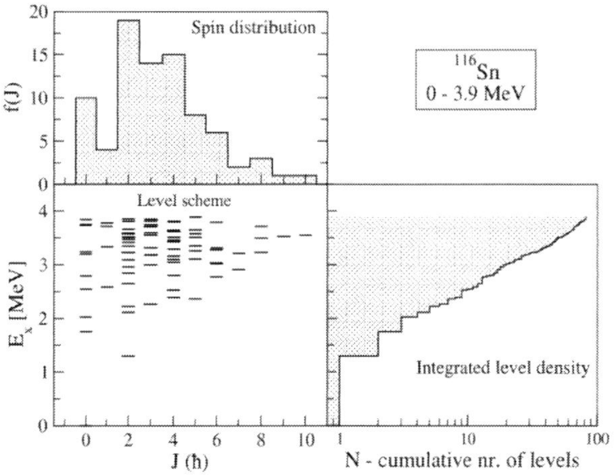

Figure 4. Example of the complete level scheme of ^{116}Sn up to $E_x = 3.9$ MeV, shown in two dimensions: spin and excitation energy. Projections on the two axes illustrates the spin distribution function and the level density, respectively.

Namely, we calculated individual moments in the (E, J) plane for each nucleus with the known levels (E_i, J_i) in the given energy and spin range, $M_{m,n}^{exp} = \sum_i (J_i^m \cdot E_i^n)$. The following nine moments were determined from the experimental values: J, J^2, J^3, $J \cdot E$, $J^2 \cdot E$, $J^3 \cdot E$, $J \cdot E^2$, $J^2 \cdot E^2$, $J^3 \cdot E^2$ and compared with the corresponding moments calculated with the CT model, eq. (3), for the level density. In the fit procedure, the dependence on both mass number A and excitation energy E were taken into account. The following result was obtained:

$$\sigma^2 = 0.391 \cdot A^{0.675}(E - 0.5 \cdot Pa')^{0.312}. \quad (7)$$

It is found that a backshift of the energy gives the best results. For this backshift we found that Pa', the so-called deuteron pairing energy, is very useful; this quantity will be discussed in detail in the next section. Formula (7) gives a good description to the existing (rather sparse) experimental determinations of σ ([20] and references therein). Because it is based mainly on the known low-lying levels, we think that the use of this formula up to energies of the order of the neutron binding energy is rather realistic. Formula (7) represents a practical alternative to the more classical formulas, such as those based on statistical mechanical calculations [21] or on the assumption of the nucleus as a rigid sphere [11], because both these formulas use the theoretical level density parameter a. Our formula depends only on mass and excitation energy, and through Pa' it implicitly distinguishes between isobars (types of nucleus).

3. Level density parameters

In our most recent work [20] we have determined the parameters of the BSFG model (eq. (2)) and of the CT model (eq. (3)) by fitting experimental level spacings and average neutron resonance spacings for the 310 nuclei in our data set (level density at the neutron binding energy is not known for 14 of these nuclei, but they were kept because their level scheme was complete up to rather high excitation energy). This new determination followed the procedure of ref. [4], but used the spin distribution eq. (7) described above with its experimentally determined dependence on mass and excitation energy, and taking into account the spin staggering for the even-even nuclei given by eq. (6). The first step was to determine for each nucleus the parameters a, E_1, and T, E_0 of the BSFG and CT models, respectively. These are shown in Figs. 4 and 5, respectively. A table with these parameters is given in [20]. The next step, also similar to the approach from [4], was to study the correlations between these parameters and different other quantities to be able to propose simple formulas (parametrizations) of these values. The quantities with which these parameters were found to be strongly correlated are the shell correction and the deuteron pairing [4, 20]. The correlation of a with the shell correction was observed in many works, starting with ref. [1]. The backshift parameters E_1 and E_0 turned out to be very similar to the deuteron pairing Pa'. The definitions of these two quantities are as follows. The shell correction S is the difference between the mass calculated with a liquid-drop Weizsäcker-type formula and the experimental atomic mass. For the liquid drop formula we use the parametrization of ref. [22], while the experimental masses are taken from [23]. The deuteron pairing energy is calculated from mass or mass excess values $M(A, Z)$ of the mass tables [23]:

$$Pa'_k = \frac{1}{2}[M(A+2, Z+1) - 2 \cdot M(A, Z) + M(A-2, Z-1)]. \quad (8)$$

It should be noted that this quantity has a different sign convention than Pa, the one used in our first work [4], which is also published in tables [24]: $Pa' = -(-1)^Z Pa$.

Based on these correlations, we were able to propose simple formulas that describe the empirical values of the parameters of the two level density models [20] as function of mass, shell correction S, and deuteron pairing Pa'. Thus, for the BSFG model parameters, we propose the following formulas:

$$a = (p_1 + p_2 S')A^{p_3} \;=\; (0.199 + 0.0096 S')A^{0.869}, \quad (9)$$

with $S' = S + 0.5 \cdot Pa'$, and $p_1 = 0.199(7)$; $p_2 = 0.0096(4)$; $p_3 = 0.869(7)$.

$$E_1 = q_1 + 0.5 \cdot Pa' \;=\; -0.381 + 0.5 \cdot Pa', \quad (10)$$

with $q_1 = -0.381(14)$. The experimental values of a and E_1 and their fits with formulas (9) and (10) are shown in Fig. 5.

For the CT model parameters we obtain the following formulas:

$$T = A^{-2/3}/(p_1 + p_2 S') \;=\; A^{-2/3}/(0.0597 + 0.00198 S'), \quad (11)$$

with $S' = S + 0.5 \cdot Pa'$ and $p_1 = 0.0597(2)$; $p_2 = 0.00198(10)$.

$$E_0 = q_1 + 0.5 \cdot Pa' \;=\; -1.004 + 0.5 \cdot Pa', \quad (12)$$

with $q_1 = -1.004(21)$. The experimental values of T and E_0 and their fits with formulas (11) and (12) are shown in Fig. 6.

In ref. [25] it was observed that there are very good correlations between the parameters of the BSFG and CT models. In the case of the present values, these very compact correlations (see Fig. 13 of ref. [20]) are:

$$T = 5.164 \cdot a^{-0.791}, \quad (13)$$

while for the energy backshifts a linear relationship results from eqs. (10) and (12):

$$E_0 = E_1 - 0.623, \quad (14)$$

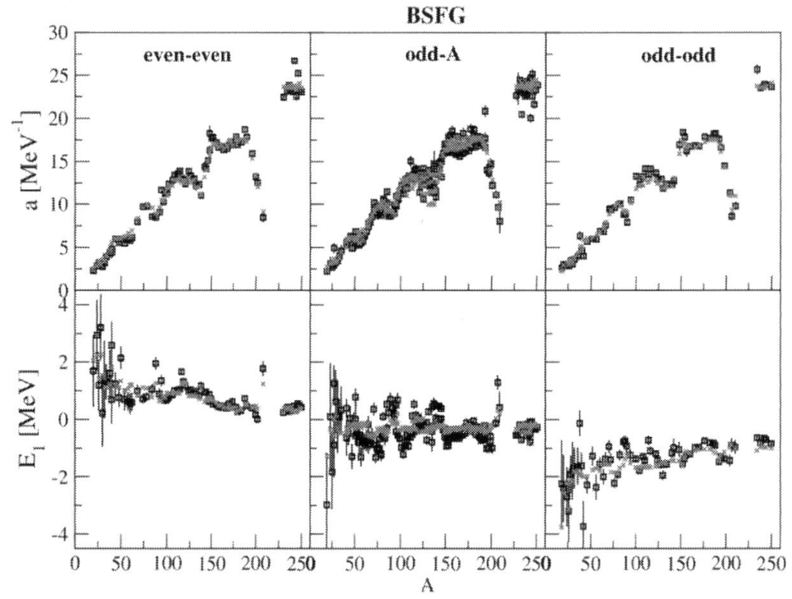

Figure 5. Level density parameters for the BSFG model. Rectangles with error bars: experimentally determined values; red x symbols: values provided by formulas (9), (10).

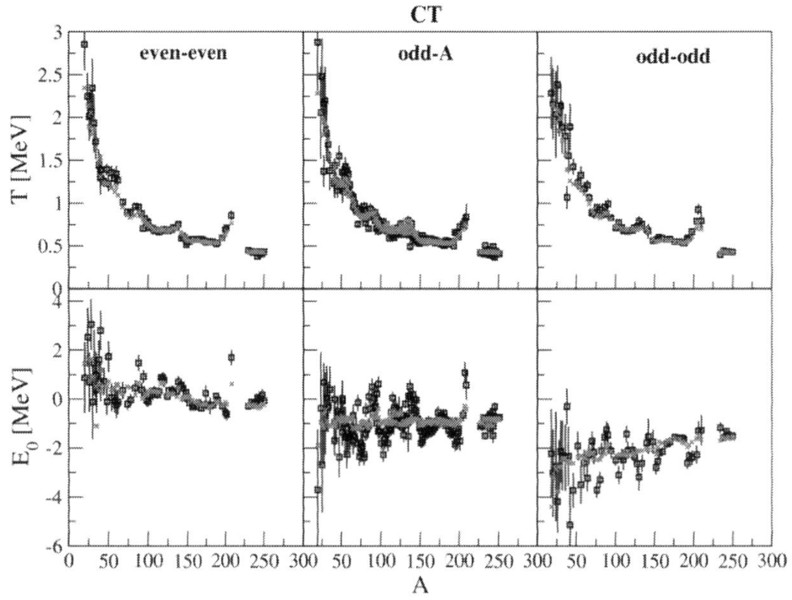

Figure 6. Level density parameters for the CT model. Rectangles with error bars: experimentally determined values; red x symbols: values provided by formulas (11), (12).

4. Constant temperature in nuclei

As remarked in our previous papers [4, 20, 25], both the BSFG model and the CT model can be successfully applied in an empirical way to low-energy levels and neutron resonance spacings, therefore both appear as convenient ways to parametrize the level densities at least up to the neutron binding energy. Figure 7 illustrates, by several cases of nuclei from light to heavy, and of different types (even-even, odd-A, and odd-odd) that the two models offer an equally good description to the cumulative number of levels.

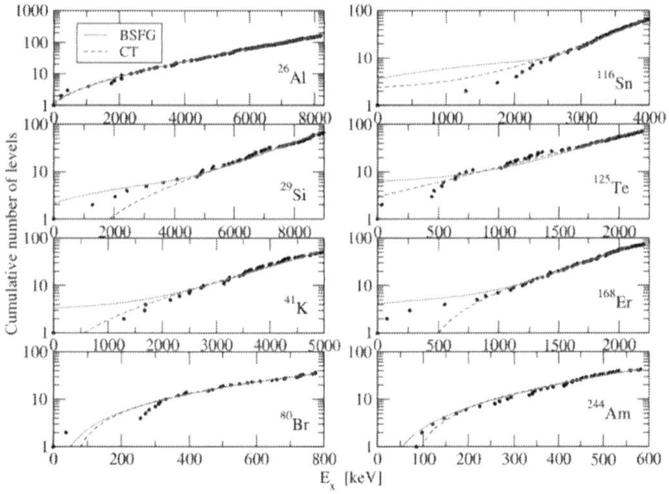

Figure 7. Examples of fits of the experimental cumulative number of levels of different nuclei from our set with the BSFG and CT model formulas. The fit also includes the level spacing at the neutron resonance energy, which is not shown in the graphs.

The BSFG model, based on the original Fermi-gas model of Bethe [26], calculates the level densities by counting the number of possibilities in which the excitation energy can be distributed among the single-particle states, a combinatorial problem that can use methods of statistical mechanics [27]. On the other hand, the much simpler CT model is based on the definition of the nuclear temperature T as [6, 27]

$$\frac{1}{T} = \frac{d[\ln \rho(E)]}{dE}$$

where $\rho(E)$ is the level density at the excitation energy E. If one assumes that T is constant, one obtains the CT model expression of eq. (3). Therefore, according to the CT model in a certain energy range, if we increase the excitation energy of the nuclei their temperature remains constant. This behavior is similar to that of systems with a first-order critical phase transition (like solid-liquid or liquid-gas). The nuclei behave like a water/ice mixture what means that their number of degrees of freedom increases with the excitation energy in such a way that the temperature stays constant; this effect may be explained by the destruction of the nuclear Cooper pairs with the excitation energy ("melting" of the Cooper pairs)[28], thus leaving the mean energy constant per excited nucleon according to the formula $T = E/n_{ex}$ [27], where n_{ex} is the number of excited nucleons. The early model of Gilbert and Cameron [1] used a composite model, with a CT level density below the neutron binding energy, and a BSFG model above that.

There is increasing experimental evidence that the CT model is the right description of the nuclei at even higher excitation energies. A strong such evidence was borne out by a Hauser-Feshbach approach analysis of neutron and proton evaporation spectra from the^{6}Li + ^{55}Mn and d + ^{59}Co reactions [29]. Using a Fermi-gas like temperature dependence failed to reproduce the experimental data, while assuming the CT model described the spectra up to about 20 MeV excitation, indicating that for mass ≈ 60 nuclei the CT model works up to that energy. This agrees with earlier experiments that revealed also that the CT formula represents well data up to about 10 MeV [30]. Another evidence came from a different field: the behavior of fragments from low-energy neutron induced fission [31]. It concerns the average number of neutrons emitted by the fission fragments as a function of their mass measured in neutron-induced fission of ^{237}Np at two neutron energies, 0.8 and 5.55 MeV, where a very interesting feature has been noted: higher incident neutron energy, that is higher excitation energy of the system, leads to an increase of the number of evaporated neutrons only for the heavy fragments. Thus, it takes place a peculiar sharing of the excitation energy between the two fragments, the heavy fragments taking practically the whole energy. This is inconsistent with a Fermi-gas behaviour, that would lead to the division of the excitation energy between the two fragments in their mass ratio (as it happens in many binary reactions involving high excitation energies). On the other hand, it is fully consistent with the Constant Temperature model. Indeed, according to eq. (11), the temperature of the heavy fragment is generally lower than that of the light fragment (except for special cases, as an effect of shell closures). If the two fragments at the scission point, therefore in thermal contact, are in the pairing regime where they have different temperatures, then the cooler heavy fragments will absorb the entire available excitation energy and therefore will emit more neutrons. This particular sorting of the available energy is a consequence of the two different temperatures of the fragments, and shows that a correct description of the behavior of the nuclei at low and moderate excitation energies by the CT model leads to an understanding of these features of the low-energy neutron induced fission.

5. Conclusions

We reviewed the description of the nuclear level density by simple models whose parameters were empirically determined from existing experimental data.

An empirical mass and energy description has been proposed for the spin-cutoff parameter of the spin distribution function. Also, for the observed spin staggering of the spin distribution function of the even-even nuclei a simple formula was proposed, and comparison with predictions of two structure models was presented.

Using this description for the spin distribution function, the parameters of the BSFG and CT models for the level density were determined for 310 nuclei between ^{19}F and ^{251}Cf by fitting both the completely known low-energy discrete levels and the level spacings at the neutron binding energy. Simple formulas were proposed that reasonably describe these sets of parameters as a function of mass number. The formulas proposed for both spin-cutoff paramater and the BSFG and CT model parameters, respectively, contain only quantities that can be extracted from mass tables, and are thought to be reasonable approximations up to at least the neutron binding energy. They can also be used as a practical way to extrapolate level density estimations for nuclei far from stability, for which such quantities cannot be experimentally determined.

Both the BSFG and CT models give equally good description to experimental level densities at excitation energies at least up to the neutron binding energy. Recent experimental evidences are reviewed that at least up to this energy (but, in ceratin mass regions, even at higher energies) the correct description of nuclei is given by the constant temperature model.

Acknowledgments
We thank Dr. S. Pascu for performing the *spdf*-IBM calculations. This work was partly supported by the Romanian project PN-II-ID-PCE-2011-3-0140 (UEFSCDI).

6. References

[1] Gilbert A and Cameron A G W 1965 *Can. J. Phys.* **43** 1446; Brancazio P J and Cameron A G W 1969 *ibidem* **47** 1029
[2] von Egidy T, Behkami A N and Schmidt H H 1986 *Nucl. Phys.* A **454** 109
[3] von Egidy T, Schmidt H H and Behkami A N 1988 *Nucl. Phys.* A **481** 189
[4] von Egidy T and Bucurescu D 2005 *Phys. Rev.* C **72** 044311; 2006 *ibidem* **73**, 049901(E)
[5] Ignatyuk A V, Smirenkin G N and Tishin A S, *Sov. J. Nucl. Phys.* 1975 **21** 255
[6] Ericson T *Adv. Phys.* 1960 **9** 425
[7] Dilg W, Schantl W, Vonach H and Uhl M 1973 *Nucl. Phys.* A **217** 267
[8] Iljinov A S, Mebel M V, Bianchi N, De Sanctis E, Guaraldo C, Lucherini V, Muccifora E, Polli E, Reolon A R and Rossi P 1979 *Nucl. Phys.* A **543** 450
[9] Rauscher T, Tielemann F -K and Kratz K -L, 1997 *Phys. Rev.* C **56** 1613
[10] Zhongfu H, Ping H, Zongdi S and Chunmei Z 1991 *Chin. J. Nucl. Phys.* **13** 147
[11] Grimes S M, Anderson J D, McClure J W, Pohl B A and Wong C 1974 *Phys. Rev.* C **10** 2373
[12] Huang P L, Grimes S M and Massey T 2000 *Phys. Rev.* C **62** 024002
[13] Al-Quraishi S I, Grimes S M, Massey T N and Ressler D A 2003 *Phys. Rev.* C **67** 015803
[14] Agrawal B K, Samaddar S K, Ansari A and De J N 1999 *Phys. Rev.* **59** 3109
[15] Alhassid Y, Liu S and Nakada H *Phys. Rev. Lett.* 2007 **99** 044304; Alhassid Y, private communication
[16] Kaneko K and Schiller A 2007 *Phys. Rev.* C **75** 044304
[17] von Egidy T and Bucurescu D 2008 *Phys. Rev.* C **C78** 051301(R)
[18] Arima A and Iachello F 1975 *Phys. Rev. Lett.* **35** 1069; Iachello F and Arima A 1987 *The interacting boson model* (Cambridge University Press)
[19] Kuznezov D F and Iachello F 1988 *Phys. Lett.* B **209** 420
[20] von Egidy T and Bucurescu D 2009 *Phys. Rev.* C **80** 0054310
[21] Faccini U and Saetta-Menichella A 1967 *Energ. Nucl.* (Milan) **15** 54
[22] Pearson J M 2001 *Hyperfine Interact.* **132** 59
[23] Audi G, Wapstra A H and Thibault C 2003 *Nucl. Phys.* A **729** 337
[24] Audi G 2003 http://amdc.in2p3.fr/masstables/Ame2003/rct.masso3
[25] Bucurescu D and von Egidy T 2005 *Phys. Rev.* C **72** 067304
[26] Bethe A H 1937 *Rev. Mod. Phys.* **9** 69
[27] Bohr A and Mottelson B R 1969 *Nuclear Structure* (Benjamin, New York/Amsterdam), vol. I
[28] Guttormsen M, Hjorth-Jensen M, Melby E, Rekstad J, Schiller A and Siem S 2001 *Phys. Rev.* C **63** 044301
[29] Voinov A V, Oginni B M, Grimes S M, Brune C R, Guttormsen M, Larsen A C, Massey T N, Schiller A and Siem S 2009 *Phys. Rev.* C **79** 031301(R)
[30] Guttormsen M, Chankova R, Hjorth-Jensen M, Rekstad J, Siem S, Schiller A and Dean D J 2003 *Phys. Rev.* C **68** 034311
[31] Schmidt K-H and Jurado B 2010 *Phys. Rev. Lett.* **104** 212501 ; 2011 *Phys. Rev.* C **83** 061601(R)

Multistep shell model in the complex energy plane

Z.X. Xu, C. Qi, R. J. Liotta, R. Wyss
KTH, Alba Nova University Center, SE-10691 Stockholm, Sweden
E-mail: chongq@kth.se

Abstract. We have adopted the multistep shell model in the complex energy plane to study nuclear excitations occurring in the continuum part of the spectrum. In this method one proceeds by solving the shell model equations in a successive manner. That is, in each step one constructs the building blocks to be used in future steps. We applied this formalism to analyze the unbound nuclei ^{12,13}Li starting from the one-particle states in ^{10}Li and two-particle states in ^{11}Li. In the former case the excitations correspond to the motion of three particles partitioned as the product of a one-particle and two-particle systems. The ground state of ^{12}Li is thus calculated to be an antibound (virtual) state. In the four-particle system ^{13}Li the states can be constructed as the coupling of two correlated pairs. We found that there is no bound or antibound state in ^{13}Li.

1. Introduction

Experimental facilities allow one nowadays to measure systems living a very short time. To describe these processes one has to consider the decaying character of the system. On the other hand, the theoretical study of unstable nuclei is a difficult undertaking. However, the system may be considered stationary if it lives a long time. In this case the time dependence can be circumvented. Of the various theories that have been conceived to analyze unbound systems, we mention an extension of the shell model to the complex energy plane [1]. The basic assumption of this theory is that resonances can be described in terms of states lying in the complex energy plane. The real parts of the corresponding energies are the positions of the resonances while the imaginary parts are minus twice the corresponding widths, as it was proposed by Gamow at the beginning of quantum mechanics [2]. These complex states correspond to solutions of the Schrödinger equation with outgoing boundary conditions. Details of the formalism and its application to two-nucleon systems can be found, e.g., in Refs. [3, 4, 5].

Our aim is to develop a suitable formalism to treat unstable nuclei involving many valence nucleons in the continuum [6]. This formalism is an extension of the shell model in the complex energy plane [1]. The correlations induced by the pairing force acting upon particles moving in decaying single-particle states is taken into account by using the multistep shell model (MSM) [7]. The formalism is presented in Section 2. Applications are in Section 3 and a summary and conclusions are in Section 4.

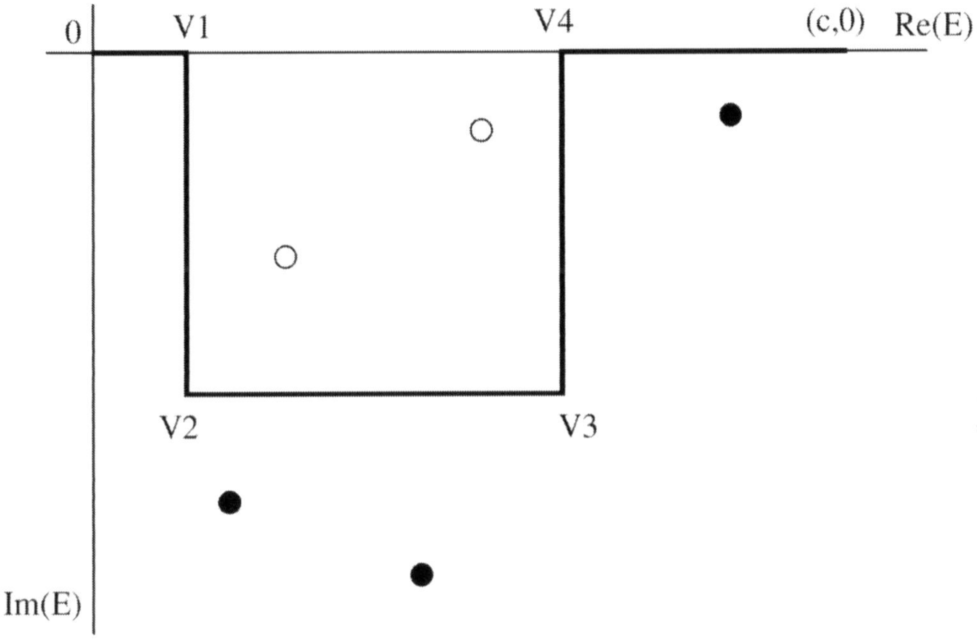

Figure 1. Integration contour L^+ in the complex energy plane. The open circles denote the resonances included in the sum of Eq. (2), while the solid circles are those excluded. The vertex $(c, 0)$ corresponds to the energy cutoff point c.

2. The formalism

The eigenstates of a central potential obtained as outgoing solutions of the Schrödinger equation can be used to express the Dirac δ-function as [8]

$$\delta(r - r') = \sum_n w_n(r) w_n(r') + \int_{L^+} dE u(r, E) u(r', E), \tag{1}$$

where the sum runs over all the bound and antibound states plus the complex states (resonances) which lie between the real energy axis and the integration contour L^+ (see, Fig. 1). The antibound states are virtual states with negative scattering length. The wave function of a state n in these discrete set is $w_n(r)$ and $u(r, E)$ is the scattering function at energy E.

Discretizing the integral of Eq. (1) one obtains the set of orthonormal vectors $|\varphi_j\rangle$ forming the Berggren representation [9]. Since this discretization provides an approximate value of the integral, the Berggren vectors fulfill the relation $I \approx \sum_j |\varphi_j\rangle\langle\varphi_j|$. One has,

$$\int_{L^+} dE u(r, E) u(r', E) = \sum_p h_p u(r, E_p) u(r', E_p), \tag{2}$$

where the quantities E_p and h_p are defined by the procedure that uses to perform the integration. Therefore the orthonormal vectors $|\varphi_j\rangle$ are given by the set of bound, antibound and Gamow states, i.e., $\langle r|\varphi_n\rangle = \{w_n(r, E_n)\}$, and the discretized scattering states, i.e.,

Journal of Physics: Conference Series **338** (2012) 012029 doi:10.1088/1742-6596/338/1/012029

$\langle r|\varphi_p\rangle = \{\sqrt{h_p}u_p(r, E_p)\}$. In Ref. [9] it was also found that few discretized scattering states in the basis are enough to obtain convergence. Using the Berggren representation one readily gets the two-particle shell-model equations in the complex energy plane (CXSM) [4], i.e.,

$$(W(\alpha_2) - \epsilon_i - \epsilon_j)X(ij;\alpha_2) = \sum_{k\leq l}\langle\tilde{k}\tilde{l};\alpha_2|V|ij;\alpha_2\rangle X(kl;\alpha_2), \tag{3}$$

where V is the residual interaction and i, j, k, l label single-particle states. The two-particle states are labeled by α_2. ϵ and W denote the energy of the single-particle and two-particle states, respectively. The tilde in the interaction matrix element denotes mirror states so that in the corresponding radial integral there is not any complex conjugate, as required by the Berggren metric.

The two-particle wave function is given by

$$|\alpha_2\rangle = P^+(\alpha_2)|0\rangle, \tag{4}$$

where the two-particle creation operator is

$$P^+(\alpha_2) = \sum_{i\leq j} X(ij;\alpha_2)\frac{(c_i^+c_j^+)_{\lambda_{\alpha_2}}}{\sqrt{1+\delta_{ij}}}, \tag{5}$$

and λ_{α_2} is the angular momentum of the two-particle state.

2.1. Separable interaction

For a separable interaction the matrix can be written as

$$\langle\widetilde{kl};\alpha|V|ij;\alpha\rangle = -G_\alpha f_\alpha(kl)f_\alpha(ij), \tag{6}$$

where G_α is the interaction strength and $f_\alpha(ij)$ is the matrix element of the field defining the interaction. The shell model equation becomes

$$-\frac{X(ij;\alpha)}{G_\alpha} = \sum_{k\leqslant l}\frac{f_\alpha(kl)f_\alpha(ij)X(kl;\alpha)}{(\omega_\alpha - \epsilon_i - \epsilon_j)}. \tag{7}$$

Multiplying $\sum_{i\leqslant j} f_\alpha(ij)$ on both sides of Eq. (7) one gets the so-called dispersion relation

$$-\frac{1}{G_\alpha} = \sum_{i\leqslant j}\frac{f_\alpha^2(ij)}{\omega_\alpha - \epsilon_i - \epsilon_j}. \tag{8}$$

The two-particle wave function amplitudes are given by

$$X(ij;\alpha) = N_\alpha\frac{f_\alpha(ij)}{\omega_\alpha - \epsilon_i - \epsilon_j}, \tag{9}$$

where N_α is the normalization constant determined by requiring

$$\sum_{i\leqslant j} X^2(ij;\alpha) = 1. \tag{10}$$

2.2. The Multistep Shell Model Method

The Multistep Shell Model Method (MSM) solves the shell model equations in several steps. In the first step the single-particle representation is chosen. In the second step the energies and wave functions of the two-particle system are evaluated with a given two-particle interaction. The three-particle states are evaluated in terms of a basis consisting of the tensorial product of the one- and two-particle states previously obtained. The MSM basis is overcomplete and non-orthogonal. To correct this one needs to evaluate the overlap matrix among the basis states also. A general description of the formalism is in Ref. [10]. The particular system that is of our interest here, i.e., the three-particle case, can be found in Ref. [7]. Below we refer to this formalism as CXMSM [6].

The three-particle energies $W(\alpha_3)$ are given by [7]

$$(W(\alpha_3) - \varepsilon_i - W(\alpha_2))\langle\alpha_3|(c_i^+ P^+(\alpha_2))_{\alpha_3}|0\rangle$$
$$= \sum_{j\beta_2}\left\{\sum_k (W(\beta_2) - \varepsilon_i - \varepsilon_k)A(i\alpha_2, j\beta_2; k)\right\}\langle\alpha_3|(c_j^+ P^+(\beta_2))_{\alpha_3}|0\rangle, \tag{11}$$

where

$$A(i\alpha_2, j\beta_2; k) = \hat{\alpha}_2\hat{\beta}_2\left\{\begin{array}{ccc} i & k & \beta_2 \\ j & \alpha_3 & \alpha_2 \end{array}\right\} Y(kj;\alpha_2)Y(ki;\beta_2), \tag{12}$$

and

$$Y(ij;\alpha_2) = (1+\delta(i,j))^{1/2}X(ij;\alpha_2). \tag{13}$$

The rest of the notation is standard.

The matrix defined in Eq. (11) is not hermitian and the dimension may be larger than the corresponding shell-model dimension. This is due to the violations of the Pauli principle as well as over-counting of states in the CXMSM basis. Therefore the direct diagonalization of Eq. (11) is not convenient. One needs to calculate the overlap matrix in order to transform the CXMSM basis into an orthonormal set. In this three-particle case the overlap matrix is

$$\left\langle 0|(c_i^+ P^+(\alpha_2))^\dagger_{\alpha_3}(c_j^+ P^+(\beta_2))_{\alpha_3}|0\right\rangle = \delta_{ij}\delta_{\alpha_2\beta_2} + \sum_k A(i\alpha_2, j\beta_2; k). \tag{14}$$

Using this matrix (14) one can transform the matrix determined by Eq. (11) into a hermitian matrix T which has the right dimension. The diagonalization of T provides the three-particle energies. The corresponding wave function amplitudes can be readily evaluated to obtain

$$|\alpha_3\rangle \;=\; P^+(\alpha_3)|0\rangle, \tag{15}$$
$$P^+(\alpha_3) \;=\; \sum_{i\alpha_2} X(i\alpha_2;\alpha_3)(c_i^+ P^+(\alpha_2))_{\alpha_3}, \tag{16}$$

where $P^+(\alpha_3)$ is the three-particle creation operator. It has to be pointed out that in cases where the basis is overcomplete the amplitudes X are not well defined. But this is no hinder to evaluate the physical quantities. For details see Ref. [7].

The CXMSM allows one to choose in the basis states a limited number of excitations. This is because in the continuum the vast majority of basis states consists of scattering functions. That is, the majority of the two-particle states provided by the CXSM are complex states which form a part of the continuum background. Only a few of those calculated states correspond to physically meaningful resonances, i.e., resonances which can be observed. Below we call a "resonance" only to a complex state which is meaningful. These resonances are mainly built upon single-particle states which are either bound or narrow resonances. Yet, one cannot ignore the continuum when evaluating the resonances. The continuum configurations in the resonance wave function are small but many, and they affect the two-particle resonance significantly [4]. Therefore, the great advantage of the CXMSM is that one can include in the basis only two-particle resonances, while neglecting the background continuum states, which form the vast majority of complex two-particle states.

2.3. *The neutron-proton correlation*

The influence of the neutron-proton correlation can be considered in a straightforward way in our CXMSM. For example, we consider a system with three neutrons and one proton , say ^{12}Li by assuming ^{8}He as the inert core [6]. With the basis denoted as

$$|p\alpha_3;\alpha_4\rangle = (c_p^+ P^+(\alpha_3))_{\alpha_4}|0\rangle, \tag{17}$$

where p labels the proton state and α_4 are the three-neutron one-proton state, the four-particle energies $W(\alpha_4)$ in ^{12}Li are given by

$$\begin{aligned}(W(\alpha_4) - \varepsilon_p - W(\alpha_3))\langle\alpha_4|(c_p^+ P^+(\alpha_3))_{\alpha_4}|0\rangle \\ = \sum_{q\beta_3}\left\{\sum_{kl\lambda\alpha_2}\langle pk;\lambda|V|ql;\lambda\rangle B_1 + \sum_{ijkl\lambda\alpha_2\beta_2}\langle pi;\lambda|V|ql;\lambda\rangle B_2\right\}\langle\alpha_4|(c_q^+ P^+(\beta_3))_{\alpha_4}|0\rangle,\end{aligned} \tag{18}$$

where,

$$B_1 = (-1)^{p+q+k+l} X(k\alpha_2;\alpha_3) F(l\alpha_2;\beta_3)\hat{\alpha}_3\hat{\beta}_3\hat{\lambda}^2 \left\{\begin{array}{ccc} p & k & \lambda \\ \alpha_2 & \alpha_4 & \alpha_3\end{array}\right\}\left\{\begin{array}{ccc} q & l & \lambda \\ \alpha_2 & \alpha_4 & \beta_3\end{array}\right\}, \tag{19}$$

and

$$\begin{aligned}B_2 = (-1)^{p+q+i+l} Y(ji;\alpha_2)Y(jk;\beta_2)X(k\alpha_2;\alpha_3)F(l\beta_2;\beta_3) \\ \times\hat{\alpha}_2\hat{\alpha}_3\hat{\beta}_2\hat{\beta}_3\hat{\lambda}^2 \left\{\begin{array}{ccc} p & i & \lambda \\ \beta_2 & \alpha_4 & \alpha_3\end{array}\right\}\left\{\begin{array}{ccc} q & l & \lambda \\ \beta_2 & \alpha_4 & \beta_3\end{array}\right\}\left\{\begin{array}{ccc} i & j & \alpha_2 \\ k & \alpha_3 & \beta_2\end{array}\right\}.\end{aligned} \tag{20}$$

Here p and q label proton states, while i, j, k, l label neutron states. $\langle pk;\lambda|V|ql;\lambda\rangle$ denotes the corresponding proton-neutron interaction matrix element. The wave function amplitudes X and the projected quantities F have been evaluated in previous steps of the CXMSM. Notice that in this case the overlap matrix is the unit matrix, i.e.,

$$\langle 0|(c_{p'}^+ P^+(\alpha_3'))^\dagger_{\alpha_4}(c_p^+ P^+(\alpha_3))_{\alpha_4}|0\rangle = \delta_{pp'}\delta_{\alpha_3\alpha_3'}. \tag{21}$$

2.4. *Four-particle states*

Systems with four like particles can be described in a correlated two-particle basis as

$$|\alpha_2\beta_2;\alpha_4\rangle = (P^\dagger(\alpha_2)P^\dagger(\beta_2))_{\alpha_4}|0\rangle, \tag{22}$$

where $\alpha_2 \leqslant \beta_2$. One can obtain the four-particle MSM equation as

$$\begin{aligned}(W(\alpha_4) - W(\alpha_2) - W(\beta_2))\langle\alpha_4|(P^\dagger(\alpha_2)P^\dagger(\beta_2))_{\alpha_4}|0\rangle = -\sum_{\gamma_2\leqslant\delta_2}\sum_{ijkl}(1+\delta_{\gamma_2\delta_2})^{-1} \\ \times(W(\gamma_2) + W(\delta_2) - \epsilon_i - \epsilon_j - \epsilon_k - \epsilon_l)B(ijkl,\alpha_2\beta_2\gamma_2\delta_2;\alpha_4)\langle\alpha_4|(P^\dagger(\gamma_2)P^\dagger(\delta_2))_{\alpha_4}|0\rangle,\end{aligned} \tag{23}$$

where

$$B(ijkl,\alpha_2\beta_2\gamma_2\delta_2;\alpha_4)=\hat{\alpha}_2\hat{\beta}_2\hat{\gamma}_2\hat{\delta}_2 Y(ij;\alpha_2)Y(kl;\beta_2)Y(ik;\gamma_2)Y(jl;\delta_2)\begin{Bmatrix} i & j & \alpha_2 \\ k & l & \beta_2 \\ \gamma_2 & \delta_2 & \alpha_4 \end{Bmatrix}. \quad (24)$$

The four-particle overlap matrix can be obtained in a similar way as

$$\begin{aligned}\langle 0|(P^\dagger(\alpha_2)P^\dagger(\beta_2))^\dagger_{\alpha_4}(P^\dagger(\gamma_2)P^\dagger(\delta_2))_{\alpha_4}|0\rangle &= \delta_{\alpha_2\gamma_2}\delta_{\beta_2\delta_2} \\ +(-1)^{\alpha_2+\beta_2-\alpha_4}\delta_{\alpha_2\delta_2}\delta_{\beta_2\gamma_2} &- \sum_{ijkl} B(ijkl;\alpha_2\beta_2\gamma_2\delta_2). \end{aligned} \quad (25)$$

To restore the Pauli principle, one can use the Schmidt procedure to generate a set of orthonormal basis $|u_i\rangle$ (which has the right dimensions) from the overlap matrix. The overlap matrix is thus transformed into an identity matrix

$$\langle u_m|u_n\rangle = I_{mn} = \sum_{ij} \xi_m(i)O(i,j)\xi_n(j), \quad (26)$$

where $O(i,j)$ is the overlap matrix, and the wave function in the Schmidt basis is

$$|u_m\rangle = \sum_i \xi_m(i)|i\rangle. \quad (27)$$

One can transform the dynamical matrix into a hermitian matrix T in the right dimensions if all possible basis states are included. The dynamical matrix $M(ij)$ in the MSM basis is given by

$$W(\alpha)\langle i|\alpha\rangle = \sum_j M(i,j)\langle j|\alpha\rangle. \quad (28)$$

To transform it into the orthonormal basis, one can use Eq. (27) and the identity equation. One thus obtains

$$W(\alpha)\sum_i \xi_m(i)\langle i|\alpha\rangle = \sum_i \xi_m(i)\sum_j M(i,j)\sum_n \langle j|u_n\rangle\langle u_n|\alpha\rangle, \quad (29)$$

$$W(\alpha)\langle u_m|\alpha\rangle = \sum_n \left\{\sum_{ij}\xi_m(i)M(i,j)\sum_k \xi_n(k)\langle j|k\rangle\right\}\langle u_n|\alpha\rangle. \quad (30)$$

The expression in the bracket is therefore the hermitian matrix T, which can be written as

$$T(m,n) = \sum_{ijk} \xi_m(i)M(i,j)O(j,k)\xi_n(k). \quad (31)$$

3. Applications

Using the Berggren single-particle representation described above, we will evaluate the complex energies and wave functions of the unbound nuclei ^{12,13}Li using the MSM basis states consisting of the Berggren one-particular states, which are states in ^{10}Li, times the two-particle states corresponding to ^{11}Li. The spectrum of ^{11}Li was already evaluated within the CXSM including antibound states [13]. Here we will repeat that calculation in order to determine the two-particle states to be used in the calculation of the three- and four-particle systems.

To define the Berggren single-particle representation we still have to choose the integration contour L^+ (see Eq. (1)). The valence shells are the low lying resonances $0p_{1/2}$ at (0.195,-0.047)

MeV and $0d_{5/2}$ at (2.731, -0.545) MeV. Besides, the state $1s_{1/2}$ appears as an antibound state. To include in the representation the antibound $1s_{1/2}$ state as well as the Gamow resonances $0p_{1/2}$ and $0d_{5/2}$ we will use two different contours. The number of points on each contour defines the energies of the scattering functions in the Berggren representation, i.e., the number of basis states corresponding to the continuum background. This number is not uniformity distributed, since in segments of the contour which are close to the antibound state or to a resonance the scattering functions increase strongly.

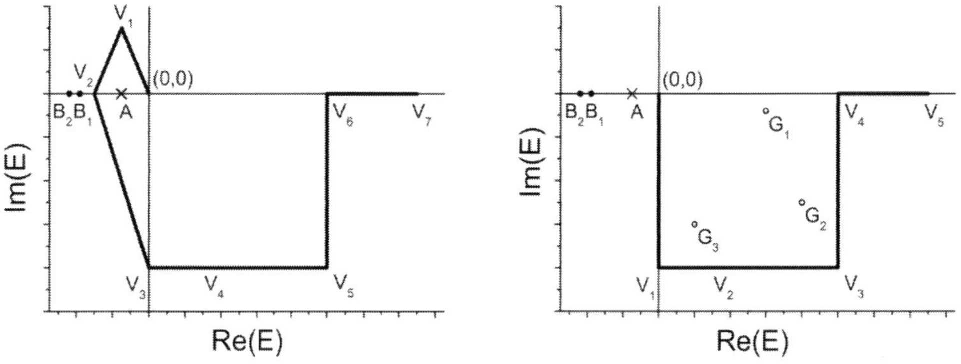

Figure 2. Left: Contour used to include the antibound state (see, also, Ref. [12]). The points B_i denote bound state energies while A denotes the antiboud state. Right: Contour used to include the Gamow resonances represented by the points G_i.

The dynamics of ^{11}Li is determined by the pairing force acting upon the two neutrons coupled to a state 0^+, which behaves as a normal even-even ground state [11, 13]. Besides the energy, this state has been measured to have an angular momentum contain of about 60 % of s-waves and 40 % of p-waves, although small components of other angular momenta are not excluded.

We will use a separable interaction as in Ref. [13]. The strength G_{λ_2}, corresponding to the states with angular momentum λ_2 and parity $(-1)^{\lambda_2}$, will be determined by fitting the experimental energy of the lowest of these states, as usual. It is worthwhile to point out that G_{λ_2} defines the Hamiltonian and, therefore, is a real quantity. The energies are thus obtained by solving the corresponding dispersion relation. The two-particle wave function amplitudes are given by

$$X(ij;\alpha_2) = N_{\alpha_2}\frac{f(ij,\alpha_2)}{\omega_{\alpha_2} - (\epsilon_i + \epsilon_j)}, \tag{32}$$

where $f(ij,\alpha_2)$ is the single particle matrix element of the field defining the separable interaction and N_{α_2} is the normalization constant determined by the condition $\sum_{i\leq j} X(ij;\alpha_2)^2 = 1$.

Due to the large number of scattering states included in the single-particle representation the dimension of three-particle basis is also large. The scattering states are needed in order to describe these unstable states. In the calculations we took into account all the possibilities described above regarding the energies of the single-particle state $0p_{1/2}$ as well as the binding energy of the state ^{11}Li(gs).

With the single-particle states and the two-particle states ^{11}Li(gs) and ^{11}Li(2_1^+) discussed above, we formed all the possible three-particle basis states. We found that the only physically relevant states are those which are mainly determined by the bound state ^{11}Li(0_1^+). The corresponding spins and parities are $1/2^+$, $1/2^-$ and $5/2^+$. States like $3/2^+$, which arises from the CXMSM configuration $|1s_{1/2} \otimes 2_1^+; 3/2^+\rangle$, is not a meaningful state. The corresponding calculated energies are shown in Fig. 3.

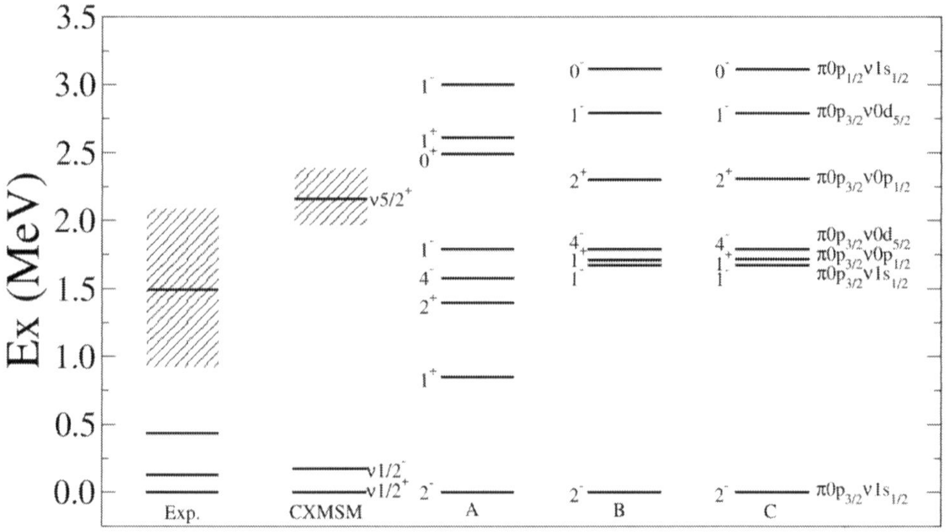

Figure 3. Experimental level scheme in ^{12}Li. The three lowest levels are from [16], while the one at 1.5 MeV is from [15]. In the second column are the three-neutron CXMSM results. In the columns A-C are the shell-model calculations corresponding to different truncation schemes: A) Maximum of 1p excitation from p to sd shell, B) maximum of 3p excitations and C) full psd space. Dashed lines indicate the widths of the resonances.

In Ref. [16] it was also found that ^{12}Li(gs) is an antibound state but, in addition, two other low-lying states were observed at 0.250 MeV and 0.555 MeV by using two-proton removal reactions. In this case the $0p_{3/2}$ proton in the core may interfere with the neutron excitations evaluated above. In particular, the antibound $1/2^+$ ground state would provide, through the proton excitation, a state 1^- and a 2^-. This is the situation encountered in the shell model calculation [17]. One sees in this Figure that the full calculation predicts all excited states to lie well above the corresponding experimental values. It is worthwhile to point out that the calculated states exhibit rather pure shell model configurations. For instance the states 2_1^- (ground state) and 1_1^- are mainly composed of the configuration $|\pi\left[0p_{3/2}\right]\nu\left[(0p_{1/2})^2 1s_{1/2}\right]\rangle$. This does not fully agree with our CXMSM calculation, since in our case this wave function is mainly of the form $|1s_{1/2} \otimes^{11} \mathrm{Li(gs)}\rangle$. This differs from the shell model case in two ways. First, the state ^{11}Li(gs) contains nearly as much of $1s_{1/2}$ as of $0p_{1/2}$. Second the continuum states contribute much in the building up of the antibound ^{12}Li(gs) wave function, as discussed above. In our representation it is straightforward to discern the antibound character of this state, which is not the case when using harmonic oscillator bases.

Using the two-particle states of ^{11}Li within the Berggren single-particle representation, we can calculate the four-particle system ^{13}Li. The four-particle CXMSM basis is constructed by two correlated two-particle states. The two-particle states we included here are the $^{11}\mathrm{Li}(0_1^+)$, $^{11}\mathrm{Li}(1_1^-)$ and $^{11}\mathrm{Li}(2_1^+)$, which are the same as the three-particle case. In this basis set we formed

Table 1. Calculated energies (in MeV) of the four-particle states in ^{13}Li corresponding to different single-particle $\epsilon_{p_{1/2}}$ energies. The two-particle states $W(0_1^+) = -0.295$MeV, $W(1_1^-) = (0.084, -0.002)$ and $W(2_1^+) = (2.300, -0.372)$MeV are included in the calculation.

	$\epsilon_{p_{1/2}} = (0.195, -0.047)$	$\epsilon_{p_{1/2}} = (0.563, -0.252)$
0^+	(0.868,-0.059)	(1.505,-0.041)
	(5.127,-0.964)	(5.274,-1.009)
1^-	(0.836,-0.117)	(1.722,-0.166)
	(3.244,-0.593)	(3.921,-0.721)
2^+	(0.715,-0.114)	(1.802,-0.257)
	(2.907,-0.445)	(3.373,-0.478)
	(5.131,-0.910)	(5.205,-0.966)
3^-	(2.541,-0.391)	(2.674,-0.413)
4^+	(5.715,-1.119)	(5.715,-1.119)

Table 2. Same as Table 1 but for $W(0^+) = -0.369$MeV.

	$\epsilon_{p_{1/2}} = (0.195, -0.047)$	$\epsilon_{p_{1/2}} = (0.563, -0.252)$
0^+	(0.855,-0.057)	(1.527,-0.033)
	(5.142,-0.970)	(5.303,-1.020)
1^-	(0.828,-0.122)	(1.744,-0.169)
	(3.244,-0.593)	(3.921,-0.721)
2^+	(0.715,-0.114)	(1.802,-0.257)
	(2.903,-0.450)	(3.367,-0.477)
	(5.137,-0.916)	(5.213,-0.973)
3^-	(2.541,-0.391)	(2.674,-0.413)
4^+	(5.715,-1.119)	(5.715,-1.119)

the symmetric Hamiltonian matrix by evaluating the dynamical matrix Eq. (23) and the overlap Eq. (25).

The only four-particle states we can get are 0^+, 1^-, 2^+, 3^- and 4^+ states. The calculated results refer to different single-particle $\epsilon_{p_{1/2}}$ and two-particle ^{11}Li(gs) energies are all listed in Tables 1 and 2. We found that there is no bound or antibound state in ^{13}Li, which agrees with the experiment that the ^{13}Li is unbound [14]. The ground state is a resonance at around 1 MeV, however the configuration and the energy strongly depend on which single-particle energy $\epsilon_{p_{1/2}}$ we choose. For $\epsilon_{p_{1/2}} = 0.195$MeV case, the ground state is $|^{11}\mathrm{Li}(1_1^-)\otimes{}^{11}\mathrm{Li}(1_1^-);2^+\rangle$ at the energy of $(0.715, -0.114)$MeV, while for $\epsilon_{p_{1/2}} = 0.563$MeV case, the ground state is 0^+ which at the energy of $(1.505, -0.041)$MeV and with the components $|^{11}\mathrm{Li}(1_1^-)\otimes{}^{11}\mathrm{Li}(1_1^-);0^+\rangle$ and $|^{11}\mathrm{Li}(0_1^+)\otimes{}^{11}\mathrm{Li}(0_1^+);0^+\rangle$ about half and half. Comparing between tables one can find that the four-particle states are not much affected by the energy of ^{11}Li(gs).

In our calculation we only included the resonances of ^{11}Li which is considered physically meaningful for building the whole four-particle basis. However as we mentioned before, there are lots of other continuum states which belong to the background and have a large component of the poles. Although these states are not in themselves physically meaningful resonances, they might have some influence on the spectrum of ^{13}Li. Therefore for a more precise calculation, one should include some of these two-particle states.

4. Summary and conclusions

We have studied excitations occurring in the continuum part of the nuclear spectrum which are at the limit of what can be observed within present experimental facilities. These states are very unstable but yet live a time long enough to be amenable to be treated within stationary formalisms. We have thus adopted the shell model in the complex energy plane for this purpose. In addition we performed the shell model calculation by using the multistep shell model. In this method of solving the shell model equations one proceeds in several steps. In each step one constructs building blocks to be used in future steps [10]. We applied this formalism to analyze ^{12,13}Li as determined by the neutron degrees of freedom.

By using single-particle energies (i.e., states in ^{10}Li) as provided by experimental data when available or as provided by our calculation, we found that the only physically meaningful two-particle states are ^{11}Li(gs), which is a bound state, and ^{11}Li(2_1^+), which is a resonance. As a result there are only three physically meaningful states in ^{12}Li which, besides the antibound ground state, it is predicted that there is a resonance $1/2^-$ lying at about 1 MeV and about 800 keV wide and another resonance which is $5/2^+$ lying at about 1.1 MeV and 500 keV wide. That the ground state is an antibound (or virtual) state was confirmed by a number of experiments [14, 15, 16] and the state $5/2^+$ has probably been observed in [15]. However, in [16] two additional states, lying at rather low energies, have been observed which do not seem to correspond to the calculated levels. It has to be mentioned that neither a shell model calculation, performed within an harmonic oscillator basis, provides satisfactory results in this case. Yet, we found that this shell model calculation works better than one would assume given the unstable character of the states involved. In four-particle system we found that there is no bound or antibound state in ^{13}Li, which agrees with the experiment that the ^{13}Li is unbound.

Acknowledgments

This work has been supported by the Swedish Research Council (VR) under grant Nos. 623-2009-7340 and 2010-4723.

References

[1] Betan R Id, Liotta R J, Sandulescu N and Vertse T 2002 *Phys. Rev. Lett.* **89** 042501.
[2] Gamow G 1928 *Z. Phys.* **51** 204.
[3] Michel N, Nazarewicz W, Ploszajczak M and Bennaceur K 2002 *Phys. Rev. Lett.* **89** 042502.
[4] Betan R Id, Liotta R J, Sandulescu N and Vertse T 2003 *Phys. Rev. C* **67** 014322.
[5] Michel N, Nazarewicz W, Płoszajczak M and Vertse T 2009 *J. Phys. G* **36** 013101.
[6] Xu Z X, Liotta R J, Qi C, Roger T, Roussel-Chomaz P, Savajols H and Wyss R 2011 *Nucl. Phys. A* **850** 53 (preprint: arXiv:1012.2303).
[7] Blomqvist J, Liotta R J, Rydstrom L and Pomar C 1984 *Nucl. Phys. A* **423** 253.
[8] Berggren T 1968 *Nucl. Phys. A* **109** 265.
[9] Liotta R J, Maglione E, Sandulescu N and Vertse T 1996 *Phys. Lett. B* **367** 1.
[10] Liotta R J and Pomar C 1982 *Nucl. Phys. A* **382** 1.
[11] Esbensen H, BertschG F and Hencken K 1997 *Phys. Rev. C* **56** 3054, and references therein.
[12] Betan R Id, Liotta R J, Sandulescu N, and Vertse T 2004 *Phys. Lett. B* **584** 48.
[13] Betan R Id, Liotta R J, Sandulescu N, Vertse T and Wyss R 2005 *Phys. Rev. C* **72** 054322.
[14] Aksyutina Yu et al. 2008 *Phys. Lett. B* **666** 430.
[15] Roger T, PhD thesis; Roussel Chomaz P et al., to be published.
[16] Hall C C et al. 2010 *Phys. Rev. C* **81** 021302(R).
[17] Qi C and Xu F R 2008 *Chin. Phys. C* **32** (S2) 112.

Physics of Nuclear Processes Triggered by the Interplay of Strong and Weak Interactions

Jouni Suhonen
Department of Physics, P.O. Box 35 (YFL), FI-40014 University of Jyväskylä, Finland
E-mail: `jouni.suhonen@phys.jyu.fi`

Abstract. Neutrinoless double beta ($0\nu\beta\beta$) decay of nuclei is a process that requires the neutrino to be a massive Majorana particle and thus cannot proceed in the standard model of electro-weak interactions. Recent results of the neutrino-oscillation experiments have produced accurate information on the mixing of neutrinos and their squared mass differences. The $0\nu\beta\beta$ decay takes place in atomic nuclei where it can be observed, at least in principle, by underground neutrino experiments. The information about the weak-interaction observables, like the neutrino mass, has to be filtered from the data through the nuclear matrix elements (NMEs). In this article recent work of the Jyväskylä group on the NMEs related to double beta decays is reviewed. This work concerns (1) the relevance to neutrinoless double beta decay of occupancies of single-particle orbitals close to the Fermi surface and (2) an example of the resonant neutrinoless double electron-capture decay of an atomic nucleus.

1. Introduction

Thanks to the present-day accurate neutrino-oscillation experiments our knowledge about the relative neutrino masses and neutrino mixing is quite accurate. Unlike in the quark sector the mixing in the neutrino sector turns out to be strong, roughly bi-maximal [1]. Unfortunately it is impossible for the oscillation experiments to access the absolute mass scale and fundamental character (Majorana or Dirac) of the neutrinos. Instead, the neutrinoless double beta ($0\nu\beta\beta$) decay, if detected, can shed ligt on these issues. From the measured $0\nu\beta\beta$-decay half-lives the *effective mass* of the neutrino

$$\langle m_\nu \rangle = \sum_j \lambda_j^{\rm CP} m_j |U_{{\rm e}j}|^2 \qquad (1)$$

can be extracted. Above the m_j are the values of the mass eigenstates of the neutrino, $\lambda_j^{\rm CP}$ are the Majorana CP phases and $U_{{\rm e}j}$ are the components of the electron-neutrino row of the neutrino-mixing matrix. The mass $\langle m_\nu \rangle$ can be extracted [2] provided that the decay goes mainly via the Majorana-neutrino mass and the associated nuclear matrix elements (NMEs) can be evaluated in a reliable way. Once $\langle m_\nu \rangle$ is extracted it can be analyzed by the use of the measured data on the elements of the neutrino-mixing matrix to produce information on e.g. the possible mass hierarchies of neutrinos.

2. Neutrinoless double beta decay: orbital occupancies and spin-orbit partners

2.1. Decay half-life and structure of final states

The $0\nu\beta\beta$ decay can proceed via the exchange of a virtual light Majorana neutrino. By assuming the neutrino mass mechanism to be the dominant one over the other possible mechanisms [1], the inverse of the $0\nu\beta\beta$ half-life can be written as

$$t_{1/2}^{(0\nu)} = g^{(0\nu)} \left| M^{(0\nu)\prime} \right|^{-2} \left(|\langle m_\nu \rangle| \right)^{-2} , \tag{2}$$

where the involved nuclear matrix element can be expressed as

$$M^{(0\nu)\prime} = \left(\frac{g_{\rm A}}{g_{\rm A}^{\rm b}} \right)^2 \left[M_{\rm GT}^{(0\nu)} - \left(\frac{g_{\rm V}}{g_{\rm A}} \right)^2 M_{\rm F}^{(0\nu)} + M_{\rm T}^{(0\nu)} \right] , \tag{3}$$

where $g_{\rm A}^{\rm b} = 1.25$ is the bare-nucleon value of the axial-vector coupling constant. The quantity $\langle m_\nu \rangle$ is the effective mass defined in (1), $g^{(0\nu)}$ is a suitably defined leptonic phase-space factor and $g_{\rm V}$ ($g_{\rm A}$) is the vector (axial-vector) coupling constant of the weak interaction. In the present study the tensor matrix element $M_{\rm T}^{(0\nu)}$ is dropped as its magnitude is quite small [3, 4, 5].

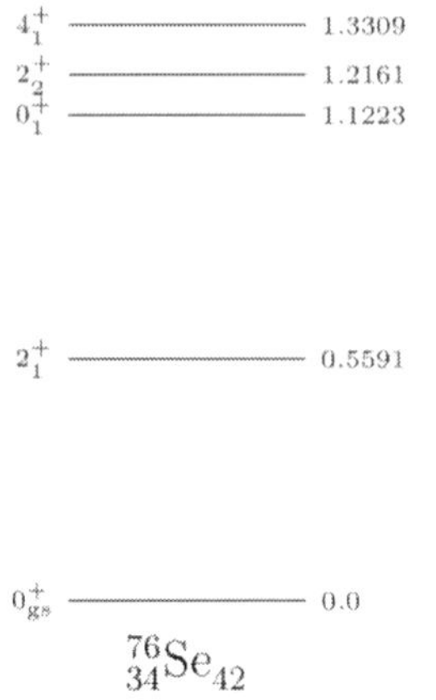

Figure 1. Experimental low-energy spectrum of ^{76}Se. The numbers to the right of the energy levels are excitation energies.

The half-life formula (2) applies to both the ground-state and excited-state transitions. Here it is of interest to address the decay to the first excited 0^+ state in addition to the decay to the ground state. The virtual transitions between this 0^+ state and the intemediate states can be described by the multiple-commutator model (MCM) [6, 7] and it is designed to connect excited states of an even-even reference nucleus to states of the neighboring odd-odd nucleus. The states of the odd-odd nucleus are given by the pnQRPA. The excited states of the even-even nucleus are generated by the (charge conserving) quasiparticle random-phase approximation (QRPA) described in detail in [8]. In the present work the first excited 0^+ state, 0_1^+ (see Fig. 1), is assumed to be a two-phonon state of the form

$$|0_{\rm 2-ph}^+\rangle = \frac{1}{\sqrt{2}} \left[Q^\dagger(2_1^+) Q^\dagger(2_1^+) \right]_0 |{\rm QRPA}\rangle , \tag{4}$$

where $Q^\dagger(2_1^+)$ creates the collective lowest-lying 2^+ state, 2_1^+, in an even-even nucleus and $|{\rm QRPA}\rangle$ is the QRPA vacuum. A multiplet of ideal two-phonon states consists of partner states

$J^\pi = 0^+, 2^+, 4^+$ that are degenerate in energy, and exactly at an energy twice the excitation energy of the 2_1^+ state. In practice this degeneracy is always lifted by the residual interaction between the one- and two-phonon states [9]. In Fig. 1 the proposed two-phonon triplet of states is shown for ^{76}Se, the $0\nu\beta\beta$ grand-daughter nucleus of ^{76}Ge. One can see that indeed the centroid of the energies of the two-phonon triplet $J^\pi = 0^+, 2^+, 4^+$ is roughly twice the energy of the first 2^+ state, whereas the energy spread is some 200 keV. Based on these considerations the wave function of the first excited 0^+ state in ^{76}Se is in this work assumed to be reasonably well described by the ansatz wave function (4).

2.2. Role of orbital occupancies and spin-orbit partners

The size of the adopted single-particle model space is an important issue in nuclear-model calculations as evidenced by the recent analysis of [10]. The various steps in including the spin-orbit partners on top of the minimal proton and neutron space (Toy-1 basis) for ^{76}Ge are described as

$$\begin{aligned}
&A = 76 \\
&\text{Toy-1 basis:} \quad 1\mathrm{p}_{1/2}, 1\mathrm{p}_{3/2}, 0\mathrm{f}_{5/2}, 0\mathrm{g}_{9/2} \\
&\text{Toy-2 basis:} \quad \text{Toy 1 basis} + 0\mathrm{f}_{7/2} \\
&\text{Toy-3 basis:} \quad \text{Toy 1 basis} + 0\mathrm{g}_{7/2} \\
&\text{Toy-4 basis:} \quad \text{Toy 1 basis} + 0\mathrm{f}_{7/2} + 0\mathrm{g}_{7/2} \\
&\text{Large basis:} \quad \text{10 single-particle levels.}
\end{aligned} \tag{5}$$

The Toy-1 basis corresponds to the smallest space that makes any sense in practical calculations. The spin-orbit partners are added to the Toy-1 basis from below (Toy 2), from above (Toy 3) and from both below and above (Toy 4). The large basis includes enough single-particle orbitals to enable a realistic calculation in the pnQRPA framework.

The measured occupancies of the neutron [11] and proton [12] 1p-$0\mathrm{f}_{5/2}$-$0\mathrm{g}_{9/2}$ orbitals for ^{76}Ge and ^{76}Se have inspired theoretical studies of the related effects on the $0\nu\beta\beta$ NMEs. In [13] it was found that the magnitude of the pnQRPA-calculated NME reduces some 30% for the Jastrow SRC and 23% for the UCOM SRC when going from the Woods-Saxon mean-field-based orbital occupancies to occupancies stemming from a kind of hybrid approach. In this hybrid approach the measured neutron occupancies were adopted and the proton occupancies were derived from an adjusted mean field that reproduces the low-energy spectra of the neighboring proton-odd nuclei.

In [10] various possibilities for the orbital occupancies were studied for the $0\nu\beta\beta$ decays to the ground states in the $A = 76, 82, 128, 130, 136$ double-beta systems. It was found that the different occupancies did not play a decisive role in the magnitude of the NME, except in the case of the decay of ^{76}Ge with the hybrid occupations of [13] as visible in Fig. 2. A more important effect comes from the variation of the values of g_A as shown in Fig. 3.

In the case of the excited-state transitions a careful study was performed in [14, 15]. As an example the case of ^{76}Ge decay is taken up here. The computed values of the NME (3) of ^{76}Ge are given in Fig. 4 for the five model spaces quoted in (5) and depicted in the left panel of Fig. 2. The UCOM short-range correlations are used together with the above-mentioned occupancy schemes. If one compares Fig. 4 with Fig. 2 one notices a sharp difference between the behaviors of the NMEs: The NMEs for the decays to the 0_1^+ excited state are much less dependent on the size of the model space than those for the decays to the ground state. This means that the inclusion of all the spin-orbit partners in the model space is not so essential in producing a reliable estimate for $M^{(0\nu)'}(0^+_{2-\mathrm{ph}})$, as long as at least one major shell for protons and neutrons is included in the calculations. Furthermore, the NME with the "PLB" occupancies is now the largest one whereas for the ground-state transitions it is by far the smallest one. In general the

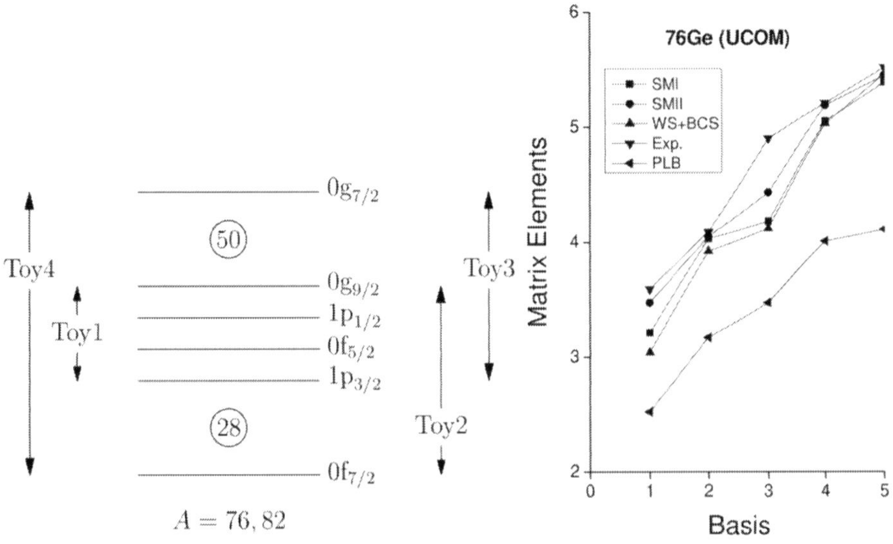

Figure 2. The NMEs (3) for the ground-state-to-ground-state decay of ^{76}Ge with different occupancies and UCOM short-range correlations for four different sizes (Toy1-Toy4) of model bases indicated in the left figure. Number 5 denotes a realistic basis with two oscillator major shells.

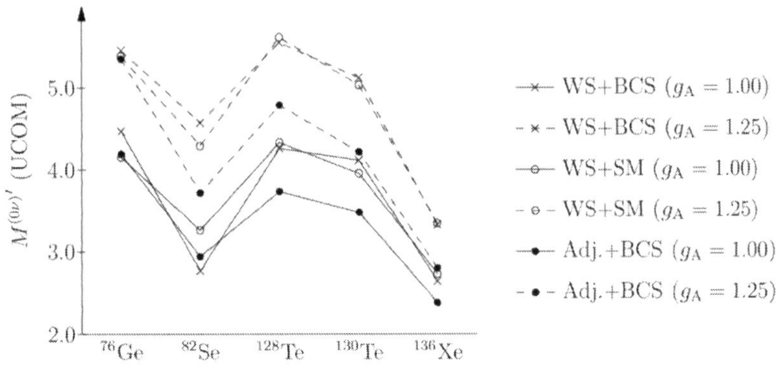

Figure 3. The NMEs (3) corresponding to the ground-state-to-ground-state transitions in five $0\nu\beta\beta$ systems. Three different orbital occupancies and two different values of the axial-vector coupling constant are used. The UCOM SRC are assumed.

relative magnitudes of the NMEs with the different occupancy schemes are reshuffled relative to the order of the NMEs of the ground-state transitions.

From the left panel of Fig. 5 one observes that the effect of different values of the axial-vector coupling constant g_A on the values of the $0\nu\beta\beta$ NMEs is considerable. This difference stems from two sources: The fitting of the values of the experimental $2\nu\beta\beta$ NMEs produces a range of g_{pp} values in the calculations. This effect is small since the g_{pp} dependence of the $M^{(0\nu)'}(0^+_{2-ph})$ NMEs is very weak. The larger effect is the explicit g_A dependence of the NMEs as evident from the expression (3). This strong g_A dependence is also typical of the $0\nu\beta\beta$ NMEs of the ground-

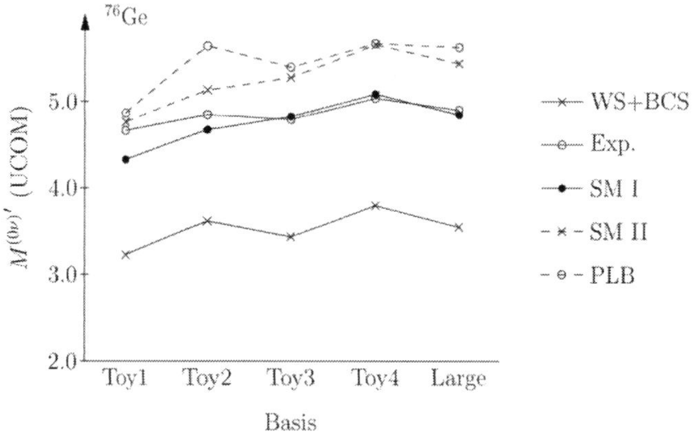

Figure 4. Values of the $0\nu\beta\beta$ NME (3) of the ground-state-to-0_1^+-excited-state transition in ^{76}Ge for different model spaces (5) and for $g_A = 1.25$. Results for five different orbital occupancies and the UCOM short-range correlations are given.

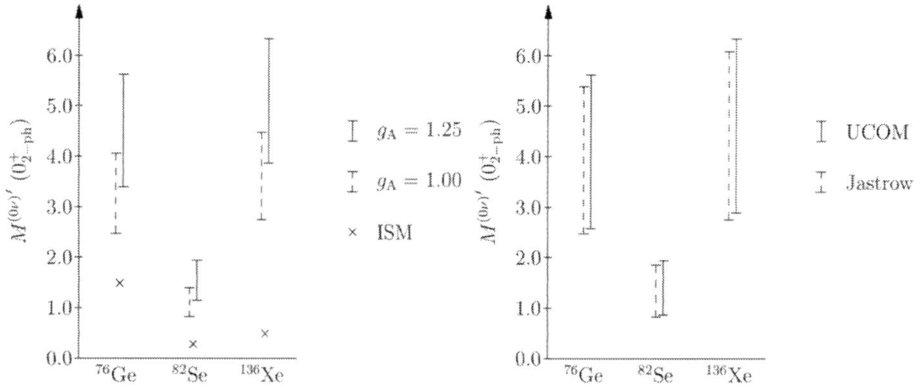

Figure 5. Calculated NMEs for the $0\nu\beta\beta$ decays of ^{76}Ge, ^{82}Se and ^{136}Xe to the first excited 0^+ states in ^{76}Se, ^{82}Kr and ^{136}Ba. Both Jastrow and UCOM short-range correlations have been used. The left panel shows the NMEs separately for $g_A = 1.25$ and $g_A = 1.00$, whereas the right panel displays the NMEs separately for the UCOM and Jastrow short-range correlations. The crosses in the left panel denote the ISM results with Jastrow short-range correlations and $g_A = 1.25$ [5].

state transitions as seen in Fig. 3. Thus attempts to pinpoint the value of g_A in medium-heavy and heavy nuclei are welcomed. The NMEs of the ISM are smaller than the presently calculated, especially for the decay of ^{136}Xe. In the case of the ground-state transition of ^{136}Xe the ISM and pnQRPA results are rather close for both the UCOM and Jastrow short-range correlations. It seems that the differences between the QRPA framework and the ISM framework are enhanced in calculations trying to describe $0\nu\beta\beta$ decays to (presumably) collective excited 0^+ states.

In the right panel of Fig. 5 the final values of the $0\nu\beta\beta$ NMEs are displayed separately for the UCOM and Jastrow short-range correlations. From this panel one observes that the spread in the calculated values of the NMEs is large, the biggest values being roughly twice the smallest ones. This spread is of the same order of magnitude or even slightly bigger than for the corresponding NMEs of the ground-state transitions. The UCOM and Jastrow results are almost the same for the excited-state transitions contrary to the rather large differences encountered for the ground-state transitions (see Ref. [10]).

Starting from the expression (2) the $0\nu\beta\beta$ half-lives can be cast in the form

$$t_{1/2}^{(0\nu)}(0_i^+ \rightarrow 0_f^+) = \frac{C^{(0\nu)}}{(|\langle m_\nu \rangle|[\mathrm{eV}])^2} \times 10^{25}\,\mathrm{yr}\,, \tag{6}$$

where the effective neutrino mass should be given in units of eV. The computed ranges of the factors $C^{(0\nu)}$ are listed in Table 1 for both the ground-state and excited-state transitions. It is seen that for ^{76}Ge and ^{136}Xe the $0\nu\beta\beta$ half-lives for the transitions to the first excited 0^+ states are one order of magnitude longer than for the ground-state transitions, whereas for ^{82}Se the difference is two orders of magnitude owing to the small NME. Detection of the excited-state transitions could be hard because of this at least one-order-of-magnitude suppression relative to the ground-state decays but at the same time the background could be cleaned up considerably by the coincident gammas emitted when the 0_1^+ state decays to the ground state through the 2_1^+ state.

$C^{(0\nu)}$	^{76}Ge	^{82}Se	^{136}Xe
0_{gs}^+	0.14 – 0.40	0.046 – 0.12	0.053 – 0.11
0_1^+	1.53 – 7.92	2.76 – 15.3	0.316 – 1.68

Table 1. Calculated values of the nuclear-structure coefficients of the half-life expression (6).

3. Resonant neutrinoless double electron-capture of ^{106}Cd

A lot of work has been done in experimental [16] and theoretical [1, 2] investigations of the double β^- decays of nuclei due to their favorable decay Q values. The positron-emitting modes of decays, $\beta^+\beta^+$, β^+EC and ECEC are much less studied. Theoretical studies of these modes include [17] for the general, nuclear-model independent frameworks of $2\nu2\beta$ $\beta^+\beta^+$, β^+EC and ECEC decays and [18] for the general frameworks of the $0\nu2\beta$ $\beta^+\beta^+$ and β^+EC decays. The formalism for the resonant neutrinoless double electron capture (R0νECEC) was first developed in [19] and later discussed and extended to its radiative variant ($0\nu\gamma$ECEC) in [20].

The first nuclear-structure calculations of the nuclear matrix elements (NMEs) involved in the above decays were performed in [21] and soon after in [22] and [23]. All of these calculations considered transitions to the final ground states only. Later the $2\nu2\beta$ $\beta^+\beta^+$, β^+EC and ECEC decays of ^{78}Kr ^{92}Mo, ^{96}Ru, ^{106}Cd, ^{124}Xe and ^{130}Ba were examined in [24] for both the ground states and first excited 0^+ states. This study was complemented by a joint theoretical and experimental investigation for the decay of ^{106}Cd in [25]. The in 1995 introduced renormalised quasiparticle random-phase approximation [26] was used in [27] to calculate the NMEs of the $2\nu2\beta$ $\beta^+\beta^+$ decays of ^{78}Kr and ^{106}Cd to the ground and first excited 0^+ states of the final nuclei. In [28, 29] the single-state-dominance hypothesis (SSDH) was examined and the NMEs related to the $2\nu2\beta$ ECEC decays of ^{106}Cd and ^{136}Ce to the final ground state and two lowest excited 0^+ states were derived. More recently in [30] the $2\nu2\beta$ $\beta^+\beta^+$, β^+EC and ECEC modes of decay were discussed under the SSDH, without a quantitative nuclear-structure calculation, for several nuclei and for several 0^+ and 2^+ final states. A more refined NME and half-life calculation of the $2\nu2\beta$ $\beta^+\beta^+$, β^+EC and ECEC decays of ^{106}Cd to the ground state and first excited 0^+ state in ^{106}Pd was carried out in [31]. The $2\nu2\beta$ $\beta^+\beta^+$, β^+EC and ECEC decays of ^{106}Cd to the final ground state was also examined within the Hartree-Fock-Bogoliubov model in [32].

The calculation of the NMEs related to the neutrinoless positron emission modes was started, as mentioned earlier, in [21, 23]. Later, in [33] the ground-state $0\nu2\beta$ $\beta^+\beta^+$ decays of ^{124}Xe and ^{136}Ce were compared with several $\beta^-\beta^-$ decays and in [34] a systematic study of the $0\nu2\beta$ $\beta^+\beta^+$ and β^+EC decays to excited 0^+ states in ^{92}Mo, ^{96}Ru, ^{106}Cd, ^{124}Xe, ^{130}Ba and ^{136}Ce was performed. In the latter two calculations the computational scheme was based on the relativistic harmonic confinement model (RHCM) of quarks and the resulting nucleon form factors [35, 36, 37]. In this framework a folding of the free nucleon current with the confined quark degrees of freedom was done resulting in a nucleonic current that differed from that of the standard formulation [38, 39]. In addition, no short-range correlations were taken into account beyond the RHCM-predicted nucleon form factors. In [40] the $2\nu2\beta$ $\beta^+\beta^+$, β^+EC and ECEC decays as well as the $0\nu2\beta$ $\beta^+\beta^+$ and β^+EC decays of ^{106}Cd to the ground state were treated within the second quasi random phase approximation framework.

Nuclear-structure calculations associated with the R0νECEC processes were performed for ^{112}Sn in [19, 41], for ^{74}Se in [42] and for ^{136}Ce in [43]. In these calculations the non-relativistic Schrödinger wave functions of the involved electron orbitals were used.

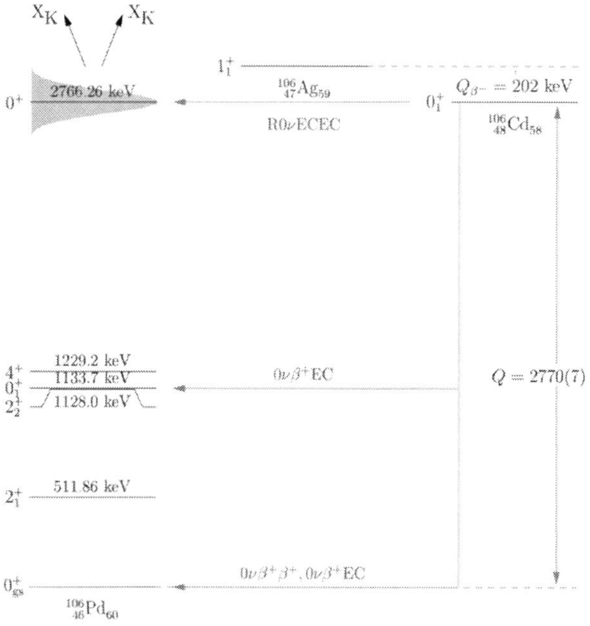

Figure 6. Experimental low-energy spectrum of ^{106}Pd and the resonant 0^+ state at 2766.26 keV of excitation (the atomic two-K-hole energy has been added). Also the Q value of the resonant 0νECEC transition has been given. All the $0\nu2\beta$ transition modes under consideration in this article are indicated by horizontal arrows.

In the present article the work of [31] is extended to description of various $0\nu2\beta$ decay modes of ^{106}Cd. These include the $\beta^+\beta^+$ and β^+EC transitions to the ground state, $0^+_{\rm gs}$, and to the first excited 0^+ state, 0^+_1. The $0\nu2\beta$ decays to only the 0^+ states are considered due to the large suppression of the mass mode for the $0\nu2\beta$ decays to 2^+ states [44]. In addition, the R0νECEC transition to the 0^+ state at excitation energy 2766.26 keV is considered where the energy of the two electron K holes in the palladium atom has been added to the nuclear excitation energy 2717.56 keV of the 0^+ state. The relativistic Dirac wave functions of the K-electron orbitals are used. All these transitions are displayed in Fig. 6.

The half-lives of the different $0\nu 2\beta$ modes can be compactly written as

$$T_{1/2}^{\beta^+\beta^+} = T_0^{\beta^+\beta^+}(\langle m_\nu\rangle[\text{eV}])^{-2}, \quad (7)$$

$$T_{1/2}^{\beta^+\text{EC}} = T_0^{\beta^+\text{EC}}(\langle m_\nu\rangle[\text{eV}])^{-2}, \quad (8)$$

$$T_{1/2}^{\text{ECEC}} = T_0^{\text{ECEC}}\frac{x^2+26.21}{(\langle m_\nu\rangle[\text{eV}])^2}, \quad (9)$$

where the effective neutrino mass should be inserted in units of eV and the variable x is defined as $|Q-E| = x\,\text{eV}$. Here Q is the atomic mass difference of the mother and daughter atoms and E the excitation energy of the resonant state in the final atom, including both the nuclear and atomic contributions. In Table 2 the auxiliary factors of the above half-life expressions are given for both the Jastrow and UCOM short-range correlations.

State	s.r.c.	T_0		
		$\beta^+\beta^+(10^{27}\text{yr})$	$\beta^+\text{EC}\ (10^{26}\text{yr})$	R0νECEC (10^{22}yr)
0^+_{gs}	UCOM	1.93 – 2.20	1.35 – 1.54	
	Jastrow	3.11 – 3.24	2.18 – 2.28	
0^+_1	UCOM		981 – 1220	
	Jastrow		832 – 1070	
0^+_{res}	UCOM			3.20 – 8.08
	Jastrow			3.77 – 9.69

Table 2. The auxiliary factors of Eqs. (7)-(9) for the decays of ^{106}Cd. The symbol 0^+_{res} refers to the R0νECEC 0^+ resonance.

From Eq. (9) and Table 2 one can infer that for $\langle m_\nu\rangle = 0.3\,\text{eV}$, i.e. for a value of $\langle m_\nu\rangle$ consistent with the range predicted by the analysis of [45], the favorable case of close degeneracy (values of x below few eV) would yield a R0νECEC half-life around 10^{25} years which is not discouragingly large and within reach of the next-generation $\beta\beta$ experiments. In the 100 eV range of x the R0νECEC half-life would already exceed 10^{27} years.

The most recent experimental half-life limit [46] for the resonant decay of ^{106}Cd is $T_{1/2}^{\text{ECEC}} > 1.1\times 10^{20}\,\text{yr}$ and is far from the theoretically predicted range of detectability. The quoted numbers in Fig. 6 suggest that an accurate prediction of the R0νECEC half-life is only possible if the energy (and multipolarity) of the resonant state is verified and the atomic mass difference of the Cd and Pd atoms is measured accurately, say, in Penning-trap experiments. Such experiments have already been performed, e.g. for the ^{74}Se $\rightarrow$ ^{74}Ge transition in [42] and for the ^{136}Ce $\rightarrow$ ^{136}Ba transition in [43].

4. Conclusions

The rare weak-interaction processes in nuclei are of high current interest. Their theoretical and experimental studies are in a position to shed light on fundamental properties of neutrinos and their mixing. Nuclear structure is intimately intertwined with the decay processes and therefore the calculation of the associated nuclear matrix element is an issue of high relevance. In the present work it is pointed out that the orbital occupancies and inclusion of all spin-orbit partners in the model space can affect significally the magnitudes of the nuclear matrix elements for the double beta-minus decays. At the same time the various neutrinoless double beta-plus processes attract more and more attention in the scientific community of neutrino physics. Here the example case of ^{106}Cd is treated in order to gain insight in the orders of magnitude of the associated decay half-lives.

Acknowledgments
This work was supported by the Academy of Finland under the Finnish Center of Excellence Program 2006-2011 (Nuclear and Accelerator Based Program at JYFL).

References

[1] Avignone III F T, Elliott S R and Engel J 2008 *Rev. Mod. Phys.* **80** 481–516
[2] Suhonen J and Civitarese O 1998 *Phys. Rep.* **300** 123–214
[3] Kortelainen M and Suhonen J 2007 *Phys. Rev. C* **75** 051303(R)
[4] Kortelainen M and Suhonen J 2007 *Phys. Rev. C* **76** 024315
[5] Menéndez J, Poves A, Caurier E and Nowacki F 2009 *Nucl. Phys. A* **818** 139–151
[6] Suhonen J 1993 *Nucl. Phys. A* **563** 205–224
[7] Civitarese O and Suhonen J 1994 *Nucl. Phys. A* **575** 251–268
[8] Suhonen J 2007 *From Nucleons to Nucleus: Concepts of Microscopic Nuclear Theory* (Berlin: Springer)
[9] Delion D S and Suhonen J 2003 *Phys. Rev. C* **67** 034301
[10] Suhonen J and Civitarese O 2010 *Nucl. Phys. A* **847** 207–232
[11] Schiffer *et al* J P 2008 *Phys. Rev. Lett.* **100** 112501
[12] Kay *et al* B P 2009 *Phys. Rev. C* **79** 021301(R)
[13] Suhonen J and Civitarese O 2008 *Phys. Lett. B* **668** 277–281
[14] Suhonen J 2011 *Int. J. Mod. Phys. E* **20** 451–458
[15] Suhonen J 2011 *Nucl. Phys. A* **853** 36–60
[16] Barabash A S 2010 *Phys. Rev. C* **81** 035501
[17] Doi M and Kotani T 1992 *Prog. Theor. Phys.* **87** 1207–1231
[18] Doi M and Kotani T 1993 *Prog. Theor. Phys.* **89** 139–159
[19] Bernabeu J, De Rujula A and Jarlskog C 1983 *Nucl. Phys. B* **223** 15–28
[20] Sujkowski Z and Wycech S 2004 *Phys. Rev. C* **70** 052501
[21] Staudt A, Muto K and Klapdor-Kleingrothaus H V 1991 *Phys. Lett. B* **268** 312–316
[22] Suhonen J 1993 *Phys. Rev. C* **48** 574–576
[23] Hirsch M, Muto K, Oda T and Klapdor-Kleingrothaus H 1994 *Z. Phys. A* **347** 151–160
[24] Aunola M and Suhonen J 1996 *Nucl. Phys. A* **602** 133–166
[25] Barabash A S, Umatov V I, Gurriarán R, Hubert F, Hubert P, Aunola M and Suhonen J 1996 *Nucl. Phys. A* **604** 115–128
[26] Toivanen J and Suhonen J 1995 *Phys. Rev. Lett.* **75** 410–413
[27] Toivanen J and Suhonen J 1997 *Phys. Rev. C* **55** 2314–2323
[28] Civitarese O and Suhonen J 1998 *Phys. Rev. C* **58** 1535–1538
[29] Civitarese O and Suhonen J 1999 *Nucl. Phys. A* **653** 321–337
[30] Domin P, Kovalenko S, Šimkovic F and Semenov S V 2005 *Nucl. Phys. A* **753** 337–363
[31] Suhonen J and Civitarese O 2001 *Phys. Lett. B* **497** 221–227
[32] Shukla A, Raina P K, Chandra R, Rath P K and Hirsch J G 2005 *Eur. Phys. J. A* **23** 235–242
[33] Aunola M and Suhonen J 1998 *Nucl. Phys. A* **643** 207–221
[34] Suhonen J and Aunola M 2003 *Nucl. Phys. A* **723** 271–288
[35] Suhonen J, Khadkikar S B and Faessler A 1990 *Phys. Lett. B* **237** 8–13
[36] Suhonen J, Khadkikar S B and Faessler A 1991 *Nucl. Phys. A* **529** 727–740
[37] Suhonen J, Khadkikar S B and Faessler A 1991 *Nucl. Phys. A* **535** 509–547
[38] Doi M, Kotani T, Nishiura H and Takasugi E 1983 *Prog. Theor. Phys.* **69** 602–635
[39] Doi M, Kotani T and Takasugi E 1985 *Prog. Theor. Phys. Suppl.* **83** 1–175
[40] Stoica S and Klapdor-Kleingrothaus H V 2003 *Eur. Phys. J. A* **17** 529–536
[41] Rahaman S, Elomaa V, Eronen T, Hakala J, Jokinen A, Julin J, Kankainen A, Saastamoinen A, Suhonen J, Weber C and Äystö J 2008 *Phys. Lett. B* **662** 111–116
[42] Kolhinen V, Elomaa V, Eronen T, Hakala J, Jokinen A, Kortelainen M, Suhonen J and Äystö J 2010 *Phys. Lett. B* **684** 17–21
[43] Kolhinen V S, Eronen T, Gorelov D, Hakala J, Jokinen A, Kankainen A, Rissanen J, Suhonen J and Äystö J 2011 *Phys. Lett. B* **697** 116–120
[44] Tomoda T 2000 *Phys. Lett. B* **474** 245–250
[45] Klapdor-Kleingrothaus H V and Krivosheina I V 2006 *Mod. Phys. Lett. A* **21** 1547–1566
[46] Rukhadze *et al* N I 2011 *Nucl. Phys. A* **852** 197–206

Institute of Physics
Dirac House, Temple Back
Bristol BS1 6BE UK

proceedings
.com